T0073935

Henry Enfield Roscoe

Henry Roscoe circa 1872. Copyright of the University of Manchester.

Henry Enfield Roscoe

The Campaigning Chemist

PETER J.T. MORRIS AND PETER REED

OXFORD
UNIVERSITY PRESS

OXFORD
UNIVERSITY PRESS

Oxford University Press is a department of the University of Oxford. It furthers
the University's objective of excellence in research, scholarship, and education
by publishing worldwide. Oxford is a registered trade mark of Oxford University
Press in the UK and certain other countries.

Published in the United States of America by Oxford University Press
198 Madison Avenue, New York, NY 10016, United States of America.

© Oxford University Press 2024

All rights reserved. No part of this publication may be reproduced, stored in
a retrieval system, or transmitted, in any form or by any means, without the
prior permission in writing of Oxford University Press, or as expressly permitted
by law, by license, or under terms agreed with the appropriate reproduction
rights organization. Inquiries concerning reproduction outside the scope of the
above should be sent to the Rights Department, Oxford University Press, at the
address above.

You must not circulate this work in any other form
and you must impose this same condition on any acquirer.

Library of Congress Cataloging-in-Publication Data
Names: Morris, Peter J. T. (Peter John Turnbull), 1956– author. |
Reed, Peter, 1942– author.
Title: Henry Enfield Roscoe : the campaigning chemist /
Peter J.T. Morris and Peter Reed.
Description: New York, NY : Oxford University Press, [2024] |
Includes bibliographical references and index.
Identifiers: LCCN 2023043421 (print) | LCCN 2023043422 (ebook) |
ISBN 9780190844257 (hardback) | ISBN 9780197637616 (epub)
Subjects: LCSH: Roscoe, Henry E. (Henry Enfield), 1833-1915. |
Chemists—Great Britain—Biography.
Classification: LCC QD22.R8 M67 2023 (print) | LCC QD22.R8 (ebook) |
DDC 540.92 [B]—dc23/eng/20231023
LC record available at https://lccn.loc.gov/2023043421
LC ebook record available at https://lccn.loc.gov/2023043422

DOI: 10.1093/oso/9780190844257.001.0001

Printed by Integrated Books International, United States of America

MIX
Paper
FSC FSC® C183721

This book is dedicated to John Pickstone (1944–2014) and Jeff Hughes (1965–2018) whose work at the Centre for the History of Science, Technology and Medicine at the University of Manchester inspired both authors.

Contents

List of Illustrations

Figures

Unless credited otherwise, the figures have been taken from H. E. Roscoe, *Life and Experiences*, from the library of Peter Morris.

Tables

Preface

We have written this book to rescue Henry Enfield Roscoe, one of the foremost chemists and reformers of the Victorian period, from his undeserved obscurity in the historical literature. There has been no biography of Roscoe since his former student Sir Edward Thorpe expanded his obituary into a relatively short biography (accurately called "a biographical sketch") in 1916 soon after Roscoe's death. While there have been books that have dealt with aspects of Roscoe's activities, a complete scholarly account of his life has so far been lacking. What is even more striking is Roscoe's absence from works in which one would have expected him to feature, including biographies of his contemporaries and histories of chemistry, on one hand, and histories of the Liberal Party, on the other.

One of us (Peter Morris) first thought of writing a biography of Roscoe after reading in a journal in the early 1980s the colorful statement that in the 1870s one could have spotted the portly figure of Roscoe running along a platform to catch a connecting train as he strove to maintain the links between academia and industry (unfortunately the publication in which this appeared cannot now be found). He was also interested in Roscoe as one of the relatively few academic chemists who became a Member of Parliament. Hence Roscoe's career seemed the ideal medium for exploring the links between chemistry, industry, and politics in the second half of the nineteenth century. After having the initial inspiration, he was encouraged to write this biography by John Pickstone; it is a matter of great personal regret that neither he nor his colleague at the University of Manchester, Jeff Hughes, are alive to read the end result.

For the coauthor (Peter Reed), it was while working for the museum service in Liverpool that he first became aware of the Roscoe family and the prominent role played by William Roscoe (Henry's grandfather) in the cultural life of Liverpool. Later researching the lives of Robert Angus Smith and Alfred Evans Fletcher, both of whom had strong links with Manchester and worked for the Alkali Inspectorate, and a following discussion with Jeff Hughes revealed Henry Roscoe's role as professor of chemistry in the development of Owens College and its attainment of university status. But it soon became apparent that no definitive biography had been written placing his work and achievements in a broad contextual framework. During a subsequent conversation with Peter Morris in which we discussed our shared interest in Henry Roscoe, we agreed to pool our knowledge and expertise to fill this gap in scientific biography.

When one writes a biography of a multifaceted person such as Roscoe, the issue arises of what to include and what should be left out. In the case of Roscoe, one can envisage a biography which would be entirely about Roscoe the chemist, Roscoe the educational reformer, or Roscoe the Liberal politician. However, we have chosen to deal with Roscoe in the round, and to explore how these different aspects of his career influenced each other. We are very much aware that in doing so, we risk making chemists, educational historians, and political historians unhappy, as they would all wish to have a greater emphasis (and more space given over) to the phase of Roscoe's career that is of most interest to them. In particular we have not sought to write a scientific biography of Roscoe as this would not be accessible to the majority of our readers. There is another reason why we have combined all three elements of Roscoe's life. As a chemical researcher, he was not one of the most important chemists of his era, and hence a scientific biography would be of limited value. Similarly, as an MP, he was not one of the most significant politicians in the late Victorian Parliament as he was never a government minister. As an educational reformer, he played an important role in the promotion of scientific and technical education, but he was one of many such reformers. It is only when one combines all three elements of Roscoe's life that his uniqueness is evident and his contribution to Victorian Britain comes to the fore. The golden thread that runs through his whole career was Roscoe's eagerness to campaign for what he believed was essential for the progress of British society and its economy, be it the improvement of chemical education, the expansion of Owens College, the reform of technical instruction, the promotion of the metric system, the opening of museums and libraries on Sunday, or the introduction of new vaccines. Roscoe was a tireless reformer and above all, a lifelong campaigner.

Peter J.T. Morris and Peter Reed
March 2023

Acknowledgments

For his initial support, when this project was just an idea, Peter Morris (PM) wishes to thank the late John Pickstone of the University of Manchester who believed PM was the right person to write this book; John is sorely missed by all. PM has known Peter Reed (PR) for many years, but it was at a meeting of the Society for the History of Alchemy and Chemistry at the Royal Institution in November 2015 that we first discussed the idea of writing this biography together. PR has brought to this project his deep knowledge of chemistry in the nineteenth century and especially his familiarity with Liverpool and Manchester, stemming from his time as Assistant Director at Merseyside County Museums (later National Museums and Galleries on Merseyside). His enthusiasm for the history of educational reform perfectly balanced PM's interest in the history of laboratories and political history. Without PR's involvement, this volume would be much poorer and may have never been written at all. PM thanks him most heartily for accepting the invitation to write the book together.

PR wishes to say that it has been a pleasure working with PM over the past five years or so on the Roscoe book. We must thank Magda Wheatley for her important contribution to the checking of the final manuscript. As a result of her meticulous checking, many improvements have been made to the text and numerous errors have been removed.

It is always vital to have a fresh pair of eyes to read material one has written, and we are grateful to everyone who read our drafts. Roland Jackson read all the chapters in the book and offered many judicious comments. Bill Brock read several chapters and gave excellent advice based on his profound understanding of the history of chemistry. Robert Bud also read several chapters, and made many helpful suggestions. Robert Fox read Chapter 1, while Mark Curthoys and the former MP Brian Iddon read Chapter 10. Frank James has read the scientific material in Chapters 4 and 7, as well as answering a myriad of questions as the research and writing progressed. Their detailed comments and feedback have improved the contents of the book. Nevertheless, as always, any errors that remain are the responsibility of the authors.

Scholars invariably owe a major debt to librarians and archivists. Here the practice of the two authors has diverged. PR has visited many libraries across the United Kingdom and the United States, while PM has remained for the most part in his study and used the internet to advantage. First and foremost, we wish to thank David Allen (Royal Society of Chemistry Library) and his staff, who provided copies of several articles from the Society's journals and arranged our access to the Roscoe correspondence manuscripts before they were available online. We are both grateful to James Peters (Curator of the University Archives, John Rylands Library, Manchester) for his support over many years both in person and online, and for giving PR access to the Roscoe letterbooks. The staff of the John Rylands Library were very helpful in tracking down books and documents, often at very short notice. PM benefited from the library's remote photographing service and would like to particularly thank Ian

Graham for his expertise in handling the fragile and bulky class registers. The staff of the Manchester "Lit and Phil" were very helpful in providing access to the printed catalog of papers published in the *Proceedings* and *Memoirs*, which were of particular value for writing the bibliography of Roscoe's own publications.

PR is grateful to the staff of the Liverpool Record Office for their help in making the archives of William Roscoe available on a regular basis, and to the staff of the Harold Cohen Library at the University of Liverpool. The University of Heidelberg Library has been exceptionally helpful at a distance, providing copies of the letters by Roscoe to Bunsen, and PM would like to thank Clemens Rohfleisch and Sabrina Zinke for their help. Both authors have benefited from membership of the London Library and their prompt service in dispatching books to our respective home addresses.

PM would also like to thank Matthias Röschner (Deutsches Museum archives); Alexandra Fisher (Parliamentary archives); Anne Barrett and Lucy Shepherd (Imperial College archives); Matt Dunne (University of Leeds archives); Virginia Mills (Royal Society archives); Nick Wyatt (Science Museum Library); Sian Prosser (Royal Astronomical Society Library); and Richard Temple (Archivist Senate House Library, University of London).

For his part, PR is grateful for the help received from the staff of the following libraries: British Library, Bodleian Library, and the Cheshire and Chester Archives and Local Studies Centre. He moved to California in January 2017 and wishes to thank the staff of Sacramento Public Library, the Mechanics Institute Library in San Francisco, and especially the University of California Library at Davis, whose collections in both the history of science and history in general are outstanding.

One cannot know everything, and inevitably one has to rely on the help and advice of others. We are both grateful to several historians who tried to help us find the evidence for scientific biography becoming unfashionable in the 1970s, namely: Helge Kragh; Roy MacLeod; Mary Jo Nye; Lewis Pyenson; Thomas Söderqvist; and Jeffrey Sturchio. PM is especially grateful for the expert assistance of Christoph Meinel (chemistry in Heidelberg); Arnold Thackray (Dalton); Yoshi Kikuchi (Japanese students at Owens College); chemists Richard Wayne and David Phillips (photochemistry); and economic historian Steve Broadberry. He also wishes to thank (in no particular order!): Kate Alderson-Smith, John Brooke, and Alan Ruston (history of Unitarianism); Stuart Anderson and Anna Simmons (history of the Pharmacy bills); Ruth Barton (Roscoe-Huxley correspondence); James Donnelly (history of technical education); Alan J. Rocke (Dalton and Karlsruhe conference); Jeffrey A. Johnson (German academic chemistry); Lee Macdonald and Ileana Chinnici (history of astronomy); Christine Nawa (chemistry in Heidelberg); Alwyn Davies (chemistry at UCL); John Burrows (photochemistry); Ernst Homburg (German chemists and other questions); Rudolf Werner Soukup (chemists in Heidelberg); Franco Calascibetta (Italian chemists); David Lewis (Russian chemists); Isabel Malaquias (Portuguese chemists); Gerrylynn K. Roberts (Owens College students); Gregory S. Girolami (whisky money and for obtaining hard-to-get materials); William Griffith (vanadium chemistry); Jan Trofast (Berzelius and vanadium); Tony Travis (the Fast Blackley Red case); Fyaz Ismail (chemical papers and questions relating to Manchester); Diana Leitch (Dalton papers); Yona Siderer (translations of Roscoe's books); Neil Rankin (economic history); and Alex May (biographical questions).

For his part, PR is grateful to Mark Fletcher and Mike Morley Fletcher for access to the diaries of Alfred Fletcher; and Jonathan Aylen for sourcing and then helping access several important documents relating to the history of Manchester. He also wishes to thank Jon Hall for his assistance in revising Figure 8.1.

On a personal level, Peter Reed would like to thank his family in the United Kingdom, Australia, and the United States for their support, encouragement, and forbearance over many years, and hopes the narrative of the published book will provide a better account of Henry Roscoe's life and work than his often long answers to their inquiries. Peter Reed's wife Margaret has provided constant support and encouragement over a long period; he hopes when she reads the book she can finally understand how all the topics we discussed fit together in a more coherent manner.

Finally, we would like to express our gratitude to our senior editor at Oxford University Press, Jeremy Lewis, for accepting our book in the first place, and for his forbearance and helpfulness over the past few years as we have sought to write it. We also wish to thank Madeline Hoverkamp (OUP) and Dharuman Bheeman (Newgen) for their help in the final stages of this book's publication.

Peter Morris (Essex) and Peter Reed (California), March 2023

Abbreviations

We have used two online sources for biographical information by their abbreviations without further citation, namely: *Oxford Dictionary of National Biography* (*ODNB*): https://www.oxforddnb.com/; and *Deutsche Biographie* (*DB*): https://www.deutsche-biographie.de/.

1

Who Was Henry Enfield Roscoe?

The chemist Henry Enfield Roscoe (hereafter Harry Roscoe; he was known as Harry because his father was Henry) was born in 1833, four years before the accession of Queen Victoria to the throne of Great Britain and Ireland, and he died in 1916, 15 years after Queen Victoria. When she died in January 1901, he had not known any other monarch, although he was now 68 years old. He was thus a product of the Victorian era and considered himself to be such. If he perceived a key characteristic of this period, it would have been the relative success of moderate and peaceful reform, much of which he had campaigned for himself. His grandfather William Roscoe had been a campaigner for the abolition of slavery and parliamentary reforms at the be-ginning of the nineteenth century, and Roscoe followed in his footsteps as a progres-sive Liberal. Roscoe's lifetime also marked an important period in the relationship between Britain and Germany. When Roscoe was born, there was a personal union between Britain and the German kingdom of Hanover. Germany was a patchwork of largely small states and still largely rural. By the time of Roscoe's death, Germany had been unified under Prussia, the new country was highly industrialized, and the two nations were now fighting each other in an increasingly desperate struggle. Roscoe had seen Germany as a model for Britain to follow, at least in the field of education and its links with industrial development, but thanks to the war the idea of adopting German methods was now anathema. This book will examine the life of Roscoe, in the context of the Victorian period on one hand, and his relationship with Germany on the other, as a chemist, an educational reformer, and a political reformer.

Our Use of the Biographical Approach

We have taken the biographical form to enable us to analyze key aspects of Victorian chemistry and education using the life of Roscoe as a framework. Some readers may object to the idea of writing a scientific biography as an outmoded and inevi-tably hagiographic way of writing the history of science. As a means for exploring a scientist's life and work, scientific biography has received a good deal of scholarly criticism; Elizabeth Garber has suggested that, "no genre of history fell under more odium than that of biography."[1] In the past, many biographies have focused solely on the personal life of the person or have explored their scientific work without endeav-oring to bring the two together as an integrated life story; John Herivel's biography of the French physicist and mathematician Joseph Fourier is an example of the latter.[2]

---

[1] Elizabeth Garber, ed., *Beyond History of Science: Essays in Honor of Robert E. Schofield* (Bethlehem, PA: Lehigh University Press, 1990), 9.
[2] John Herivel, *Joseph Fourier, the Man and the Physicist* (Oxford: Clarendon Press, 1975).

Henry Enfield Roscoe. Peter J.T. Morris and Peter Reed, Oxford University Press. © Oxford University Press 2024.
DOI: 10.1093/oso/9780190844257.003.0001

Thomas Hankins has drawn attention to the dilemmas facing scientific biographers and how the genre was exploited and then criticized to undermine its potential value.[3] He goes on to stake out the requirements for a successful outcome for the biographical approach: concentrating on the science; integrating the various component parts of the person's life into a coherent whole; and making it readable since the genre is a literary form requiring an emphasis on the subject's character and personality. Thomas Söderqvist, in his introduction to a compilation of essays on scientific biography, draws attention to how "public understanding of science and its practices may have been significantly shaped by the genre," and suggests that "there has been a growing scholarly interest in discussing scientific biography in the last two decades—a tradition which apparently reflects the renaissance of the genre in the 1980s and 1990s."[4] Söderqvist sees Hankins's article "as a minor citation classic" for the analysis of the genre and for setting out a content framework for scientific biography if it is to be successful.[5]

The subject's life and their scientific work do not occur in isolation. Both are influenced by a multiplicity of interactions. However, there is also a wider intellectual context—the social, political, and economic conditions of the time—and the changes these undergo during the lifetime of the subject. This is likely to provide some difficult questions to answer and involve conflicting evidence, but this is not a reason to abandon the genre. As Mary Jo Nye has concluded, "the most compelling scientific biographies are ones that portray the ambitions, passions, disappointments, and moral choices that characterize a scientist's life."[6]

Many recent scientific biographies have embraced the wider intellectual context of their subject to varying degrees, but there is nearly always a balance between the themes around which the book is written and the extent to which events are presented in chronological order. One of the most acclaimed scientific biographies of recent times is Crosbie Smith and Norton Wise's *Energy and Empire*, which concentrates on themes relating to Lord Kelvin's work in physics with few contextual chapters.[7] Its use of themes as the framework for the narrative structure has influenced many other scientific biographers. Adrian Desmond's biography of Thomas Huxley adopts a strict chronological approach but with an overarching theme of science and religion, while Roy MacLeod's biography of the Anglo-Australian chemist Archibald Liversidge is chronological but with each chapter having a specific theme.[8] William H. Brock, in his biographies of Liebig and William Crookes, follows the thematic approach but

---

[3] Thomas Hankins, "In Defence of Biography: The Use of Biography in the History of Science," *History of Science* 17 (1979): 1–16, on 4–5.

[4] Thomas Söderqvist, "Introduction," in *The History and Poetics of Scientific Biography*, ed. Thomas Söderqvist (Aldershot, Hants: Ashgate, 2007), 9.

[5] Söderqvist, "Introduction," 4 and 8.

[6] Mary Jo Nye, "Scientific Biography: History of Science by Another Means," *Isis* 97 (2006): 322–329, on 322.

[7] Crosbie Smith and M. Norton Wise, *Energy and Empire: A Biographical Study of Lord Kelvin* (Cambridge: Cambridge University Press, 1989).

[8] Adrian Desmond, *Huxley: The Devil's Disciple to Evolution's High Priest* (London: Penguin, 1997) and Roy MacLeod, *Archibald Liversidge, FRS: Imperial Science under the Southern Cross* (Sydney: Sydney University Press, 2009).

with a narrative driven by chronology.[9] By contrast, Ursula DeYoung's biography of John Tyndall is entirely thematic.[10] One biography that emphasizes the scientific context is Colin Russell's biography of Edward Frankland.[11] Any author has to find the right balance between chronology and themes which not only will appease a reviewer skeptical about the biographical approach, but also will appeal to the reader. However, while most of these biographies have an introduction that sets out the thesis of the book and the importance of the subject's life and work, few have a prologue that attempts to draw together the numerous strands developed through the chapters of the book and assess the impact of the subject's contribution to debates and discoveries in their lifetime and how these have influenced later times.

There is no scholarly biography of Harry Roscoe, nor has any book been published that brings together the many facets of his life and analyzes them within the context of his period. Roscoe published an autobiography late in life, which is a selective collection of reminiscences about many of the events in his life and personal recollections of people he met.[12] However, as we will note from time to time below, this autobiography was seemingly written from memory and is not entirely accurate. The two best-known anecdotes in the *Life and Experiences* are Roscoe meeting a tramp outside Owens College soon after his arrival there, and the cheating student who was later lynched in America. The former is impossible to verify, but it is now possible to show that Roscoe's account of the cheating student is partly incorrect. According to Roscoe, his name was Pearson, who joined Roscoe in his early years at Owens and undertook work on the atomic weight of uranium. Roscoe caught him cheating and let him go without publicizing his misdoings. He then joined the Congregationalists, got into trouble with them, and then became a nonconformist minister near Manchester. After robbing a house there, he fled to America and was eventually lynched for horse-rustling![13] Roscoe was clearly amused by this story about a rogue and his just deserts, but is it accurate? Remarkably there is a lengthy online biography of Robert West Pearson, the author of which appears to be unaware of the account in Roscoe's autobiography.[14] Pearson was born in Ancoats, Manchester, in 1838 to a relatively poor family. Contrary to Roscoe's account, his education took place entirely at Owens College. He won two chemistry prizes, but he was more interested in logic and philosophy. After a relatively brief training at Cavendish College, Manchester, under the Joseph Parker mentioned by Roscoe, he applied for a post at the Great George Street Chapel in Liverpool. He was accused of plagiarizing his sermons and then it was

[9] W. H. Brock, *Justus von Liebig: The Chemical Gatekeeper* (Cambridge: Cambridge University Press, 1997); Brock, *William Crookes (1832–1919) and the Commercialisation of Science* (Aldershot, Hants: Ashgate, 2008).

[10] Ursula DeYoung, *A Vision of Modern Science: John Tyndall and the Role of the Scientist in Victorian Culture* (New York: Palgrave Macmillan, 2011).

[11] Colin A. Russell, *Edward Frankland: Chemistry, Controversy and Conspiracy in Victorian England* (Cambridge: Cambridge University Press, 1996).

[12] Henry Enfield Roscoe, *The Life and Experiences of Sir Henry Enfield Roscoe DCL, LLD, FRS, Written by Himself* (London: Macmillan, 1906)

[13] Roscoe, *Life and Experiences*, 105–106.

[14] David Hughes, "Dr Robert West Pearson: The Scandalous Life of the Rev. Robert West Pearson: Overreaching Ambition or 'a Bad Egg'?" (June 2018), https://www.cottontown.org/Names%20of%20Note/Pages/Rev.-Dr.-Robert-West-Pearson.aspx.

discovered that the two doctorates he claimed to have (a PhD and an MD) were fake. Pearson then went to Park Road Church in Blackburn, where he survived until 1864 despite considerable controversy. He left Blackburn and went to America, where he became an attorney in Massachusetts. Then, after more scandals, he became a minister in the Baptist Church until yet another scandal forced his departure in 1879. After an unsuccessful attempt to take up the practice of law again, he became a preacher in San Francisco and then an Episcopalian priest in Arizona. Pearson's colorful and scandal-ridden life came to an end in Los Angeles in 1890, not through being lynched for horse-stealing, but by a stroke.

Roscoe's former student and colleague in Manchester, Sir (Thomas) Edward Thorpe, wrote a relatively brief biography which is effectively an expanded obituary, as Thorpe admitted in his "advertisement" at the front.[15] His account draws heavily on Roscoe's autobiography, supplemented by personal recollections and Roscoe's correspondence, but in the fashion of the day, he does not always reveal the identity of the other party. Both books provide a starting point for our work, and we have included the personal insights that both volumes provide when they have proved to be both apposite and accurate.

The close examination of one person's life in an important period of history allows us to see how much is owed to chance and opportunities, and how much is achieved by the seizing of these opportunities by the "prepared mind," to use the phrase of Roscoe's friend Louis Pasteur. Finally, by restricting ourselves to a specific time frame, the lifetime of Roscoe, we can examine more clearly how the activities of the late Victorian period have shaped our own period and how their concerns are reflected in our own time. By bringing all these different aspects together, we can better understand both Harry Roscoe and the world he inhabited, which he did so much to change.

Roscoe the Lancastrian

Roscoe was born in London, but his family came from Liverpool and several members of his extended family, notably the Jevons, were still living in Liverpool. Furthermore, after his father's premature death, Roscoe moved to Liverpool and went to school there. However, we have called him a Lancastrian rather than a Liverpudlian for good reasons. We want to emphasize his connection with the northwest generally, rather than specifically with Liverpool (which was still part of Lancashire at the time) and to link his background with Manchester. His family originally came from Charnock Richard, a small village southwest of Chorley, which emphasizes his Lancastrian background. Several important chemists in the nineteenth century came from Lancashire, including James Sheridan Muspratt and Edward Frankland, as well as the industrial chemists John Mercer, Edmund Knowles Muspratt, and James Hargreaves,

---

[15] Sir Edward Thorpe, *The Right Honourable Sir Henry Enfield Roscoe, PC, DCL, FRS: A Biographical Sketch* (London: Longman, Green, 1916); T.E.T[horpe], "The Right Honourable Sir Henry Enfield Roscoe," *Journal of the Chemical Society, Transactions* 109 (1916): 395–424; T.E.T[horpe], "Sir Henry Roscoe," *Proceedings of the Royal Society of London* A93 (1917): i–xxi,

among others.[16] We cannot fully understand Roscoe unless we consider his love of Lancashire and its people. His grandfather William Roscoe, the son of a Liverpool market gardener and publican, become a Radical MP for Liverpool, despite his opposition to the slave trade—the slave trade and the wider "triangular trade" being the major source of Liverpool's wealth up to that time. The Roscoe family were members of the (English) Presbyterian church and stayed in the church when it adopted Unitarianism in the late eighteenth century.[17] Roscoe embraced his Unitarian heritage and remained a member of the church throughout his life, but it was never a large part of his outlook, nor was his social network largely Unitarian. He was good friends with Bernhard Samuelson, but was not close to other Unitarian politicians, most notably Joseph Chamberlain.

Roscoe was a member of the last generation whose education was affected by their non-adherence to the Church of England.[18] He was not able to take a degree at Oxford or Cambridge Universities because of his inability to subscribe to the 39 Articles of the Church of England. Oxford had allowed nonconformists to take the Bachelor of Arts degree in 1854, and Cambridge followed two years later. However, Roscoe would not have been able to gain an academic position at Oxford or Cambridge until the Universities Test Acts (which abolished all religious tests at Oxford, Cambridge, and Durham) was passed in 1871, and after the act was passed, he was almost immediately offered a chair at Oxford, which he declined.[19] Roscoe had little time for the tutorial methods used at Oxford and Cambridge and he rightly believed that the teaching at University College London was far superior to the new natural sciences course at Oxford. Similarly in later years, he had no desire to take up a chair of chemistry at Oxford, although he was a good friend of Benjamin Collins Brodie and he could have greatly influenced the teaching of the subject at Oxford by making research part of the undergraduate degree, as another Mancunian professor, William Henry Perkin Jr., finally achieved in the 1910s.[20] It is thus curious that his only son Edmund was an undergraduate at Oxford, and his early death while he was there doubtless did little to improve Roscoe's view of that ancient university.

Roscoe the Germanophile

With the accession of Queen Victoria, the personal union with Hanover was broken and Germany became just another foreign country. In 1837 it was still a loose federation of states and Austria's power over the German Federation was waning. After the hope of a democratic German republic was raised and then crushed in 1848–1850,

[16] For the careers of James Sheridan and Edmund Knowles Muspratt, see Peter Reed, *Entrepreneurial Ventures in Chemistry: The Muspratts of Liverpool, 1793–1934* (Farnham, Surrey: Ashgate, 2015).

[17] Anne Holt, *Walking Together: A Study in Liverpool Nonconformity, 1688–1938* (London: George Allen & Unwin, 1938).

[18] Valerie K. Lund, "The Admission of Religious Nonconformists to the Universities of Oxford and Cambridge, and to Degrees in Those Universities, 1828–1871," MA thesis, College of William and Mary, Williamsburg, Virginia, 1978.

[19] Roscoe, *Life and Experiences*, 42–43, 147.

[20] Jack Morrell, "W.H. Perkin, Jr., at Manchester and Oxford: From Irwell to Isis," *Osiris* 8, *Research Schools: Historical Reappraisals* (1993): 104–126.

Prussia became the nucleus for a different kind of German unification. Hanover was annexed by Prussia in 1866 and the German Empire was formed in 1871. British hopes lay in the accession of Friedrich, the son-in-law of Queen Victoria, to the German throne, but his father lived to be 90 and Friedrich reigned for only 98 days before dying of throat cancer in 1888. By this time, Germany was industrializing rapidly and the two countries were keen trade rivals.[21] Many reformers, including Roscoe, believed that England had to adopt the best features of the German educational system in order to compete with Germany and other countries.

Thanks to his education in Heidelberg in the mid-1850s, Roscoe was a great admirer of German education and science. His chemical researches were all closely linked to Robert Bunsen's own research. The laboratory building he constructed in the early 1870s at Owens College, which became a model for other British university laboratories, was almost entirely based on the laboratory building Bunsen erected in Heidelberg between 1853 and 1855, which Roscoe saw being built and was one of its first occupants.[22] His interest in the practical applications of chemistry and his links with industry reflected Bunsen's own interest in the economic value of chemistry and his investigation of blast furnaces gases in particular. Roscoe's own teaching and promotion of original research as the best way to advance chemistry faithfully followed the German model he encountered in Heidelberg. His deep knowledge of the German education system, his companionship with German scientists, and his ardent conviction that it was probably the best educational system in the world made him immensely influential in Britain in the 1870s and 1880s when there was a general belief that the British educational system had to be completely overhauled.

Furthermore, Roscoe's political views can be placed in a German context. Coming as he did from a liberal nonconformist English background, Roscoe encountered many liberals in Germany in the tolerant state of Baden, including his original sponsor Robert von Mohl, who briefly had been minister of justice in the revolutionary German government of 1848. As a result of his stay in the country, Roscoe had a deep and abiding love of Germany, while being less enthusiastic about Prussia or Berlin. The outbreak of World War I and the strident support for Germany's actions by his German academic friends—the so-called Manifesto of the Ninety-Three—must have been devastating to the elderly Roscoe.

Roscoe the Mancunian

Although he was a Londoner by birth (and university education), and Liverpudlian by ancestry (and secondary education), Roscoe became an important figure in Manchester. Roscoe arrived at Owens College, Manchester, to prepare for the start of the new academic year in October 1857 as the new Professor of Chemistry in succession to Edward Frankland, who had resigned and taken a position at St.

[21] Ross J. S. Hoffman, *Great Britain and the German Trade Rivalry, 1875–1914* (Philadelphia: University of Pennsylvania Press, 1933).

[22] For the laboratory at Heidelberg and the laboratory at Owens College, see Peter J.T. Morris, *The Matter Factory: A History of the Chemistry Laboratory* (Reaktion: London, 2015).

Bartholomew's Hospital, London. Roscoe's arrival in Manchester was not a foregone conclusion. Although he came from a well-known Liverpool family, he had no connections with Manchester and had no particular reason to go there. Had things turned out differently, Roscoe may well have become a London-based entrepreneurial freelance chemist like William Crookes, who was six months older. Even his appointment was by no means certain. At the interviews he faced strong competition from two local men, Frederick Crace Calvert and Robert Angus Smith, both of whom—like Frankland—had worked with Lyon Playfair.

The Manchester that Roscoe moved to was very different from Heidelberg which was a university town set in hilly countryside, with a population of around 15,000 in the mid-1850s, of which perhaps about a thousand were students.[23] Manchester had once been a small country town with a population of 9,000 in 1717, although still twice the size of Heidelberg in the same year.[24] But in stark contrast to Heidelberg, Manchester was transformed by the industrial revolution, and the population of the municipal borough had reached 303,043 by 1851.[25] The growth and prosperity of Manchester was largely a result of the cotton industry. Manchester had become known as "Cottonopolis": importing raw cotton from India and the United States; spinning and weaving cotton fabrics; followed by dyeing and printing; and then exporting fabrics for making up as clothing or furnishings. The cotton industry had slowly metamorphosed from a domestic trade into industrial-scale production in large factory buildings accommodating innovative machinery (the spinning jenny, the spinning frame, and the spinning mule), all driven by steam power, and with a dramatic increase in output.[26] The import of raw cotton fluctuated in cycles during the 1800s in step with periods of war with France. The total imports of raw cotton into Britain in 1792 were 34,907,497 pounds (98,054 bales), but by 1850 these imports had grown to 685,600,00 pounds (1,749,300 bales); this represented about 80% of world cotton.[27] By 1815 cotton textiles were Britain's largest export, most of it produced in Manchester and the surrounding area. The number of mills in Manchester itself peaked at 108, just before Roscoe's arrival, in 1853. The cotton industry in turn boosted the machine-making industry, which then diversified away from textile machinery. Textile manufacture also stimulated industries with a more direct chemical connection, such as bleaching, dyeing, and calico-printing. One of the largest calico-printing firms in Europe, Edmund Potter, was based in nearby Glossop. Edmund Potter himself would become Roscoe's father-in-law in 1863. Manchester's industrialization was accompanied by vibrant commercial enterprise. The leading industrial and commercial figures were active in Manchester's cultural, scientific, and political institutions, and many became national luminaries.

[23] Wikipedia "Heidelberg" (Census results) and the University Calendars of the period.

[24] Peter Arrowsmith, "The Population of Manchester from c AD79 to 1801," *Greater Manchester Archaeological Journal* 1, 99–102.

[25] Charles Richson, *Educational Facts and Statistics of Manchester and Salford* (London: Longman, Brown, Green and Longmans, 1852), 8.

[26] The spinning jenny was invented by James Hargreaves in 1764, the spinning frame by Richard Arkwright in 1768, and the spinning mule by Samuel Crompton in 1799. S. D. Chapman, *Cotton Industry in the Industrial Revolution* (London: Macmillan, 1972).

[27] Richard Burn, *Statistics of the Cotton Trade* (London: Simpkin, Marshall & Co., 1847), 14; and Thomas Ellison, *The Cotton Trade of Great Britain* (London: Effingham Wilson, 1886), appendix, Table 1.

Turning to the chemical industry itself, the most interesting chemical firm in Manchester in the 1850s was the dyestuffs company of Roberts, Dale & Company of Cornbrook, founded in 1852.[28] They started to make the first synthetic dye, picric acid, with the help of Calvert, and then oxalic acid. At the end of the 1850s the firm began to employ German chemists, initially Rudolph Koepp (who soon left), then Heinrich Caro and Carl Alexander Martius, followed by other German chemists. This group of German chemists created several important dyes, most notably Manchester Yellow, before returning to Germany in the mid-1860s. Martius was the co-founder of Agfa in Berlin, and Caro became the research leader of BASF in Ludwigshafen. In this way, two of the major German dyestuff companies had their origins in Manchester. In the mid-1850s, Potter, working with Robert Rumney of the Ardwick Chemical Works, developed a process for the manufacture of the purple dye murexide from guano, which has a strong claim to be the second synthetic dye after picric acid.

By the 1830s Manchester had also become a thriving scientific and cultural center, with institutions of national and international renown. The Manchester Literary and Philosophical Society, founded in 1781, built "on the wealth and prestige of the 18th-century town [that] lay with its clergy, professional men and merchant manufacturers."[29] In 1823 the Royal Manchester Institution was founded to provide gallery space for fine art displays and to organize programs of scientific lectures. In 1843 Lyon Playfair was appointed the first Professor of Chemistry with Robert Angus Smith as his assistant; both had trained with the eminent chemist and educator Justus Liebig in Giessen and were awarded a PhD there. By the 1840s these institutions had been joined by the Manchester Natural History Society (founded 1821), the Manchester Mechanics' Institution (1824), the Manchester Statistical Society (1833), the Manchester Athenaeum (1835), and the Manchester Geological Society (1838). Many of the members of these societies were also members of London-based institutions, "thus providing a useful bridgehead between Manchester and London, and bringing mutual benefits."[30] The Manchester Statistical Society was one of the earliest statistical societies in Britain and took a lead in the use of statistics for advancing the social sciences and social policies. It was a paper read at one of their meetings in 1836 that first drew attention to the benefits of Manchester having a university.

Roscoe the University Builder

When Owens College was founded in 1851, it began a long and at times tortuous journey on the path to university status that required government and parliamentary approval and finally a Royal Charter. In the 1850s there were several models aspiring universities could adopt based on existing institutions: broadly, Oxford and

[28] W. V. Farrar, "Synthetic Dyes before 1860," *Endeavour* 32 (September 1974): 149–154, reprinted in the Variorum series (Aldershot, Hants: Ashgate, 1997); Carsten Reinhardt and Anthony S. Travis, *Heinrich Caro and the Creation of Modern Chemical Industry* (Dordrecht: Kluwer, 2000), passim.

[29] Robert H. Kargon, *Science in Victorian Manchester* (Manchester: Manchester University Press, 1977), 3.

[30] Peter Reed, *Acid Rain and the Rise of the Environmental Chemist in Nineteenth-Century Britain* (Farnham, Surrey: Ashgate, 2014), 46.

Cambridge with their tutorial approach to learning, or University College London with its reliance on lectures and laboratory instruction (for the sciences). But at the core of all universities were instruction at the higher education level, examinations, and the awarding of degrees.

There was much debate about the purpose of university education and the university curriculum. For John Henry Newman in the 1850s, the purpose of university education was to develop lifelong skills without any focus on specific occupations and that no subject should be excluded from the curriculum (including theology and the sciences); in reviewing the current methods of instruction, Newman favored a self-learning approach for students through interactions with fellow students, and to the exclusion of professors and examinations.[31] By the 1850s there was broad support among the newer universities for a general or liberal education that ranged across the humanities and the sciences, rather than focusing on a narrow range of subjects or on specific occupations. Later, with local communities providing much of the finance (as in the case of Manchester), there was a shift toward specific careers and professions to enhance local labor markets. This created a sharper dichotomy of purpose for universities that remains a contentious issue for government departments, educational institutions, policy planners, and the public even today.

Although Roscoe joined Owens College after it was founded, he was central to its expansion and its elevation to university status as Victoria University. Robert Kargon portrays the foundation of Owens College in 1851 as developing alongside other cultural institutions because of the concerns of local people for institutions that reflected the standing of Manchester as the major industrial and commercial town in Britain, while Anna Guagnini stresses the part played by the "rich scientific community" in Manchester.[32] Roscoe became a key figure after the College's foundation, taking the lead in securing the future for the College after its initially shaky start, by relocating it to the Oxford Road site in 1873 and in the establishment of Victoria University as a federal university in 1880.

Although he was not head of the College, it was his determination and tenacity that proved crucial to the success of its expansion. He came to realize that in order for Owens College to have a stable future, it was necessary to garner support from Manchester industrialists and merchants and convince them of the need for well-qualified managers and foremen. Manchester had already become the leading industrial center in Britain, but industrialists supported Owens College to ensure continued economic growth and prosperity as new technical advances emerged.[33] Other major towns and cities created universities to secure the same outcomes. Overall, this theme allows us to explore, through the prism of Roscoe, the connection between Owens College and industry and how far this formed a paradigm for other civic universities

[31] John Henry Newman, *The Idea of a University* (London: Longmans, Green, 1919), 19–20, 107, 131, and 145–149.

[32] Kargon, *Science in Victorian Manchester*, 153–155; and Anna Guagnini, "The Fashioning of Higher Technical Education in Britain: The Case of Manchester, 1851–1914," in *Industrial Training and Technical Innovation: A Comparative and Historical Study*, ed. Howard F. Gospel (London: Routledge, 2010), 70.

[33] Harold Perkin, "The Historical Perspective," in *Perspectives of Higher Education*, ed. Burton R. Clark (Berkeley: University of California Press, 1984), 17–55, on 46.

in their relationship with industry. To fully examine this duality, it is important to point out that no theme stands alone, but is linked to several others.

In forging these links, Roscoe campaigned on several fronts to demonstrate the benefits of universities and industry working together, of which his proactive consultancy work formed a crucial part. Besides his role in the chemistry department, Roscoe was active in creating other links between Owens College and industry. Roscoe drew on his wide network of associates in engineering and in the chemical industry for financial support in developing Owens College: an especially influential group of engineers in Manchester that included Charles Beyer of Messrs. Beyer, Peacock & Company; Joseph Whitworth and William Fairbairn raised funding to create a chair in engineering, as Guagnini has highlighted; while a prominent group of chemical manufacturers that included Messrs. Brooke, Simpson & Spiller, Messrs. Gaskell, Deacon, & Company, Messrs. Muspratt & Company, and Messrs. W. Gossage & Company contributed to funds for chemical laboratories.[34] But the influence of these industrialists went beyond just funding, to advising on appointments and the content of courses. Several were appointed trustees of the College.

For Roscoe, the relationship between a university and its locality was crucial for Owens College, but this duality was important for the civic university movement as a whole, since each was founded within their locality, receiving support (and funding) while also, like Owens, having to serve the interests of industry and commerce. Michael Sanderson concluded that Owens College became a model for the civic university movement because of its careful balancing of the sciences and the arts; most civic universities followed Owens but with some variations due to their localities.[35] There is then the role of a university's buildings beyond accommodating the different academic functions. Sophie Forgan draws attention to the role of architecture in defining a distinct alternative to Oxbridge, while the presence of a university could bestow status on a town.[36]

Other issues to emerge for Owens College (and other aspiring universities) were the role of research in the search for new knowledge and understanding and the inclusion of a medical faculty. Under the influence of scientists such as Liebig and Bunsen, research had become an important part of German university education, and the many British students (including Roscoe) who had studied in Germany became strong advocates of research when taking up posts in British universities.[37] Research was a critical element of university education for Roscoe, not just for the sciences but for the humanities as well; the embrace of research became a driving force for Owens College and the new civic universities.[38] The inclusion of a medical faculty at Owens College was supported by Roscoe, but serious organizational and financial

---

[34] Guagnini, "The Fashioning of Higher Technical Education," 76; and Joseph Thompson, *The Owens College: Its Foundation and Growth* (Manchester: J. E. Cornish, 1886), 639 and 643.

[35] Michael Sanderson, *The Universities and British Industry, 1850–1970* (London: Routledge and Kegan Paul, 1972), 105.

[36] Sophie Forgan, "The Architecture of Science and the Idea of a University," *Studies in the History and Philosophy of Science* 20 (1989): 405–434, on 410.

[37] Michael Sanderson, *The Universities in the Nineteenth Century* (London: Routledge & Kegan Paul, 1975), 6–7.

[38] Henry E. Roscoe, "Original Research as a Means of Education," in *Essays and Address by Professors and Lecturers of the Owens College, Manchester* (London: Macmillan, 1874), 21–57.

Figure 1.1 Harry Roscoe sitting at his desk at Woodcote Lodge, 1906.

issues in the context of national concern over the proliferation of medical schools and medical qualification (where no national system existed) took many years to resolve (Figure 1.1).[39]

Roscoe the Chemical Educator

Roscoe was respected by his contemporaries as a chemical educator, rather than as a leading researcher (despite his commitment to research). He transformed Owens College into one of the major centers of academic chemistry in Britain, and his students founded schools at other new universities, including Leeds and Newcastle. At Owens College, Roscoe taught chemistry through a series of graduated practical exercises, followed by original research. He argued that this kind of teaching promoted "freedom of enquiry, independence of thought, disinterested and steadfast labour, habits of exact and truthful observation, and of clear perception."[40] As Roscoe acknowledged, this method of chemical training was introduced in the laboratory of

---

[39] Stella V.F. Butler, "A Transformation in Training: The Formation of University Medical Faculties in Manchester, Leeds and Liverpool, 1870–84," *Medical History* 30 (1986): 115–132, on 115 and 120.
[40] Roscoe, "Original Research as a Means of Education," 57.

Justus Liebig at Giessen.[41] Liebig's laboratory was closely examined by Jack Morrell in his now legendary paper on "the chemist breeders."[42] He introduced the concept of "research school" and compared Liebig's research school with Thomas Thomson's research school at Glasgow (although arguably Thomson never had one). He identified several factors as being crucial, including institutional and financial support, the scientific reputation of the school's leader, and the leader's charisma. One of the dangers of Morrell's paper was the priority it gave to Giessen. This has been contested by Homburg, who argued a good case that Friedrich Stromeyer at Göttingen (Bunsen's teacher and hence Roscoe's intellectual "grandfather") could be considered the birthplace of modern chemical training, although as Rocke has pointed out, Stromeyer taught inorganic analysis rather than research.[43] Roscoe used a combination of inorganic analysis, practical organic chemistry, and in the case of the most able students, original research in the fourth and fifth years to teach chemistry.

Given the paucity of students at Owens when he arrived, Roscoe had to develop a persuasive case for the value of the chemistry course (degrees were not obligatory at Owens).[44] He argued that the course was suitable for men going into professional or commercial life, especially chemical manufacturing. He pointed out that German chemical manufacturers did not take chemists unless they had done original research.[45] Another important career for chemistry students was school teaching, and this was obviously connected to the introduction of science teaching in schools in this period. This symbiotic relationship with industry and business was very different from the situation at Oxbridge. Gerrylynn Roberts has studied the teaching of chemistry at Cambridge during the long tenure (1861–1908) of Roscoe's contemporary George Downing Liveing.[46] The focus of the Cambridge chemistry course was on agriculture and medicine, reflecting its clientele of landowners and physicians rather than businessmen and industrialists.[47]

Thanks to his training in Germany, Roscoe had a specifically Germanic vision for the creation of what became an outstanding chemical school at Owens College. He was influenced at different stages of his life by his schoolteacher William Balmain, Thomas Graham and Alexander Williamson at University College London (UCL), and, as we have seen, Bunsen at Heidelberg. In a series of papers, the Israeli sociologist Joseph Ben-David analyzed the factors for the supremacy of German science in the second half of the nineteenth century. He identified the innovative and flexible nature of German science as the main reason and argued that this stemmed from the

---

[41] Ibid., 50.

[42] Jack B. Morrell, "The Chemist Breeders: The Research Schools of Liebig and Thomas Thomson," *Ambix* 19 (1972): 1–46.

[43] Ernst Homburg, "The Rise of Analytical Chemistry and Its Consequences for the Development of the German Chemical Profession (1780–1860)," *Ambix* 46 (1999): 1–32; Alan J. Rocke, "Origins and Spread of the 'Giessen Model' in University Science," *Ambix* 50 (2003): 90–115, on 92.

[44] Roscoe, "Original Research as a Means."

[45] Ibid., 37.

[46] Gerrylynn K. Roberts, "The Liberally-Educated Chemist: Chemistry in the Cambridge Natural Sciences Tripos, 1851–1914," *Historical Studies in the Physical Sciences* 11 (1980): 157–183.

[47] There is an underlying assumption here that the students will follow in their father's footsteps in terms of their career. This has been demonstrated to be the case for public schoolboys by W. D. Rubinstein, *Capitalism, Culture, and Decline in Britain, 1750–1990* (London: Routledge, 1993).

competition between German universities and their lack of centralization.[48] However, Ben-David also came to realize that the hierarchical nature of German science, with its all-powerful institute director, suppressed innovation and new disciplines.[49] Ben-David argued that the foreign students in Germany had a misleading impression about German education, having gone through neither the secondary education system nor the undergraduate system with all their drawbacks. They therefore took an idealized view of a research-based education back to their own countries.[50] We will study how Roscoe applied what he saw at first hand in Germany and how he modified it to meet the local conditions, in particular the creation of the specifically English concept of a department, rather than the German institute. The design of the laboratories and its debt to Heidelberg has been discussed by one of the authors (PJTM).[51] He concludes that the most important aspect of the laboratories at Owens College was the way they established a template for other universities and the architectural partnership of Alfred Waterhouse (a Unitarian like Roscoe) as the preeminent designer of British university laboratories.

The contrast and competition between the English (specifically the UCL) model of chemistry teaching and the hierarchical German model is the theme of the pioneering work of Yoshiyuki Kikuchi on the setting up of the chemistry school at the Imperial University in Tokyo.[52] He examines how the final character of the department was shaped by this competition, which the Anglophile group eventually won. He presents the interaction of the English influence, the German influence, and the Japanese milieu in terms of "contact zones," a concept introduced by Mary Louise Pratt in the field of transcultural studies.[53]

Roscoe was adamant about the importance of studying the fundamental principles of chemistry as they underpinned the applied or technological aspects, in his view, as Robert Bud and Gerrylynn Roberts have analyzed.[54] In this Roscoe followed in the footsteps of German chemists such as Liebig and Bunsen, who, as Christoph Meinel points out, adopted the Humboldtian concept of university education that focused on pure chemistry rather than applied chemistry.[55] At Owens College, Watson Smith was given the title of Lecturer of Technological Chemistry in 1884, an attempt to keep industrial chemistry within the fold of pure chemistry while acknowledging

[48] Joseph Ben-David, "Scientific Productivity and Academic Organization in Nineteenth Century Medicine," *American Sociological Review* 25 (1960): 828–843.

[49] Joseph Ben-David and Awraham Zloczower, "Universities and Academic Systems in Modern Societies," *European Journal of Sociology* 3 (1962): 45–84. This point has since been developed in a more sophisticated manner by Jeffrey A. Johnson, "Academic Chemistry in Imperial Germany," *Isis* 76 (1985): 500–524.

[50] Joseph Ben-David, "The Universities and the Growth of Science in Germany and the United States," *Minerva* 7 (1968): 1–35, on 7–8.

[51] Morris, *Matter Factory*, 180.

[52] Yoshiyuki Kikuchi, *Anglo-American Connections in Japanese Chemistry: The Lab as Contact Zone* (New York: Palgrave Macmillan, 2013).

[53] Mary Louise Pratt, "Arts of the Contact Zone," *Profession* 91 (1991): 33–40; Pratt, *Imperial Eyes: Travel Writing and Transculturation*, 2nd ed. (New York: Routledge, 2008).

[54] Robert Bud and Gerrylynn K. Roberts, *Science versus Practice: Chemistry in Victorian Britain* (Manchester: Manchester University Press, 1984), 85–86.

[55] Christoph Meinel, "Artibus Academicis Inserenda: Chemistry's Place in Eighteenth and Early Nineteenth Century Universities," *History of Universities* 7 (1988): 89–115, on 107.

the importance of the industrial aspects of chemistry.[56] For Roscoe, teaching technologists meant, "science, science, science," as he expressed in a talk to the Society of Chemical Industry in 1884.[57] Roscoe's firm stance on pure chemistry brought him into conflict with some manufacturers. Martin Saltzman has mapped the lengthy debates between Roscoe and Ivan Levinstein, the Manchester dyestuffs manufacturer, whether academic laboratories should serve as research laboratories for local companies.[58] Levinstein demanded more support from institutions such as Victoria University (the former Owens College).

Roscoe the Educational Reformer

Partly because of his experience in expanding Owens College and partly because of his belief that the English education system was flawed, Roscoe became an ardent educational reformer. At the time of the Great Exhibition in 1851, Britain appeared to be on the crest of the wave. Not only did it have the largest economy in the world, its products were widely admired, as evidenced by the awards won by British manufacturers at the Great Exhibition.[59] Furthermore, despite Charles Babbage's "Decline of Science" campaign in the 1830s, British science was still world-leading with the work of Michael Faraday in electricity, James Joule in thermodynamics, John Herschel in astronomy, Charles Darwin in biology, Charles Lyell in geology, and James Clerk Maxwell about to unify electricity and magnetism in the 1860s. Even in chemistry, Britain held a respectable position thanks to Thomas Graham, Alexander Williamson, and Edward Frankland, and by having the leading German chemist August Wilhelm Hofmann (who even Anglicized his name to William Hofmann) in its midst.[60] Not unnaturally, this apparent economic and scientific superiority created the belief—especially among industrialists—that Britain must be doing things right.[61]

The British system of technical training in the 1850s and for many decades thereafter was based on the time-honored apprentice system, or in industries lacking an

---

[56] H. E. Roscoe, *Record of Work Done in the Chemistry Department of Owens College* (London: Macmillan, 1887), 19–20.

[57] Roscoe, "Remarks on the Teaching of Chemical Technology," *Journal of the Society of Chemical Industry* 3 (1884): 592–594. A similar view was taken by Alexander Williamson in his inaugural speech as dean of the new science faculty at UCL; see Alexander W. Williamson, *A Plea for Pure Science: Being the Inaugural Lecture at the Opening of the Faculty of Science in University College, London* (London: Taylor & Francis, 1870). The issue is discussed with reference to Williamson and more generally in Gerrylynn K. Roberts, "'A Plea for Pure Science': The Ascendancy of Academia in the Making of the British Chemist, 1841–1914," in *The Making of the Chemist: The Social History of Chemistry in Europe, 1789–1914*, ed. David Knight and Helge Kragh (Cambridge: Cambridge University Press, 1998), 107–120.

[58] Martin Saltzman, "Academia and Industry: What Should Their Relationship Be? The Levinstein-Roscoe Dialog," *Bulletin of the History of Chemistry* 23 (1999): 34–41.

[59] Michael Leapman, *The World for a Shilling: How the Great Exhibition of 1851 Shaped a Nation* (London: Headline, 2001).

[60] For the Anglicizing of Hofmann's name, see the cover of the Calendars for the Royal College of Chemistry, Imperial College archives.

[61] For an excellent survey of the issue of technical education in the nineteenth century and beyond, see Michael Sanderson, *Education and Economic Decline in Britain, 1870 to the 1990s* (Cambridge: Cambridge University Press, 1999).

apprentice system, by learning on the job, which was called "picking up."[62] By contrast, both the owners of industrial firms and their workers regarded university education with suspicion. The English (as opposed to the Scottish) university system was based on Oxford and Cambridge, which were open only to those accepting the tenets of the Church of England. In the early nineteenth century, new universities appeared, but two of them (Durham and King's College London) were also Anglican, leaving just University College London open to nonconformists. Many businessmen and industrialists in this period were nonconformists. Few of them had any university education or saw any need for one—it is perhaps a truism that one has to have had a university education to appreciate the value of it. The Unitarian textile industrialist George Courtauld remarked in 1856 that "I have had some little experience now, and my feeling is very strongly that, for the great majority of young men in our position in life, a college course is not fit preparation for business life."[63] It is also important to understand that the apprentice system was approved by the trade unions as much as (if not more than) the management.[64] The entrant into the industry was trained by unionized master workers, and the system also limited the number of new workers, thereby maintaining higher wages.

The proposition that British industry needed more university-educated (or at least technically trained) workers came from educators, educationalists, and their political allies, not industry. One could maintain that their argument was a self-interested one, that they needed more students and money, and the best (perhaps the only) way to do this was to persuade the British government (if not British industry) that Britain was falling behind other countries, especially Germany, in economic terms.[65] One might assume, given the woeful tone of the debates about British competitiveness, that Britain was becoming an economic basket-case or at the very least was standing still.[66] Nothing could be further from the truth. The British gross domestic product (GDP) almost trebled between 1820 and 1870 (the point at which these debates were beginning), and indeed its growth was very slightly greater than Germany in this period despite being a mature economic power.[67] Between 1870 and 1913, Britain declined in economic terms, relatively speaking, as its GDP "only" increased by 2.2 times, compared with Germany's 3.3 times. Although its GDP (in purchasing power parity terms) in absolute terms grew by over 124 billion 1990 international dollars, Germany's GDP increased by 165 billion dollars, and hence Germany had overtaken Britain by 1913

---

[62] Bernand Cronin, *Technology: Industrial Conflict and the Development of Technical Education in 19th Century England* (Aldershot, Hants: Ashgate, 2001); Roderick Floud, "Technical Education and Economic Performance: Britain, 1850–1914," *Albion: A Quarterly Journal Concerned with British Studies* 14 (1982): 153–171.

[63] D. C. Coleman, "Gentlemen and Players," *Economic History Review*, 2nd series, 26 (1973): 92–116, on 107, citing *Courtauld Family Letters, 1782–1900* (Cambridge: Bowes and Bowes, 1916), VII, 3785.

[64] Cronin, *Technology, Industrial Conflict*, 178–180.

[65] Ernst Edwin Williams, *Made in Germany*, 4th ed. (London: Heinemann, 1896).

[66] David Edgerton has consistently combated the notion of British decline and technological failure; see Edgerton, "The Decline of Declinism," *Business History Review* 71 (1997): 201–206; Edgerton, *Science, Technology, and the British Industrial "Decline," 1870–1970* (Cambridge: Cambridge University Press, 1996); Edgerton, *Warfare State: Britain, 1920–1970* (Cambridge: Cambridge University Press, 2006), 299–304 and passim.

[67] All the GDP figures mentioned here are taken from Angus Maddison, *Contours of the World Economy, 1–2030 AD: Essays in Macro-Economic History* (Oxford: Oxford University Press, 2007), 379, table A.4.

by around 12 billion dollars. However, it should also be noted, the British economy grew at 1.85% a year (compared with 2.05% in the period 1820–1870), at a time when Britain supposedly went through the so-called Great Depression.[68] Furthermore, in terms of GDP per capita, Britain remained substantially ahead of Germany until the 1960s.[69]

When comparing Britain and Germany in this period, chemistry seems to be a good example to take and has often been used as such.[70] However, considering the chemical industry, it has to be remembered that Britain was predominant in the Leblanc soda industry, which did not require many educated chemists. Furthermore, Britain dominated several important industrial sectors, including textiles and shipbuilding, until World War I.[71]

By the late nineteenth century, academic German chemistry had reached world-class status, but so had history and that most characteristic of German subjects, philology. At Heidelberg, there were far more students in the 1850s taking law or theology than those taking chemistry.[72] However, these subjects did not have the economic implications of chemistry. The German system—unable to absorb all its chemistry graduates in academia or school teaching—poured its chemists into the chemical industry.[73] As the German synthetic dye industry expanded in the 1870s and especially the 1880s, these chemists were very useful and doubtlessly made the German industry world-beating. If the British industry was to employ more trained chemists, where were these chemists to come from? We should not lose sight of the fact that there were several places in Britain that offered chemical courses: by the 1860s even Oxford and Cambridge were teaching chemistry. Yet the chemical industry, on those relatively rare occasions when it sought academically trained chemists, found them hard to recruit. Reputedly this was the reason that William Henry Perkin left the dye industry in 1873.[74] As British chemistry graduates were reluctant to work in industry, firms had to employ German or Swiss chemists; for example, Ferdinand Hurter, who joined Gaskell, Deacon & Company of Widnes in 1867. After the reunification of Germany in 1871, few German chemists were willing to work in Britain. Yet teaching more British chemistry students would in the short run create a demand for more chemistry teachers. A massive expansion of chemistry teaching was effectively

[68] S. B. Saul, *The Myth of the Great Depression, 1873–1896* (London: Macmillan, 1969); Sidney Pollard, *Britain's Prime and Britain's Decline: The British Economy, 1870–1914* (London: Edward Arnold, 1989). The term "Great Depression" traditionally reserved for the long depression between the 1870s and 1890s is increasingly used, especially by American scholars, for the Depression of the 1930s.

[69] For a detailed discussion of these matters, see Steve N. Broadberry, *The Productivity Race: British Manufacturing in International Perspective* (Cambridge: Cambridge University Press, 1997).

[70] A good standard treatment of this topic is Gordon W. Roderick and Michael D. Stephens, *Education and Industry in the Nineteenth Century* (London: Longman, 1978), 122–129.

[71] Broadberry, *The Productivity Race.*

[72] Heidelberg University Calendars, https://www.ub.uni-heidelberg.de/helios/digi/unihdadressbuch. html.

[73] This was pointed out by John J. Beer in his pioneering work, *The Emergence of the German Dye Industry to 1925* (Urbana: University of Illinois Press, 1956).

[74] Roderick and Stephens, *Education and Industry in the Nineteenth Century*, 16, citing W. H. Perkin [Jr.], "Research Work at the Universities," *Manchester University Magazine*, December 1908. The reasons were actually more complex; the main reason for Perkin's departure from the dye industry was the recently renegotiated alizarin agreement with BASF, which made the firm much more sellable; see Anthony S. Travis, *The Rainbow Makers* (Bethlehem, PA: Lehigh University Press, 1993), 196–199.

impossible because of the lack of suitably qualified graduates. This is perhaps one reason why Roscoe told the Devonshire Commission that the German universities were overstaffed.

The triumph of English education in the late nineteenth century was the introduction of universal literacy (and presumably numeracy) after William Forster's Elementary Education Act of 1870 and Anthony Mundella's Elementary Education Act of 1880, which compelled children to attend school up to the age of 12.[75] By 1914 practically all men and women in Britain could read and write. For certain social groups the effect was astounding: literacy among miners grew from less than half in 1874–1879 to almost complete literacy by 1904–1909. The next step was to improve secondary education and to include more science in the secondary curriculum. This was a work in progress and would take longer, but the task was taken in hand by civil servants such as William de W. Abney, who oversaw the building of school laboratories.[76] The real problem was at the third level, beyond the age of 14 or 16. Technical education was both more labor-intensive and expensive. If there was actually a need for workers to be better trained, how were they going to receive this training? In keeping with the tradition of apprenticeships and "picking-up," the emphasis in Britain was on in-job training, which in practice meant evening classes. This was a route which remained popular until the 1960s.

In what way was the German educational system advocated by Roscoe (and others) better? Certainly, in terms of literacy, it was no better: Germany achieved universal literacy about the same time as Britain.[77] Its system of more practically oriented secondary schools (the Realschule and the Realgymnasium) was praised and in the 1870s had no counterpart in England. At the tertiary level, the Germans had the Technische Hochschule, which also had no exact English equivalent.[78] In contrast to Britain, a chemist or engineer was expected to learn the principles before starting a job, which meant that they were filled with theoretical knowledge, but had little practical experience.[79] However, there is a need to be cautious here. The standard Gymnasium was as classics-oriented as any English public school. And in the late nineteenth century, the Technische Hochschulen were not considered to be on the same level as the universities and could not award PhDs until about 1900. Furthermore, there was an inconsistency on the part of Roscoe (and other Germanophiles) in that they argued for the importance of teaching pure chemistry and hence the basic principles underlying industrial chemistry, rather than teaching applied chemistry as the Technische Hochschulen did in Germany. In the end, Roscoe and his colleagues at Owens College reached a compromise in the 1880s: applied chemistry became part of the chemical curriculum without becoming a separate subject.

[75] Sanderson, *Education and Economic Decline*, 3–13.
[76] See *ODNB*.
[77] Sanderson, *Education and Economic Decline*, 18.
[78] Wolfgang König, "Technical Education and Industrial Performance in Germany: A Triumph of Heterogeneity," in *Education, Technology and Industrial Performance in Europe, 1850–1939*, ed. Robert Fox and Anna Guagnini (Cambridge: Cambridge University Press, 1993), 65–87.
[79] Roderick and Stephens, *Education and Industry in the Nineteenth Century*, 132, citing the *Report on the Education and Status of Civil Engineers in the United Kingdom and in Foreign Countries* (London: Institution of Civil Engineers, 1870), xi.

Roscoe the Industrial Consultant

The consultant chemist had emerged as an occupational group by the mid-nineteenth century when rapid industrialization and urban expansion necessitated the analysis of air, water, and food to meet safety standards.[80] Many chemists took up consultancy work between academic appointments, not just as chemical analysts, but advising industrial enterprises how to make their processes work more efficiently, or recommending modifications to existing plants or even plant or process replacement, since such knowledge and expertise were not available within the industrial firms. Roscoe came to appreciate the close connection between chemistry and industry from his association with Bunsen, who had always sought to make chemistry useful to industry. Bunsen had demonstrated the importance and benefits of this link in collaboration with Lyon Playfair in a research project to improve iron smelting using his new analytical technique of gasometry (funded by the British Association for the Advancement of Science, BAAS).[81] Many British chemists, both academic chemists and independent chemical consultants, conducted research projects funded by the BAAS to address technical deficiencies across industrial sectors and not individual businesses. Chemical consultants drew on their chemical knowledge and often access to specialist instrumentation in carving out this occupational niche.

Roscoe, having returned from his studies with Bunsen and before his appointment at Owens College, contemplated a career as a chemical consultant since there were few academic appointments, especially for a young chemist.[82] At Owens College, however, Roscoe needed to persuade local industrialists of the value of a pure chemical education for their sons, in order to gain much-needed financial support and students.[83] Here, too, he was following in the footsteps of Bunsen, who welcomed the sons of several important English industrialists into his laboratory, as William H. Brock has examined in detail.[84] Furthermore, like some (but not all) academic chemists of the period, Roscoe was happy to take on a wide range of consultancy work, especially the regulation of air pollution, though such enterprise did not escape some criticism from outside Owens. But not all academic chemists supported the link with industry. Ira Remsen of Johns Hopkins University in Baltimore was totally opposed to the involvement of academic chemists with industry, despite having a German education similar to Roscoe's.[85]

[80] Viviane Quirke and Peter Reed, "Chemistry, Consultants, and Companies, c. 1850–2000: Introduction," *Ambix* 67 (2020): 207–213.

[81] Robert Bunsen and Lyon Playfair, "Report on the Gases Evolved from Iron Furnaces, with Reference to the Theory of the Smelting of Iron," *Report of the Fifteenth Meeting of the British Association for the Advancement of Science* (London: John Murray, 1846), 142–186. See also Robert Bunsen, *Gasometry*, trans. Henry Enfield Roscoe (London: Walton and Maberly, 1857).

[82] Roscoe, *Life and Experiences*, 100.

[83] Joris Mercelis, Gabriel Galvez-Behar, and Anna Guagnini, "Commercializing Science: Nineteenth- and Twentieth-Century Academic Scientists as Consultants, Patentees, and Entrepreneurs," *History and Technology* 33 (2017): 4–22, on 8.

[84] W. H. Brock, "Bunsen's British Students," *Ambix* 60 (2013): 203–233.

[85] William Albert Noyes and James Flack Norris, "Biographical Memoir of Ira Remsen (1846–1927)," *National Academy of Sciences of the United States Biographical Memoirs* 14 (1931): 207–257; see also Owen Hathaway, "The German Model of Chemical Education in America: Ira Remsen at Johns Hopkins (1876–1913)," *Ambix* 23 (1976): 145–164.

Roscoe was the leading founder of the Society of Chemical Industry (SCI) in 1881, which was specifically formed to strengthen the links between academia and industry, though as Wilfred Farrar points out, this was mainly an initiative of the chemical industry in response to the German example where there was "the rapid application of academic discoveries to large-scale production."[86] Through the SCI, the interlocution between academia and industry was aided by regular meetings across the country and by the publication of a journal that informed members of recent advances in chemistry likely to improve the performance of chemical industry.[87]

Toward the end of his working life, Roscoe's role with industry changed dramatically when he became a director of the Aluminium Company and a founder director of the associated Castner–Kellner Alkali Company. Roscoe was closely associated with developing this new electrolytic technology for manufacturing alkalis, replacing the heavily polluting Leblanc process.[88] That the Castner–Kellner process operates today largely unchanged is indicative of the scale of its innovation.

Roscoe the Progressive MP

Up to 1885, Roscoe had operated outside of Parliament and was in no sense a professional politician. Yet at the end of 1885, Roscoe was elected as MP for the relatively prosperous South Manchester constituency and served in Parliament for almost 10 years. This career change was partly a reaction to the recent death of his only son Edmund and his subsequent desire to leave Manchester, with all its memories of a happier time.[89] We have noted that his grandfather William Roscoe had been an MP for Liverpool. His father-in-law Edmund Potter had been the Liberal MP for Carlisle from 1861 to 1874. He was an MP in a turbulent period when the dominant issue was Irish Home Rule, a political question which divided the Liberal Party, but Roscoe remained devoted to Gladstone despite the Grand Old Man's lack of interest in or understanding of modern science.[90] Hence Roscoe provides an excellent case study for the relationship between academic science and parliamentary politics. He was a scientist who was an MP, rather than a politician with a scientific background. The major issues that arose while Roscoe was in Parliament, for example, Irish Home Rule or the rise of socialism, were not topics that he was particularly concerned about or to which he made a major contribution.[91] David Bebbington has characterized the period between 1870 and 1910 as the high water of nonconformist politics, invariably from the Liberal side of the Commons (they were barely represented in the Lords).[92]

---

[86] W. V. Farrar, "The Society for the Promotion of Scientific Industry, 1872–1876," *Annals of Science* 29 (1972): 81–86.

[87] "The Society of Chemical Industry," *Journal of the Society of Chemical Industry*, Jubilee volume (July 1931): 9–14. See also "Proceedings of the First General Meeting," *Journal of the Society of Chemical Industry* 1 (1881): 3–4.

[88] "Castner–Kellner Alkali Company Ltd," *Chemical Trade Journal*, 3 June 1899, 367.

[89] Roscoe, *Life and Experiences,* 271.

[90] Ibid., 282–283.

[91] G. R. Searle, *The Liberal Party: Triumph and Disintegration, 1886–1929* (Basingstoke: Macmillan, 1992); Alan O'Day, *Irish Home Rule, 1867–1921* (Manchester: Manchester University Press, 1998).

[92] David W. Bebbington, *The Nonconformist Conscience: Chapel and Politics, 1870–1914* (London: George Allen and Unwin, 1982).

While Roscoe as a Unitarian supported all the causes dear to the nonconformist heart, from overcrowded housing to slavery, he never played a major role in nonconformist issues. Indeed, his campaign for museums to be open on Sundays was opposed by Sabbatarian nonconformists.[93]

What can scientists bring to the political table? Can their scientific training give them a broad perspective on politics, or are they best employed by applying themselves to specific science-related issues? Roscoe did not make a major impression on the House of Commons and tended to favor speeches (and parliamentary questions) on scientific matters such as medical regulation, ventilation of the Houses of Parliament, the control of cholera in Mecca, and the introduction of the metric system.[94] He was a Vice-President of the Decimal Association and with Lord Kelvin supported a bill for compulsory metrication in 1904.[95] Roscoe associated with a like-minded circle of Liberal MPs mostly from a business or academic background with a strong radical streak, especially in the field of educational reform. One of the few political historians to study Roscoe's political career, James Moore, positions him as part of a local movement to promote radical Liberalism in Manchester in 1880s.[96] According to Moore, he was part of "an enlightened, meritocratic and moral force opposed to a self-interested and unenlightened property-owning elite."[97] Can scientists become accustomed to the grind of politics, or do they soon tire of its excessive demands? Roscoe gradually found the demands of the parliamentary life irksome, especially when Liberal MPs had to stay around in case there was an opportunity to ambush the Conservative and Liberal Unionist government. In all honesty, he was not completely unhappy to lose his seat to a Liberal Unionist in 1895.[98]

Roscoe the Institution Builder

After he lost his seat in 1895, rather than returning full-time to academia, Roscoe became a builder of new or reformed institutions, including the University of London, the Lister Institute, and the Science Museum. As Robert Bud has pointed out, the end of the nineteenth century and the first decade of the twentieth century were a period of energetic reform and the development of new or reinvigorated institutions.[99] This period encompassed the setting up of the Laboratory of the Government Chemist in 1894, headed by Roscoe's former colleague Thomas Edward Thorpe, and the National

---

[93] Roscoe, Life and Experiences, 296.

[94] As shown by Hansard online, http://hansard.millbanksystems.com/.

[95] Roscoe, Life and Experiences, 286–288. This bill fell through, as did another one in 1910. With his strong views on the subject, Roscoe would doubtlessly be horrified to discover that metrication is still not fully implemented in the United Kingdom; see Norman Biggs, "A Tale Untangled: Measuring the Fineness of Yarn," Textile History 35 (2004): 120–129, for Roscoe, see 120.

[96] James R. Moore, The Transformation of Urban Liberalism: Party Politics and Urban Governance in Late Nineteenth-Century England (Aldershot, Hants: Ashgate, 2006).

[97] Ibid., 186.

[98] Roscoe, Life and Experiences, 310–311; Moore, The Transformation of Urban Liberalism, 176–177.

[99] Robert Bud, "'Infected by the Bacillus of Science': The Explosion of South Kensington," in Science for the Nation: Perspectives on the History of the Science Museum, ed. Peter J.T. Morris (Basingstoke, Hants: Palgrave Macmillan, 2010), 11–40. Also see G. R. Searle, The Quest for National Efficiency: A Study in British Politics and Political Thought, 1899–1914 (Berkeley: University of California Press, 1971).

Physical Laboratory in 1900.[100] After the unexpected death of the previous holder of the post, Roscoe served as Vice-Chancellor of the University of London at the end of the nineteenth century and introduced reforms recommended by two commissions of inquiries.[101] By converting the University of London from being an examining body back to its original concept as a federation of London-based colleges, he thus had an important impact on London's university as well as Manchester's. Roscoe supported the formation of the Institute for Preventive Medicine (which became the Lister Institute) in 1891 and he assisted its incorporation by parliamentary act in the Commons.[102] He then became involved with its management, which eventually became a heavy burden in the years leading up to his death. Finally, in one of his last public actions, Roscoe aided his old friend Norman Lockyer and the pioneering civil servant Robert Morant with the detaching of the Science Museum in South Kensington from the Victoria & Albert Museum to become a free-standing institution in 1909.[103]

Roscoe as a Victorian Campaigner

Roscoe lived in a period of unprecedented change in Britain. He came from a middle-class background, and the Victorian era was one in which the balance of power shifted from the aristocracy to the middle class. Britain became industrialized—the urban population of England and Wales exceeded the rural population by the Census of 1891—and British cities witnessed explosive growth. Roscoe lived in the major cities of London, Liverpool, and Manchester. Science became a professional activity rather than an amateur pursuit. People began to take degrees in scientific subjects, and Roscoe was a professional scientist with academic qualifications. William Whewell famously introduced the term "scientist" the year Roscoe was born. The chemical industry also grew enormously and diversified into new fields such as dyes and pharmaceuticals, not always with happy results for Britain. With his close links to industry, Roscoe was part of that process, and he was aware of the issues it raised, from river pollution to the threat from German industry.

It might seem self-evident that a country undergoing such massive change had to change the way it did things, to move from a system suitable for a largely land-based society to one which was appropriate for an urbanized and industrialized society. Yet Britain (and England in particular) was a deeply conservative country and these

---

[100] For the Laboratory of the Government Chemist, see Peter W. Hammond and Harold Egan, *Weighed in the Balance: A History of the Laboratory of the Government Chemist* (London: HMSO, 1992); for the origins of the National Physical Laboratory, see Lee T. Macdonald, *Kew Observatory and the Evolution of Victorian Science, 1840–1910* (Pittsburgh: University of Pittsburgh Press, 2018).

[101] F.M.G. Wilson, *The University of London, 1858–1900: The Politics of Senate and Convocation* (Woodbridge: Boydell Press, 2004), Part VI, 221–461; Negley Harte, *The University of London 1836–1986: An Illustrated History* (London: Althone Press, 1986), Chapter 4, "Examining versus Teaching, 1870–1900," 119–160.

[102] Roscoe, *Life and Experiences*, 329–333; Thorpe, *Roscoe*, 171–173 and 202; Harriette Chick, Margaret Hume, and Marjorie Macfarlane, *War on Disease: A History of the Lister Institute* (London: André Deutsch, 1971).

[103] Thorpe, *Roscoe*, 170–171; Robert Bud, "Infected by the Bacillus of Science," 28–35.

changes came slowly. The main concern of the Conservatives was to maintain the institutions of the old system, such as the established churches, the monarchy, and the House of Lords. Some of the most controversial changes opposed by the Conservatives involved the established church, which pitted Anglicans against nonconformists. On the other hand, the Conservatives and their Liberal Unionist allies did as much to reform education as the Liberals. While the Conservatives were sometimes willing to extend the electoral franchise for party advantage, they were happy to use the unelected and hereditary House of Lords to block reforming measures which had passed the House of Commons, above all Irish Home Rule. Matters came to a head in 1909 when the Lords blocked Lloyd George's so-called people's budget, with its aim to place more taxes on the wealthy to pay for new social welfare programs (such as a state pension for older people). The House of Lords rejected the Parliament Bill which would remove their veto on parliamentary legislation. Roscoe's name was near the top of the list of the 250 peers that Prime Minister Herbert Asquith would have demanded from the new monarch George V if the House of Lords had continued to reject the bill.[104] In the end, the House of Lords gave way and Roscoe never became a peer (and he probably did not want to become one, as he had a low opinion of the upper chamber and he was now suffering from poor health). Had Roscoe been ennobled in this way, it would have been a hereditary peerage as there were no life peerages until 1958, but as his only son Edmund had died in 1885, there would have been no heir.

What united the rival groups within the "science reform movement" in which Roscoe was a major player was the concept that this was as much a cultural struggle as a campaign for more funding. The leading sociocultural groups in England in the mid-nineteenth century were the landed gentry and military officers, who usually sprang from this group; the established clergy, also associated with the landed gentry for the most part; the traditional professions such as law and medicine, and the financiers who worked in the City of London. The industrialists of northern England, the nonconformist clergy, and the new professions such as pharmacy and chemistry were outside this elite. It was not entirely a matter of wealth. Individual landowners and many of the established clergy were impoverished as a result of the prolonged depression in British agriculture. Nonetheless, this elite had an established position in British society which scientists hitherto lacked. The science reformers were determined to break into this charmed circle and at least partly succeeded. For example, in May 1904, several leading scientists, including Roscoe, attended a grand dinner at the Athenaeum Club graced by the presence of Prime Minister Arthur Balfour (the then President of the British Association for the Advancement of Science) and the Prince of Wales (later King George V).[105] This was a struggle for prestige, to put eminent scientists on the same social level as peers, generals, and bishops, and to have their views taken at least as seriously. It was also important for the science reformers that

---

[104] Sir John Brunner was first on the list produced by Asquith and the chief whip Alick Murray, and Roscoe was thirteenth, just behind Arthur Acland and Sir William Mather. The list is reproduced in Andrew Cook, *Cash for Honours: The Story of Maundy Gregory* (Stroud, Glos: History Press, 2008), Appendix 3. It is not clear that this list was a definitive one, and of course it was never put into effect. It is intriguing to wonder what title Roscoe would have taken, as it was then usual to use territorial titles rather than surnames.

[105] Michael Wheeler, *The Athenaeum: More than Just Another London Club* (New Haven: Yale University Press, 2020), 182.

the cultural importance of science should be acknowledged in various ways, for example by setting up a national science museum.

While Roscoe probably considered himself to be a "man of the people," he was not in any sense a populist. In fact, surprisingly often he was in conflict with public opinion. Roscoe was a member of a denomination which denied one of the central tenets of Anglicanism and most nonconformist churches. He supported temperance and tighter control over pubs, which earned him the enmity of publicans (despite having a grandfather who was an innkeeper) and Oscar Wilde's "drinking classes." He campaigned for Sunday opening of museums and libraries, which brought him into conflict with the Sabbatarian wing of the nonconformists. He opposed the apprenticeship system, which was popular with most industrial employers and trade unions. He wanted to introduce the compulsory use of the metric system and decimalization, which was anathema to many people; Lord Kelvin's bill of 1904 led to the setting up of the British Weights and Measures Association. Roscoe supported Irish Home Rule, which was opposed by many of his own constituents. In displaying his fondness for Germany and his approval of German education, Roscoe was rowing against a tide of anti-German feeling in the public at large in the 1880s.[106] Roy MacLeod has urged that there was growing opposition to science itself in the 1880s and 1890s.[107] This was partly a result of the growing strength of anti-vivisectionism in this period, which nearly derailed the foundation of the Lister Institute, and Roscoe helped to hold the line. Clearly, in Roscoe's view, being right was more important than reflecting public opinion, which he presumably considered to be ill-informed or misled.

Roscoe's Legacy

What is the relevance of Roscoe to us today? What traces have his activities left on modern British society? At first glance, one might think he had little impact. Although metrication did eventually come about, we still measure our height in feet and inches, drink pints, and drive at 30 mph. The House of Lords has not only survived, but in 2023 still contains hereditary peers and Church of England bishops. Public houses have survived as community meeting places, and apprenticeships are actively promoted by the government as an alternative to higher education. It is in the field of education that we can find his legacy. The Science Museum continues to flourish and now has sister museums in other cities including Manchester. Museums and the British Library are open on Sundays, but relatively few council-run libraries are open on Sunday, and where they are, it is only for a few hours. The number of technical colleges grew rapidly after World War I, but university student numbers remained low. The first major expansion of tertiary education began after the publication of the Robbins Report in 1963, which expanded the number of universities and converted technical colleges into polytechnics. The polytechnics were not on the same

---

[106] James Hawes, *Englanders and Huns: How Five Decades of Enmity Led to the First World War* (London: Simon & Schuster, 2014), Part Five "(1880s): Bismarck vs. the British."

[107] Roy M. Macleod, "The Support of Victorian Science: The Endowment of Research Movement in Great Britain, 1868–1900," *Minerva* 9 (1971): 197–230, on 219–226.

level as the Technische Hochschulen in Germany, but a few university-level institutes of science and technology were created, most notably UMIST in Manchester, the ultimate successor of the Manchester Mechanics Institute. The Robbins Report can be considered the culmination of the series of reports on scientific and technical education which began with the Samuelson committee in 1868. Robbins himself was an economist; there were only one scientist (the chemist Patrick Linstead) and one industrialist (Reginald Southall, managing director of BP's Llandarcy refinery) on his committee. However, university student numbers were still relatively low compared with other Western countries, notably the United States, until the early twenty-first century. While Roscoe would have been gratified by this expansion since his death, he would have probably been disappointed that this expansion was not focused on science and technology degrees, but was across the board. While policy changes to university education have addressed student numbers, the debate over the purpose of university education—whether it is to provide general lifelong skills adaptable to most occupations or a qualification for specific careers and professions—largely remains unresolved.

2

Of Lancastrian Stock and the Formative Years

Harry Roscoe (referred to as Harry for this chapter) gave an address at the Liverpool Royal Institution on 21 December 1888 and then awarded prizes. He was Liberal MP for South Manchester, having resigned his post of Professor of Chemistry at Owens College (Victoria University), Manchester, in 1886. In his address, "An Educational Parallel," he reflected on the issues raised by his paternal grandfather, William Roscoe, in a speech at the opening of the Institution on 25 November 1817.[1] In his address, *On the Origin and Vicissitudes of Literature, Science and Art and Their Influence on the Present State of Society*, William Roscoe had sought to lay out the rationale and aims of the Liverpool Royal Institution and the important role that education should play in the enlightenment of everyone in society.[2]

Harry's address lauded his grandfather's forward-looking stance on the expansion of education, abolition of slavery, and parliamentary reforms, before speculating on whether his grandfather would support his own educational reforms. These included: freeing education from religious interference; creating a minister of education of cabinet rank; expanding teacher training; including science in the school curriculum (as the Royal Liverpool Institution had done when he was younger); expanding higher education (as with the foundation of Owens College and Victoria University in Manchester); furthering evening schools; and establishing a national system for technical education. Although William would not have differentiated between the arts, science, and literature, his grandson felt confident he would have supported these educational initiatives with as much energy and commitment as he had his own reforms. It was as if the baton for educational reforms had been handed on to Harry from his grandfather.

William Roscoe was born on 8 March 1753 at the Old Bowling Green Inn, Mount Pleasant, in Liverpool to William Roscoe (William Sr.), an innkeeper, and his wife Elizabeth, daughter of William Stevenson of Allerton, near Liverpool. William Sr. had been a butler at Allerton Hall before working at the Old Bowling Green House. A move to the nearby New Bowling Green Inn at Mount Pleasant (that William Sr. had built), with its larger market garden, allowed William Sr. to more actively pursue his interest in gardening, agriculture, and botany.[3]

The Roscoe family at the time of William's birth were English Presbyterians, a nonconformist denomination which gradually became Unitarian in its theology in the latter part of the eighteenth century, in contrast to the Church of Scotland which

---

[1] Henry Enfield Roscoe, "An Educational Parallel," *Journal of Education*, new series, 11 (1 February 1889): 118–120.

[2] William Roscoe, *On the Origin and Vicissitudes of Literature, Science and Art and their Influence on the Present State of Society* (Liverpool, 1817).

[3] Arline Wilson, *William Roscoe: Commerce and Culture* (Liverpool: Liverpool University Press, 2008), 31; see also *ODNB*.

Henry Enfield Roscoe. Peter J.T. Morris and Peter Reed, Oxford University Press. © Oxford University Press 2024.
DOI: 10.1093/oso/9780190844257.003.0002

remained Trinitarian. William Roscoe was still considered to be a Presbyterian as late as 1806.[4]

The conversion of the English Presbyterians to Unitarianism was a result of the influence of two leading Presbyterian theologians, Richard Price and Joseph Priestley nationally, and at Benn's Garden Chapel in Liverpool (where William Jr. had been baptized in 1753) through the preaching of its minister William Enfield between 1761 and 1770.[5] There were three strands: skepticism; materialism, which was Priestley's strand; and not unimportantly, a biblical strand which argued that the Trinity is not mentioned in the New Testament.[6] The Presbyterians could only become Unitarians as a denomination after the passage of the Doctrine of the Trinity Act in 1813, which made it legal for the first time to deny the Trinity.

The influence of Joseph Priestley, the natural philosopher and dissenting minister, on the Presbyterians coincided with his teaching at dissenting academies across England, in particular the Warrington Academy (1761–1767) that had "opened in 1757 as a scientific and literary centre for non-conformists who were legally debarred from the older universities."[7] At Warrington, Priestley had taught a wide range of subjects, including languages, natural philosophy, and history (social, cultural, and economic). In 1770 Enfield moved to the Warrington Academy to become tutor in *belles lettres* and *rector academiae*.[8] With meetings at the Academy being open and not restricted to registered students, William Roscoe became a regular visitor, and through Enfield's teachings Unitarianism greatly influenced his intellectual and social development.[9]

William Roscoe and later generations of the Roscoe family, and their extended family, would draw support for their endeavors from among the local Unitarian community wherever they were living. In the major towns such communities usually included merchants, industrialists, bankers, and other influential members. William utilized his network not only in his daily life, but especially when founding the cultural and educational institutions with which his name is so closely associated in Liverpool. His closest associate in Liverpool was probably Dr. James Currie, a friend of Priestley and physician to the Liverpool Infirmary, who supported abolition of the slave trade and was a regular reviewer of William's writings.[10] Other associates included Joseph Brookes Yates, a merchant and antiquary, and William Rathbone IV (1757–1809), a merchant and philanthropist.[11]

---

[4] Anne Holt, *Walking Together: A Study in Liverpool Nonconformity, 1688–1938* (London: George Allen & Unwin, 1938), 158.

[5] Frederick Dunston, *Roscoeana* (privately printed, 1910), 1 and 5; and William Roscoe Baptism, https://search.ancestrylibrary.com/cgi-bin/sse.dll?indiv=1&dbid=9841&h=178150740&tid=&pid=&usePUB=true&_phsrc=DOY178&_phstart=successSource.

[6] Robert E. Schofield, *The Enlightenment of Joseph Priestley: A Study of His Life and Work from 1733 to 1773* (University Park: Pennsylvania State University Press, 1997), 167.

[7] Sheila Marriner, *The Economic and Social Development of Merseyside* (London: Croom Helm, 1982), 87. See also Schofield, *Enlightenment of Joseph Priestley*, 87–120; F. W. Gibbs, *Joseph Priestley: Adventurer in Science and Champion of Truth* (London: Nelson and Sons, 1965), 16–25.

[8] For William Enfield, see *ODNB*.

[9] Gibbs, *Joseph Priestley*, 17–19.

[10] Wilson, *William Roscoe*, 39–40; For James Currie, see *ODNB*. There are several letters between Roscoe and Currie in the William Roscoe Collection held at Liverpool Central Library.

[11] Wilson, *William Roscoe*, 41.

Liverpool in the 1750s was a rapidly expanding commercial and mercantile town, and the emerging port of Liverpool was trading with many different parts of the world. Between 1700 and 1750 the population grew from 5,000 to 20,000 as more people were drawn to seek employment there.[12] Inland trade expanded with respect to coal from St. Helens and salt from Cheshire (both commodities were exported through the port of Liverpool) and this was accompanied by improvements in river navigation and later the building of canals. Sankey Brook was important in linking St. Helens to the River Mersey (and then to Liverpool), and improvements between 1757 and 1761 allowed for the easier passage of boats carrying coal. Further improvement was made in 1782 with an extension to Fiddler's Ferry, giving access to the Mersey as tidal conditions allowed; a final extension was completed to Widnes in 1830 that provided a double-lock entrance to the Mersey in all tidal conditions. The extension to Widnes would provide the setting for the development of Widnes as an iconic chemical town of the nineteenth century, as we shall discuss later. The Weaver Navigation in Cheshire was an important part of the trading arteries, and the Act of 1721 again focused on making boat navigation easier by straightening some stretches for the movement of salt and coal (used in salt refining).[13] Later, in 1767, the Bridgewater Canal was completed for transporting coal from the Duke of Bridgewater's mines in Worsley to Manchester and then to Runcorn and the Mersey. The canal network was extended further in 1770 when the Leeds-Liverpool Canal was approved, though it was not completed until 1816. Gradually the transportation of goods switched from the unsatisfactory turnpikes to the expanding network of canals and river navigations, allowing larger tonnages to be moved in a more efficient manner.

Increasing inland trade was accompanied by an expansion in the dock facilities in Liverpool to cater to the steady growth of the import and export trade. The "Old Dock," or "Pool," as it was affectionately called, was built in 1715 as an enclosed dock able to accommodate up to 100 ships and was the first commercial dock in Britain; expansion of the salt trade was accompanied by the Salthouse Dock in 1759.[14] As Liverpool vied with London, Bristol, and Glasgow for the position of Britain's leading port and as trade with different parts of the world expanded, further docks were built, including St. George's Dock (1771), King's Dock (1788), and Queens Dock (1796).[15] Much of the trade in cotton, sugar, rum, and tobacco was associated with the African slave trade and brought considerable wealth to the town, as shown by the many buildings financed from the proceeds of the slave trade. The growth in import and export commodity trading was accompanied by expansion of banking and insurance provision, as well as businesses associated with maritime services. Banks were important to the Liverpool economy, but they proved vulnerable to fluctuations in trade and during periods of war, because banking services were provided by traders in specific commodities. What made many banks in Liverpool even more exposed (as in the rest of England) was that they did not issue bank notes, but relied on bills of exchange that

[12] Marriner, *Merseyside*, 14.
[13] Marriner, *Merseyside*, 18.
[14] Francis E. Hyde, *Liverpool and the Mersey: An Economic History of a Port 1700–1970* (Newton Abbot, Devon: David and Charles, 1971), 13–14.
[15] Marriner, *Merseyside*, 30–31.

promised to pay a specific amount on receipt or on a stated date. Serious decline in a sector of commodities regularly left banks holding bills of exchange in the affected commodities unable to meet their financial commitments. This in part was responsible for many of the collapses and failures, and why partnerships changed on a regular basis. As an indicator of the importance of Liverpool as a commercial and trading center, the Bank of England opened a branch there in 1827 that helped to stabilize banking operations by circulating Bank of England notes.[16] The Bank Charter Act of 1844 brought some stability by giving note-issuing powers exclusively to the Bank of England, allowing existing banks to continue issuing their own notes but preventing new banks from starting to issue their own notes.

As the town and its position in Britain and the world grew, there was an increasing drive by merchants, bankers, shipsowners, and industrialists to establish cultural and educational institutions so that all levels of society could better understand the changing world and adapt to it, as a counterbalance to just making money and increasing wealth. The founding of the Liverpool Royal Institution in 1817 was one outcome of this aspiration, but there were to be many others, with William Roscoe again playing a leading role alongside likeminded supporters.

Although William left school when he was 12 years old, he was an avid reader, probably due to the influence of his mother, who loved books and poetry, which brought a keen interest in the world around him. After leaving school, William helped his father in the market garden, instilling a love of nature as well as an interest in horticulture, agriculture, and botany that he pursued at different stages of his later life. At age 15, William was contracted (often referred to as "articled") as a trainee clerk to a solicitor in Liverpool, but with the solicitor's death from drinking, he transferred his articles to another of the town's solicitors. It was during this period that he had the good fortune to meet Francis Holden, a gifted young schoolmaster who gave lessons in Latin and French (and later Italian) to a small group of young people. The group included William Clarke, who was to subsequently play a major part in William's developing interest in Renaissance Italy that would lead later to publication of two outstanding books. Being part of this group, William was motivated to extend his studying and reading to many hours each day. The historian Stella Fletcher has referred to this group as "a Renaissance-style academy" and it was to have a defining role in William's life.[17]

In 1775 William Roscoe was admitted an attorney of the King's Bench, a legal jurisdiction of the legal system, and continued in this role for the next 20 years while also continuing his literary studies. Four years later, on 22 February 1781, William married Jane, daughter of William Griffies, a draper in Castle Street, Liverpool, at the Anglican Church of St. Anne (Richmond), Liverpool. In 1797, unexpectedly (and for reasons that are unclear) he decided to leave Liverpool for London and pursue a career as a barrister in Gray's Inn. However, this venture proved short-lived, for after two years he had tired of London and returned to Liverpool.

---

[16] Marriner, *Merseyside*, 42.
[17] Stella Fletcher, ed., *Roscoe and Italy: The Reception of Italian Renaissance History and Culture in the Eighteenth and Nineteenth Centuries* (Farnham, Surrey: Ashgate, 2012), 4.

Figure 2.1 William Roscoe, Harry Roscoe's grandfather.

In 1773 William and several friends founded a Society for the Encouragement of the Arts, Painting and Design, and held in Liverpool the first public exhibition of paintings held outside London.[18] It was around this time that he became acquainted with the artist Henry Fuseli, who had studied in Rome for about eight years before his appointment as Professor of Painting at the Royal Academy in London. It is out of their friendship and Fuseli's knowledge of Italy and Italian art that William decided to write a biography of Lorenzo de' Medici, the Florentine statesman, which would take 10 years of his leisure time to complete. He was helped by his friend and fellow study group member William Clarke, who having suffered ill health spent the winter of 1789 in Italy. While there Clarke acquired a fine collection of study materials for his friend, and without this assistance it is very likely William would not have completed the book. When *Life of Lorenzo de' Medici* was published in 1796 it brought William international fame as a historian and scholar, and its success is gauged by the seven editions published in his lifetime and the six after his death.[19] In 1805 William published *The Life of Pope Leo X, Son of Lorenzo de' Medici* and, although it was not as widely acclaimed as the earlier book, it still ran to six editions and was translated into French, German and Italian.[20]

[18] See the entry on William Roscoe in *ODNB*.
[19] William Roscoe, *Life of Lorenzo de' Medici* (London: A. Strahan, 1796). See also his *ODNB* entry.
[20] William Roscoe, *The Life of Pope Leo X, Son of Lorenzo de' Medici* (Liverpool: Cadell and Davies, 1805).

Soon after the publication of *Life of Lorenzo de' Medici* in 1796, William retired from his legal practice at the age of just 43. He was wealthy and was readily able to expand his outstanding collection of rare and valuable manuscripts, books, and works of art of the Italian Renaissance period. Just three years later, William purchased Allerton Hall, a fine three-story red sandstone Palladian-styled building where his father had been butler before becoming a Liverpool innkeeper.[21] A new wing was added to accommodate his extensive library and collection of paintings.

Living at Allerton Hall gave William more time for his literary and cultural activities. While working on *The Life of Pope Leo X, Son of Lorenzo de' Medici*, over a period of about 17 years, William also had a major role in establishing several of Liverpool's outstanding cultural institutions, with which his name is still closely associated. These include the Liverpool Athenaeum (1797), Liverpool Botanic Garden (1802), Liverpool Literary and Philosophical Society (1812), and Liverpool Royal Institution (1814). With Liverpool growing as a major trading and commercial center, the Liverpool Athenaeum was established so merchants and business people in general could get the latest news with access to newspapers, journals, and an extensive library.[22] William was a founder "proprietor," as members were called, because they held shares: one was Sir John Gladstone (father of the future British prime minister, William Ewart Gladstone). The library (which opened some 14 months after the official opening) became an important part of the Athenaeum, and later acquired books from William's personal library. It was judged "one of the world's most highly regarded private literary collections."[23]

William's retirement in 1796 also gave him more time to pursue his botanical interests, and soon after acquiring Allerton Hall, he purchased an extensive plot of marshy land on Chat Moss and Trafford Moss (between Liverpool and Manchester) for the purpose of reclaiming the land for agricultural purposes. When an appeal was launched in 1799 to create a botanic garden in Liverpool, it was William who became a leading supporter for not only the botanic garden, but also a library and an herbarium. He agreed to coordinate the fundraising and used the same system of shares and proprietors employed so successfully at the Liverpool Athenaeum. At the official opening in 1802, William's address took a broad view of the Botanic Garden's role, emphasizing the material contribution to Liverpool's prosperity through its efforts to introduce, grow, and acclimatize exotic plants for amenity, for science, and most of all for their economic uses.[24]

The library and especially the herbarium benefited from William's active role. In 1799 he had acquired the outstanding herbarium of Johann Reinhold Forster of Halle, Germany, who had sailed with James Cook.[25] This collection provided the foundation

---

[21] Allerton Hall was originally the home of the Lathom family. It was badly damaged by fires in 1994 and 1995, but parts of Roscoe's library survive to the present day.

[22] The Liverpool Athenaeum (which survives to the present day) pre-dates the London club of the same name by 27 years.

[23] Ken Pye, *Discover Liverpool* (Liverpool: Trinity Mirror Media, 2011), 70.

[24] John Edmondson, *William Roscoe and Liverpool's First Botanic Garden* (Liverpool: National Museums Liverpool and The University of Liverpool, 2005), 6.

[25] Edmondson, *Roscoe*, 7. The herbarium was transferred to Liverpool City Museum (now part of National Museums Liverpool) in 1909.

herbarium at Liverpool's Botanic Garden and was boosted by a fine collection of 5,000 specimens from Sir James Edward Smith, a fellow Unitarian and President of the Linnean Society of London. William had been elected a Fellow of the Linnean Society in 1804 and had met Sir Joseph Banks, President of the Royal Society, who gave William several important books with which to inaugurate the Botanic Garden's library.[26]

Against the backdrop of England's very limited educational opportunities at the beginning of the nineteenth century, a number of important initiatives took root. One of the most pervasive and active was the Literary and Philosophical Society movement, which was to serve, as the educational historian Tom Kelly has suggested, "the intellectual needs of the middle-classes and the gentry, although they not infrequently gave their aid and patronage to efforts to provide education for the 'lower orders.'"[27] William again demonstrated his commitment to cultural and educational institutions as a founding member of the Liverpool Literary and Philosophical Society in 1812, and then as its President from 1817 until his death in 1831. It was not the first "Lit and Phil" (by which the societies were often known); the Manchester Lit and Phil, founded in 1781, was the first and probably the model for other Lit and Phils established across the urban centers of Britain during the nineteenth century.[28] It was the Manchester Lit and Phil that William's grandson Harry would later serve as Secretary and President, giving many talks at its meetings and publishing several articles in its publications.

Almost inevitably, with his high-profile role in the cultural affairs of Liverpool, William Roscoe was drawn increasingly into politics, and in 1806 was adopted as an Independent candidate in the parliamentary election just two days before voting took place. He was up against Whig and Tory incumbents (both of whom supported the slave trade) and "regardless of his abolitionist views, polled more than the combined votes of the other two candidates and unseated Tarleton [the Whig candidate]."[29] During his time in Parliament, William supported Lord Grenville's government (often referred to as the "Ministry of All the Talents") when politicians of different political persuasion came together during a time of military threat from France.[30] He spoke forcibly in several important debates, including those concerning Catholic emancipation, the Poor Laws Bill, and slave abolition.[31]

During his parliamentary campaign, William Roscoe had spoken out in support of abolition of the slave trade even though the trade was widely supported in Liverpool, and brought huge wealth to the many local merchants and shipowners in the town engaged in the trade, which later paid for the many outstanding buildings that Liverpool is still renowned for today. William had been an ardent supporter of the abolition of slavery for several decades, having witnessed daily the many ships engaged in the

---

[26] Edmundson, Roscoe, 7.

[27] Thomas Kelly, A History of Adult Education (Liverpool: Liverpool University Press, 1970), 105.

[28] Gordon W. Roderick and Michael D. Stephens, Scientific and Technical Education in Nineteenth-Century England (Newton Abbot, Devon: David & Charles, 1972), 134.

[29] Fletcher, Roscoe and Italy, 15. Most parliamentary constituencies at this time elected several members rather than one.

[30] A. D. Harvey, "The Ministry of All the Talents: The Whigs in Office, February 1806 to March 1807," The Historical Journal 15 (1972): 619–648, on 619.

[31] Slave Trade Abolition Bill, HC Deb 23 February 1807, Hansard, vol. 8, cc945–995 (961); and Poor Law Bill, HC Deb 24 April 1807, Hansard, vol. 9, cc538–551.

trade entering and leaving the port of Liverpool, albeit with no slaves on board. He had assisted William Wilberforce in establishing the African Institute in London and had supported him in the parliamentary campaign against slavery, while in parallel Thomas Clarkson toured every part of Britain giving speeches in support of abolition. During the debate on the Slave Trade Abolition Bill of 1807, William spoke passionately in support of the Bill and took the opportunity, in anticipation of those who would continue to support the trade because of the economic impact, to call for trade with India to be opened up for all British merchants and not restricted to the East India Company. William concluded his speech in an elated mood:

> I have long resided in the town of Liverpool; for 30 years I have never ceased to condemn this inhuman traffic; and I consider it the greatest happiness of my existence to lift up my voice on this occasion against it, with the friends of justice and humanity.[32]

The Abolition of the Slave Trade Act 1807 was approved by Parliament on 23 February, by a vote of 265 votes for and 16 votes against, and received its Royal Assent on 25 March 1807, prohibiting British citizens from engaging in the slave trade. However, it was only with the Slavery Abolition Act 1833 that slavery became illegal in most of the British Empire. On the same day as the Royal Assent in 1807, Grenville's ministry fell, following a dispute with the king over Catholic Relief that would have allowed members of the Catholic Church to sit in Parliament.[33] Following his support for the abolition of slavery and Catholic Relief, William was met by many violent protests on his return to Liverpool and he decided not to seek re-election to Parliament.

In 1799 and before his short parliamentary career, William had been asked by the family of his close friend William Clarke to investigate the state of the family's bank (which was connected to the coal trade) after getting into financial difficulties due to fluctuations in demand for coal.[34] Feeling an obligation to help after his friend's assistance with *Life of Lorenzo de' Medici*, William was able in due course to stabilize the bank's position and agreed to become a partner. Unfortunately, misfortune destabilized the bank again in 1802, and although additional partners helped shore it up temporarily, in 1816 the bank collapsed as a result of the depression following the Napoleonic wars. William faced bankruptcy "with remarkable equanimity and courage," but Allerton Hall was sold, as was his fine collection of books, manuscripts, and works of art.[35] However, a concerted effort by William and his friends ensured that a major part of the collection of paintings was retained in Liverpool, and these were presented to the Liverpool Royal Institution. In 1893 this collection was lent to the Walker Art Gallery (part of Liverpool Corporation) and then in 1948 donated as a gift.[36] Although William was still able to view his paintings in Liverpool, he was

---

[32] Slave Trade Abolition Bill, *Hansard*, HC Deb 23 February 1807, vol. 8, cc945–95 (961).

[33] For William Grenville, see *ODNB*. The Catholic Relief Act was finally approved by Parliament and given royal assent in 1829.

[34] Marriner, *Merseyside*, 42.

[35] Roscoe took 14 days to prepare a catalog of his books and manuscripts for the sale in 1816.

[36] Many of these paintings now form an important part of the displays at the Walker Art Gallery (currently part of National Museums Liverpool). For a detailed description of the collection, see Xanthe Brooke, "Roscoe's Italian Paintings in the Walker Art Gallery, Liverpool," in *Roscoe and Italy*, ed. Fletcher, 65–93.

greatly upset by the loss of his fine collection of books—including a first folio of Shakespeare (1623), Chaucer (1561), Donne (1639), and Dryden (1687)—which had been at the center of his intellectual endeavors over many years; as he expressed in a poem, "On parting with his books":

> As one who, destined from his Friends to part,
> Regrets his loss, yet hopes again erewhile
> To share their converse and enjoy their smile,
> And tempers, as he may, Affliction's dart, —
> Thus, loved associates! Chiefs of older art!
> Teachers of wisdom! Who could once beguile
> My tedious hours, and lighten every toil,
> I now resign you; nor with fainting heart —
> For, pass a few short years, or days, or hours,
> And happier seasons may their dawn unfold,
> And all your sacred fellowship restore;
> When, freed from earth, unlimited its powers
> Mind shall with Mind direct communion hold,
> And kindred spirits meet to part no more.[37]

From an early age, William's life was bound up with different networks; most concerned cultural, educational, or political progress, but Unitarianism was very influential. William was a member of the Benn's Garden Chapel in Liverpool and came under the influence of Rev. Dr. William Enfield. Enfield was born in Suffolk in 1741 and had studied at the Daventry Academy before moving to Liverpool in 1761.[38] In 1770 Enfield moved to the Warrington Academy, where the outstanding theologian and natural philosopher Joseph Priestley had taught a wide range of subjects, including languages, natural philosophy, and history (social, cultural, and economic), between 1761 and 1767, and William was a regular visitor at their open meetings.[39]

It was through William Enfield's wide network of associates that William was reunited with some of his paintings and books. Enfield's move to the Octagon Chapel in Norwich in 1785 brought Roscoe an introduction (through mutual friends who included the botanist James Smith) to Thomas Coke of Holkham Hall, created the first Earl of Leicester of Holkham in 1837.[40] It has been suggested that it was probably their shared interest in agriculture that had brought them together, with William reclaiming land on Chat Moss and Coke cultivating tracts of Norfolk, but they also shared a keen interest in collecting books and works of art. William's visits to Holkham were interrupted by his overseeing the extension at Allerton Hall and the cataloguing of his collections. When finally visiting Holkham in 1814, William was captivated by the

---

[37] William Roscoe, *The Poetical Works of William Roscoe* (London: Ward and Lock, 1857), 94–95.
[38] For William Enfield, see *ODNB*. See also D. Thom, "Liverpool Churches and Chapels; Their Destruction, Removal, or Alteration: With Notices of Clergymen, Ministers, and Others [Pt. 2]," *Proceedings and Papers of the Historic Society of Lancashire and Cheshire* 5 (1852–1853): 3–56, on 18–19.
[39] Gibbs, *Joseph Priestley*, 16–25.
[40] Andrea M. Gáldy, "William Roscoe and Thomas Coke of Holkham," in *Roscoe and Italy*, ed. Fletcher, 204–206.

outstanding collection of books and manuscripts that complemented his own. Many of the books and manuscripts were in serious disrepair and in need of rebinding, and William was able to convince Coke to send these volumes to his bookbinder friend, John Jones, in Liverpool as a first stage in restoring the collection. It was Coke (among his many other friends) who purchased several paintings and books from William's collection following his 1816 bankruptcy, and although Roscoe refused subsequently to accept them as a gift, he was able to admire them again during his many visits while reorganizing and cataloguing the Holkham Hall collections.[41]

Throughout his life William chose verse to express his thoughts and ideas, an interest he probably inherited from his mother. Some of his most popular poems were *Mount Pleasant*, about the area of Liverpool where he spent his early life; *The Wrongs of Africa*, in which he expressed his total opposition to slavery; *Monody on the Death of Burns*, on his admiration for the poet Robert Burns; and *The Butterfly's Ball and the Grasshopper's Feast*, in which William showed his love of nature with his personification of creatures, in anticipation of Beatrix Potter (his great, great niece). In later life William returned to his literary ventures, and in 1821 he started on a new edition of the works of Alexander Pope. This, together with a new biography of Pope, occupied 10 volumes when it was published in 1824.[42] He also never neglected his botanical work, and between 1824 and 1828 William issued, in parts, his *Monandrian Plants of the Order Scitamineae*.[43] This work proved exhausting and wearying, on top of the many troubles through his life, and after suffering a series of strokes, he died on 27 June 1831 in Liverpool. William was interred in grounds adjoining the Renshaw Street chapel (where the Benn's Garden Chapel had moved in 1811), overseen by William Shepherd, the Unitarian minister and politician who was a close friend of William.[44]

William and his wife Jane, had a large family of seven sons (including a banker, a merchant, two lawyers, a doctor, and an author), and three daughters, two of whom (including Mary Anne, who later married Thomas Jevons) were poets.[45] The youngest son Henry (Harry's father) was born on 17 April 1800 at Allerton Hall.[46] His early life was disrupted by having to move home regularly, and notwithstanding his father's bankruptcy, he was privately tutored before being articled to a firm of Liverpool solicitors in 1817. Two years later, Henry moved to London to study for the bar using his literary ability to support his studies, and in 1826 was called to the bar of the Inner Temple, thereafter practicing in the northern circuit and at the Liverpool and Chester sessions. Later he was appointed judge of the Liverpool Court of Passage that heard civil court cases rather than criminal.[47] In 1831 Henry married Maria (1798–1885), daughter of Thomas Fletcher, a Liverpool merchant and granddaughter of Dr. William

[41] Wilson, *William Roscoe*, 172.
[42] Alexander Pope and William Roscoe, *Works of Alexander Pope, with notes and illustrations by himself and others. To which are added a new life of the author ... and occasional remarks by W. Roscoe* (London: C. & J. Rivington, 1824).
[43] *Roscoea* is a plant from Nepal named after William Roscoe.
[44] For William Shepherd, see *ODNB*.
[45] See the entry on William Roscoe in *ODNB*.
[46] For Henry Roscoe, see *ODNB*.
[47] The Liverpool Court of Passage was abolished by the Courts Act 1971.

Enfield (William Roscoe's Unitarian friend). They had two children, a son Henry Enfield Roscoe (born 7 January 1833) and a daughter Harriet (born 11 January 1835).

In 1833 the government set up a select committee to review the state of municipal corporations in England, Wales, and Ireland, and finding it did not have the powers to investigate the matter fully, a royal commission was appointed.[48] Because of the scale of the work, two commissioners were appointed to gather evidence for different districts or circuits (areas through which a judge conducts court cases), and Henry Roscoe was appointed to the southwestern circuit.[49] The report was presented to Parliament in 1835, and in the same year the Municipal Corporations Act was passed, leading to a radical overhaul of local government.

Henry was an outstanding legal writer, and, among many legal treatises, published *A digest of law of evidence on the trial of actions at nisi prius* (1827) and *A digest of the law of evidence in criminal cases* (1835); with his brother Thomas he published *Westminster Hall, or, Professional relics and anecdotes of the bar, bench, and woolsack* (1825).[50] He published a two-volume biography of his father in 1833.[51] Henry suffered from consumption (tuberculosis) for a number of years and died at age 36 on 25 March 1836 at Gateacre, near Liverpool. Henry's character was summed up by the journalist Henry Chorley:[52]

> His accomplishments were many and real; his solidity of judgement was as greater as his quickness of sympathy. Like all first-class peoples I have known, his patience with those inferior to himself, patience entirely clear of painful condescension, was great and genuine. Everyone was seen to the best possible advantage when beside him. He could listen and encourage, as well as talk with a natural and flowing brilliancy I have rarely heard equalled, not three times in my life excelled.[53]

Henry Enfield Roscoe (Harry Roscoe) was born in London in 1833, but when his father died the family was living in a large house in Gateacre, then a rural village with a Unitarian chapel about six miles from the Liverpool town center.[54] Facing financial constraints, the family moved into a small cottage nearby. It was close to where Harry's maternal grandfather lived, as well as Dr. William Shepherd, the Unitarian minister in Gateacre who also ran a school there to which many Liverpool nonconformist families (including Harry's uncles) sent their sons.[55] Shepherd was a close associate of William Roscoe and a fellow scholar of Italian literature. In 1803 he published his *Life of Poggio Bracciolini* which was widely translated. While living in Gateacre the

---

[48] *The Times*, 24 June 1833, 3.

[49] *The Times*, 16 September 1833, 3.

[50] Henry Roscoe, *A Digest of Law of Evidence on the Trial of Actions at nisi prius* (London: Saunders and Benning, 1827); Henry Roscoe, *A Digest of the Law of Evidence in Criminal Cases* (London: Benning and Saunders, 1835); and Henry Roscoe and Thomas Roscoe, *Westminster Hall, or, Professional Relics and Anecdotes of the Bar, Bench, and Woolsack* (London: J. Knight and H. Lacey, 1825).

[51] Henry Roscoe, *The Life of William Roscoe*, 2 vols. (London: Thomas Cadell, 1833).

[52] For Henry Chorley, see ODNB.

[53] Henry Enfield Roscoe, *The Life and Experiences of Sir Henry Enfield Roscoe DCL, LLD, FRS, Written by Himself* (London: Macmillan, 1906), 10–11.

[54] Gateacre was absorbed into Liverpool in 1913.

[55] Wilson, *William Roscoe*, 41.

Roscoe family likely attended the Unitarian Gateacre Chapel where Dr. Shepherd was the minister, but it is not known whether Harry attended Shepherd's school.[56] He did, however, attended Miss Hunt's school at Gateacre and boarded for a period while his mother and sister were away in London. Harry always remarked that it was only as a boarder that you "understand school life."[57] Another boy attending at the same time as Harry was Edmund Muspratt (youngest son of the alkali manufacturer James Muspratt and later a chemical industrialist), and they relate an amusing story about the school's rice pudding:

> We had rice pudding at dinner before meat, and on Saturday night there was a general washing of feet in tin foot-pans. It was our firm belief that the watery rice puddings which we were made to eat were cooked in the foot-pans.[58]

The family moved into Liverpool in 1842, and between 1842 and 1848 Harry attended the High School for Boys of the Liverpool Institute. This school was established by the Liverpool Mechanics' Institute in 1825, with the original aim of preparing young boys for the Institute's evening courses (especially in science subjects); later, in 1844, the High School for Girls was established with the same broad educational aims. These two schools, however, took on a more significant role in the provision of secondary education in Liverpool.[59] Liverpool, unlike most major towns, did not have a grammar school, and so the two High Schools, together with the Liverpool Collegiate School, met the provision for secondary education. The High School was deemed a "modern" school in that the curriculum did not include the "classics grind," as Harry referred to it, but rather a broad range of subjects that included English, mathematics, French, natural philosophy (physics), and chemistry.[60] The school had a good library and Harry became an avid reader, devouring novels by Walter Scott, Frederick Marryat, and Henry Fielding.

During his time at the High School, Harry was greatly influenced by several teachers, including William Hodgson, Hugo Reid, and William Balmain. Hodgson was born in 1815 in Edinburgh, and having abandoned his law studies, lectured in literature, education, and phrenology in Fife after studying for a short time at Edinburgh University. In 1839 he was appointed secretary to the Liverpool Mechanics' Institute and taught at the High School for Boys. In his autobiography, Harry relates that it was to Hodgson that "I attribute a good deal of any literary facility which I may possess to the practice in essay-writing which he gave us and to the criticisms which he made upon our work."[61] Having left the Liverpool Mechanics' Institute High School in 1847, Hodgson pursued a wide range of interests that included support for the Anti-Corn Law League and women's education; in 1858 he was appointed an assistant commissioner for the Newcastle Commission on popular education, and in 1871

---

[56] George Eyre Evans, *Vestiges of Protestant Dissent* (Liverpool: F & E Gibbons, 1897), 92–93.

[57] Roscoe, *Life and Experiences*, 14.

[58] Ibid., 14, and E. K. Muspratt, *My Life and Work* (London: Bodley Head, 1917), 48.

[59] Michael D. Stephens and Gordon W. Roderick, "Science, the Working Classes and Mechanics' Institutes," *Annals of Science* 29 (1972): 349–360, on 356–357.

[60] Roscoe, *Life and Experiences*, 14–15.

[61] Ibid., 16–17.

was appointed Professor of Commercial and Political Economy and Mercantile Law at Edinburgh University.[62]

Hugo Reid (1809–1872) was born in Edinburgh and became a classical scholar, a teacher of science, a prolific writer of textbooks, and a headmaster. During his time at the High School in Liverpool he taught Harry natural philosophy (physics, as it is now known). Harry relates in particular how Reid taught the history of the steam engine and the properties of gases, as well as the rudiments of mechanics and dynamics:

> He taught us well, and we had to keep accurate notes of the lessons. When I was ten years old I was able to write out a fairly good account of the rise and history of the steam-engine, including the atmospheric engine of Newcomen. One of these engines I actually saw at work in a Cheshire coal-pit in my early days.[63]

Probably Harry's most influential teacher at the High School was his chemistry teacher, William Balmain (1817–1880). Balmain spent his childhood on the North Sea island of Heligoland, then British, where his father was the surgeon to the British garrison. After his father's death, he returned to London and became a student of chemist Edward Turner and then his assistant, before becoming a lecturer at the Liverpool Institute and a chemistry teacher at the High School.[64] In 1844 he published a popular chemistry book, *Lessons on Chemistry*, for the use of pupils in schools.[65] He left the High School in 1847 to pursue a career as an industrial chemist in St. Helens and it was while working there that he registered two patents. The first, BP 732 in 1855, was for "the improved methods of or processes for recovering oxide of manganese after it has been used in the manufacture of chlorine" and may be linked to the work of Walter Weldon who, following research in the laboratory of Messrs J. C. Gamble & Company in St. Helens, took out a number of patents between 1866 and 1869 that resulted in the "Weldon Still," which was widely adopted in the alkali industry.[66] Balmain's second patent, BP 4152 in 1877, was for "improvements in varnishing, painting and whitewashing" with the addition of a phosphorescent substance that made them visible in the dark, by which Balmain became the inventor of luminous paint.[67] Balmain was also the discoverer of boron nitride, a compound now prepared in a number of different crystalline forms and having wide applications as a lubricant, additive to cosmetics, and for high-temperature equipment. He later made a donation to the Chemical Laboratory Fund at Owens College.[68]

---

[62] For William Hodgson, see *ODNB*.

[63] Roscoe, *Life and Experiences*, 16.

[64] For Balmain, see *Journal of the Chemical Society, Transactions* 37 (1880): 256–257. Also see the entry on Edward Turner in *ODNB*.

[65] William H. Balmain, *Lessons on Chemistry, for the Use of Pupils in Schools, etc.* (London: Longman, Brown, Green and Longmans, 1844).

[66] The "Weldon Still" used pyrolusite (manganese dioxide) to convert hydrochloric acid into chlorine (for use in bleaching powder) and the spent manganese mud was then converted into manganese dioxide again to produce more chlorine. See Peter Reed, *Entrepreneurial Ventures in Chemistry: The Muspratts of Liverpool, 1793–1934* (Farnham, Surrey: Ashgate, 2015), 107–108.

[67] See entry for "luminous" in *Oxford English Dictionary*.

[68] Joseph Thompson, *The Owens College: Its Foundation and Growth* (Manchester: J. E. Cornich, 1886), 639.

It was through his experience working in a chemistry laboratory, a rare occurrence in schools at this time, and having Balmain as an enthusiastic teacher that Harry "picked up [his] love of the subject."[69] Like many youngsters who have carried out chemical experiments in a laboratory, Harry's most vivid memory was of hydrogen sulfide with its smell of "bad eggs," as he relates mischievously:

> on one occasion a class of thirty or forty boys was ordered into the laboratory after a lecture on the method of preparing sulphuretted hydrogen [hydrogen sulfide]. There each boy was provided with a glass containing powdered sulphide of iron and with a second one containing dilute sulphuric acid. "When I give the word of command," said Balmain, "each boy will pour the acid on to the sulphide, and you must then all run away as fast as your legs can carry you." No sooner said than done! The result was such a fearful stench that each boy will carry down the recollection of that moment to his grave, and will remember to his dying day the formula which Balmain wrote on the blackboard.[70]

Harry's chemical experiments did not end at school. His mother set aside a room at home where investigations could continue, using apparatus and chemicals purchased with his limited pocket money. He undertook the analysis of chemical mixtures prepared by cousins or friends from the chemicals in his "laboratory," and these experiences allowed Harry to hone analytical skills for which he later became a leading practitioner.[71] Like many youngsters, Harry was fascinated by fireworks, and he relates a hair-raising story when, with his cousin Roscoe Jevons, in a laboratory at the Jevons family home:

> [o]n one occasion we were drying in front of a fire a quantity of rockets, squibs and Roman candles which had been packed moist in their cases when one of them ignited and set some of the others off. Our presence of mind in throwing the burning ones out of the window saved a catastrophe, for there were several pounds of blasting powder lying uncovered on the table, and if this had ignited the roof would have been blown off and we boys probably sent into eternity.[72]

Such episodes have proved a career-determining pathway for many chemists.

The Roscoes had an extended family, as well as a wide network of friends and associates, that was to play an influential role in Harry's life. The extended family included Margaret Sandbach and the Jevons family. Margaret Sandbach was the granddaughter of William Roscoe Jr., daughter of Edward Roscoe (and Harry's cousin), who continued the family's literary tradition as a poet and novelist.[73] She became a close friend of the sculptor John Gibson, who William Roscoe Jr. had supported in his early years in Rome, and her house at Hafodunnos, in Denbighshire, had an adjoining gallery

[69] Roscoe, *Life and Experiences*, 15.
[70] Ibid., 15–16. By formula, Roscoe meant what we would now call an equation.
[71] Ibid., 18.
[72] Ibid., 19.
[73] For Margaret Sandbach (née Roscoe), see *ODNB*.

displaying many of his outstanding works.[74] The member of the Jevons family who would have the closest connection with Harry was William Stanley Jevons, the ninth of eleven children born to Thomas Jevons, a Liverpool iron merchant, and Mary Anne Jevons (née Roscoe), daughter of William Roscoe.[75] Both parents were strong Unitarians, but it was probably their family connection and mutual interest in chemistry that brought about the close bond between Harry and Stanley Jevons, with their lives and careers intersecting at different times.

As Harry recorded in his autobiography, the Roscoes (and their extended family) had many close friends among the Unitarian community in Liverpool and the surrounding area, including the Booth family, Richard Vaughan Yates, and James Martineau. The Booths were a well-established and respected Liverpool family whom the Roscoes were related to through marriage, and Harry's cousin was Charles Booth, who became President of the Royal Statistical Society (1892–1894), Fellow of the Royal Society (1899), and a Privy Councillor (1904).[76] Richard Yates was a wealthy Liverpool iron merchant (and a close friend of the social commentator Harriet Martineau), who gave active support to the Liverpool Mechanics' Institute and acted as chairman of the Liverpool organizing committee for the fundraising ventures of Charles Dickens's Guild of Literature and Art.[77] Later, Yates created Princes Park in Liverpool and donated it to the borough of Liverpool for the enjoyment of others.[78] Dr. James Martineau, Harriet's brother, was a Unitarian minister who came to Liverpool in 1832 to become minister of the Paradise Street Chapel.[79] From 1840 Martineau was also Professor of Mental and Moral Philosophy and Logic at Manchester College in Manchester, but in 1853 the College moved to London, requiring Martineau to travel regularly to London to deliver his lectures. In 1857 he finally left Liverpool for London for what would become a turbulent period of his life, following publication of his many religious books.[80]

It was within these networks that Harry experienced a happy and contented childhood, all but three years shared only with his mother and sister. The impact of these two women shaping his life following his father's early death would be revealed later when Harry actively supported the rights of women in education, employment, and the professions. His intense interest in science and especially chemistry was kindled by his experiences at school under the direction of William Balmain and his own investigations at home and with his cousins. By 1848, at age 15, Harry had decided that chemistry was the subject he wanted to study further and hopefully gain a degree in, and his application to University College London (UCL) was successful.

It is not known why Harry chose UCL and dismissed other English universities that included Oxford, Cambridge, Durham, and King's College London. One of the Scottish universities was an option, but at the time chemistry was not a strong subject. Another option was the Royal College of Chemistry in London or to study abroad,

---

[74] Roscoe, *Life and Experiences*, 20–21.
[75] For William Stanley Jevons, see *ODNB*.
[76] For Charles Booth, see *ODNB*.
[77] Reed, *Entrepreneurial Ventures*, 131 and 180. See also the entry on Harriet Martineau in the *ODNB*.
[78] There is an obelisk in Princes Park as a memorial to Richard Yates.
[79] Roscoe, *Life and Experiences*, 20–21.
[80] For James Martineau, see *ODNB*.

principally in Germany where many British students were drawn to study under Justus Liebig (University of Giessen). The Royal College of Chemistry had been founded in 1845, initially as a private initiative to provide advanced study in chemistry, with the German chemist August Wilhelm Hofmann as its first Director. Hofmann had trained with Justus Liebig in Giessen and brought to London Liebig's approach to "analytical organic chemistry and systematic laboratory instruction in research methods." But as the historian Gerrylynn Roberts has pointed out, the promoters of the College sought training that would allow its students to tackle chemical issues in agriculture, manufacturing, and food and drug quality.[81] In 1848, when Harry was considering where to study chemistry, the College was in its infancy (and facing financial difficulties). Indeed, the financial constraints of the Roscoe family probably precluded Harry from attending Giessen. It was very likely William Balmain, Harry's chemistry teacher in Liverpool, who recommended UCL, since his teacher Edward Turner had taught there; more importantly, by 1848, with Thomas Graham as Professor of Chemistry and George Fownes as the first Chair in Analytical and Practical Chemistry for the Birkbeck Chemical Laboratory, UCL was becoming a major institution for studying chemistry. Whatever the reason, Harry's time at UCL confirmed his intellectual commitment to the subject, and he never regretted studying there.

---

[81] Gerrylynn K. Roberts, "The Establishment of the Royal College of Chemistry: An Investigation of the Social Context of Early-Victorian Chemistry," *Historical Studies in the Physical Sciences* 7 (1976): 437–485, on 438–439.

3

Finding His Niche

Harry Roscoe, aged 15, began his studies at University College London (UCL) in 1848, and his time there was to confirm his lifelong commitment to the study of chemistry. Harry was fortunate to have several outstanding teachers, including the chemists Thomas Graham and Alexander Williamson. Soon after he started at UCL, Roscoe was joined in London by his mother and sister so the family could retain some semblance of life among relatives there, who included the Jevons and Fletcher families. Living in London also allowed them to connect with family and Unitarian networks and develop an ever-wider circle of associates, many of whom were to remain lifelong friends of Harry.

UCL was established as London University in 1826, the first university institution in London and the first in England to admit both men and women without any religious stipulation. When in 1836 the University of London was established as a degree-awarding body, London University changed its name to University College London, which became a constituent body of the University of London (alongside King's College London, founded in 1828 with Church of England support as an alternative to the secular UCL). From its inception UCL had a chair of chemistry which, although part of the General Department (and not part of the medical school), was expected to work closely with colleagues in the medical school.[1]

From the first appointment, the post of Professor of Chemistry at UCL attracted many eminent chemists, making UCL a much sought-after institution for students wishing to study chemistry. The first Chair of Chemistry was Edward Turner (1796–1837), who had studied medicine at the University of Edinburgh and was awarded his MD in 1819, before attending the University of Göttingen to study mineral analysis and chemistry under Friedrich Stromeyer.[2] He returned to Edinburgh to become an extra-mural lecturer in chemistry before applying for the Chair of Chemistry at UCL. Turner was not UCL's first choice; Thomas Thomson, Professor of Chemistry at Glasgow University, was approached by James Mill but did not apply, and Michael Faraday, Director of the Laboratory of the Royal Institution, was offered but declined the chair because of his unfulfilled commitments at the Royal Institution.[3] From the start of Turner's tenure a laboratory was available for teaching a practical course of chemistry, and this became an important feature of chemistry teaching at UCL, especially after the Birkbeck Chemical Laboratory was built in 1845. His students spoke of Turner as a "popular and excellent teacher."[4] While at UCL, Turner published *The*

---

[1] Hugh Hale Bellot, *University College London 1826–1925* (London: University of London Press, 1929), 127.

[2] For Turner, see *ODNB*.

[3] Michael Faraday to Dionysius Lardner, 6 October 1827, in *The Correspondence of Michael Faraday*, ed. Frank James, vol. 1 (London: Institution of Electrical Engineers, 1991), 441–442.

[4] Bellot, *University College*, 127.

Henry Enfield Roscoe. Peter J.T. Morris and Peter Reed, Oxford University Press. © Oxford University Press 2024.
DOI: 10.1093/oso/9780190844257.003.0003

Elements of Chemistry, a book that gave him a wide European reputation, and he was elected a Fellow of the Royal Society of Edinburgh (1825) and a Fellow of the Royal Society (1830).[5]

When Turner died in 1837, his successor was Thomas Graham (1805–1869), who had been Professor of Chemistry at Anderson's University in Glasgow. Graham had built up a considerable reputation while in Glasgow through his outstanding research on the diffusion of gases, for his practical chemistry courses, and for his teaching, even though he spoke in a halting manner.[6] Independently of Liebig at Giessen, Graham had pursued the teaching of chemistry through practical classes and he was very keen to take the same approach at UCL, bringing his assistant James Young from Glasgow.[7] It was probably because of the prospect of being taught by Graham that Harry chose to study at UCL. Graham was a popular teacher at UCL, presenting 150 lectures each year. His textbook *Elements of Chemistry* was published in six parts between 1837 and 1841.[8]

The creation of the Birkbeck Chemical Laboratory at UCL in 1845 (following a donation from George Birkbeck, the physician and educationalist) marked an important development in chemical education, with its emphasis on laboratory-based learning rather than solely relying on formal lectures. This laboratory was not the first to adopt this approach to learning; Justus Liebig in Giessen was an early exponent, and students educated in this manner then replicated the approach in their own teaching across the world. As one of the authors (PJTM) has pointed out, the Birkbeck Laboratory was specially designed to provide safe working conditions for students to carry out practical chemical experiments involving both inorganic and organic chemistry. It was the adoption of a procedure called "group analysis" for the qualitative analysis of metallic compounds, culminating in the tables drawn up by Heinrich Will (Liebig's successor at Giessen) in 1846 that made the redesign of chemical laboratories so necessary.[9] What were the new design features?

Safety and convenient access to chemical reagents and equipment were essential components. As one of the authors (PR) has pointed out:

> Supplies of water and gas (the latter for use with Bunsen burners) were required at the bench. Fume hoods and fume chambers allowed dangerous and obnoxious fumes to be vented to the outside. Hydrogen sulfide used in qualitative inorganic analysis (or group analysis as it was often called) was generated in a Kipp's apparatus but because the gas is almost as toxic as hydrogen cyanide, the "Kipp's" was kept in a fume chamber to which students took their samples.[10]

[5] Turner was also a lecturer in geology at UCL and secretary of the Geological Society.

[6] Michael Stanley, "Thomas Graham (1805–1869)," *ODNB*.

[7] Later, Young achieved fame and wealth through the exploitation of paraffin from oil-bearing shale deposits.

[8] Stanley, "Graham," *ODNB*.

[9] Peter J.T. Morris, *The Matter Factory: A History of the Chemical Laboratory* (London: Reaktion, 2015), 109.

[10] Peter Reed, "Learning and Institutions: Emergence of Laboratory-Based Learning, Research Schools and Professionalization," in *A Cultural History of Chemistry in the Nineteenth Century*, ed. Peter Ramberg (London: Bloomsbury Academic, 2022), 191–215, on 206.

Reagent bottles containing a range of different chemicals for the analysis were stored in racks conveniently placed above the benches where the analyses were carried out.

The year after completion of the laboratory, George Fownes was appointed the first Chair in Analytical and Practical Chemistry in 1846. Fownes had studied with Justus Liebig at Giessen (as had many other aspiring British chemists of the time) and was awarded a PhD in 1841. He worked for a year with Thomas Graham at UCL on his return to Britain and then took up a series of appointments: a Lectureship in Chemistry at Charing Cross Hospital; Professor of Chemistry to the Pharmaceutical Society; and Professor of Chemistry at Middlesex Hospital. Fownes's assistant was Henry Watts, who had studied at UCL and graduated BA (London) in 1841. Sadly, Fownes died in-post in 1849.

Alexander Williamson replaced Fownes as Professor of Analytical and Practical Chemistry, and held the position for 38 years. He also became the Professor of Chemistry following the appointment of Thomas Graham as Master of the Mint in March 1855. With the agreement of his father, Williamson had gone to the University of Heidelberg in 1841 to study medicine, but soon came under the influence of the chemist Leopold Gmelin, which led Williamson in 1844 to move to Giessen to study with Justus Liebig. At Giessen, Liebig had established a laboratory that the historian Jack Morrell has termed "the chemist breeder," because:

> Liebig's was the first institutional as opposed to private laboratory in which students experienced systematic preparation for chemical research, and in which they were deliberately groomed for membership of a highly effective research school.[11]

It was while studying at Giessen that Williamson was awarded a PhD in 1845 for his work on the reactions of inorganic metal oxides and salts with chlorine. In 1846 the Williamson family moved to Paris, allowing Williamson to continue his chemical studies through friendship with a network of French chemists that included Jean-Baptiste Dumas, Auguste Laurent, and Charles Gerhardt.[12] During a trip to Paris following Fownes's death, Thomas Graham persuaded Williamson to apply for the vacant position at UCL. With his application supported by testimonials from several outstanding chemists, including Liebig and Dumas, Williamson was appointed Professor of Analytical and Practical Chemistry in 1849.

The 1840s saw the start of a remarkable transformative period for chemistry, in particular organic chemistry, and William H. Brock has drawn attention to Williamson's leading role:

> Williamson entered chemistry at one of its most exciting moments in its history. Organic chemists had moved on from solely investigating the chemistry of animal and vegetable bodies to manipulate and create a myriad of compounds that did not

[11] J. B. Morrell, "The Chemist Breeders: The Research Schools of Liebig and Thomas Thomson," *Ambix* 19 (1972): 1–46, on 2.

[12] William H. Brock, "Alexander William Williamson," *A Biographical Dictionary of 19th Century British Scientists* (Bristol: Thoemmes, 2004), 2168–2175, on 2169.

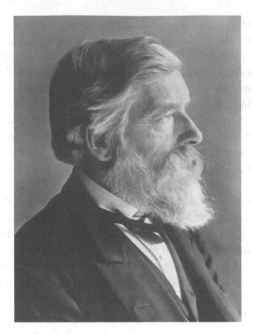

Figure 3.1 Alexander Williamson.
Image courtesy of the Royal Society of Chemistry Library.

exist in nature. The classification and determination of the relationship between these artificial substances proved a great intellectual challenge.[13]

While in Paris, Williamson was greatly influenced by Gerhardt's work on the classification of organic compounds into homologous series sharing the same functional group (i.e., alcohols, acids, and ethers). In his classic experiment, later termed the "Williamson synthesis," he investigated the relationship between alcohols, ethers, and water. This led to a better understanding of the structure of ethers and was probably the first study of how an organic chemical reaction proceeded. This work confirmed Williamson as not only an outstanding theoretical chemist, but also a chemist who appreciated the emerging importance of organic chemical analysis, using Liebig's combustion analysis, and the need for chemists trained in its practical techniques (see Figure 3.1).

We know from the diaries of Alfred Fletcher, Harry's fellow student at UCL, that Williamson was an outstanding teacher and greatly respected at UCL. He nurtured a wider interest in chemistry (and science generally) by encouraging students to attend meetings of the Chemical Society and other scientific organizations, and hosting dinner parties where students could come together and meet established chemists and

---

[13] Brock, "Williamson," 2172.

scientists in social surroundings. Harry greatly respected Williamson, as he relates in his autobiography:

> His was a mind of great originality, and his personality was a most attractive one. And despite his physical disabilities—for he lost an eye and the proper use of his left arm in early childhood—he was a diligent and accurate worker. Ardently devoted to his science, he infected all who worked under him with the same feeling. And his pupils willingly own that much of the success that they have met with in after years was due to his teaching and example. I well remember the feelings of interest he aroused as he each day came down to the laboratory brimful of new ideas.[14]

As William Tilden reflected:

> [...] he was a splendid teacher, always in the laboratory, going from one student to another, arousing and maintaining their interest in their work, and ready to discuss with them any point on which they sought his help. Now and then, when Graham was obliged to be absent, Williamson would lecture on general chemistry in his stead, and these occasions were always hailed with delight by some of the students, to whom he seemed to bring out new points of interest in the most worn subjects by the freshness of his treatment and the new light he would throw on them.[15]

Williamson's chemical instruction was not just laboratory-focused. On Saturdays in the winter session, excursions were made to local chemical manufactories, including gas works, soap works, and brass foundries, to provide students with a fuller appreciation of the large-scale production and manipulation of chemicals.[16]

The chemistry Harry studied at UCL between 1848 and 1852 was framed by the state of chemical understanding at this time. Dalton's atomic theory underpinned chemistry in general, the periodic table classification of elements was some years away, physical chemistry as a distinct subset of chemistry was in its infancy, and in organic chemistry, the homologous series of organic compounds sharing the same function group was emerging, but the classification into aliphatic and aromatic compounds would emerge only later. Harry's course had three distinct parts: general chemistry (in the main, inorganic chemistry with some physical chemistry); organic chemistry; and practical chemistry. The practical chemistry course was directed by Alexander Williamson and the rest by Thomas Graham.[17]

The general chemistry course included: the atomic theory and chemical affinity; the history and properties of many non-metallic elements and their compounds

---

[14] Henry Enfield Roscoe, *The Life and Experiences of Sir Henry Enfield Roscoe DCL, LLD, FRS, Written by Himself* (London: Macmillan, 1906), 36.

[15] William A. Tilden, *Famous Chemists: The Men and Their Work* (London: George Routledge & Sons, 1921), 231–232.

[16] "Visits to Chemical Works," *Chemical News* (3 February 1867): 71.

[17] *The University College London, Calendar for the Session 1853–54* (London: Walton and Maberly, 1853), 9–10 and 25–26. All the UCL calendars (up to 1992) can be found on JSTOR, at https://www.jstor.org/site/university-college-london/college-calendars/. I wish to thank Andrea Sella for this information.

(such as air, water, sulfuric acid, ammonia, and coal-gas); the effect of heat and light on chemical compounds; and many metallic elements, including their ores and salts. The organic chemistry included: substances derived from the animal and vegetable kingdoms, and the chemical changes they undergo. An important component of these two courses was to develop an understanding of chemical manufacture that included glassmaking, gas making, brewing, calico printing, and the preparation of commodities for pharmaceuticals. The testing of commodities, including the detection of poisons and adulterations, was also covered. While Graham lacked industrial experience, he had firsthand experience of adulterations, especially concerning tobacco, through his investigations for the Excise Board.[18]

Williamson drew on his considerable research experience when managing the practical chemistry course in the Birkbeck Laboratory. The course itself comprised practical manipulations that included: construction of apparatus, pneumatic troughs, and use of the mouth blowpipe; preparation of various gases, acids and alkalis, and organic compounds; and qualitative and quantitative analysis of inorganic and organic compounds.[19]

Harry's surviving notebook for the session 1849–1850 gives an insight into how topics were developed.[20] The general chemistry topics progressed from the latent heat of steam and ice to specific heats, catalysis, and isomerism before the inorganic chemistry addressed oxygen, hydrogen, the atmosphere, the oxides of nitrogen, chlorine, ammonia, the acids of chlorine, the acids of sulfur, phosphorus and phosphorus acids, and finally the chemistry of several metals, including copper. The organic chemistry topics reflect the infancy of the later systematic approach built around series of compounds and functional groups. The course starts with cane sugar, the analysis of sugar, glucose, fermentation of sugar before moving on to alcohol and acid series (including oxalic, tannic, citric, carbolic, picric, palmitic, benzoic, nitrobenzoic) and then to organic bases (ethylamine and diethylamine). Natural products also are featured. A notable part of Harry's notebook is the translated extracts of several papers, including Heinrich Rose on the blowpipe in chemical analysis, Williamson on etherification, and extracts from Jacob Berzelius's *Traité de chimie*.

During his studies at UCL, Harry was influenced by several outstanding teachers besides Graham and Williamson. In his autobiography, Harry mentions his English, Latin, and Greek teachers, but is particularly effusive about Augustus De Morgan, Professor of Mathematics. De Morgan had studied mathematics at Trinity College Cambridge, graduating in 1827 as fourth wrangler, but as a theist and a Unitarian decided not to study for an MA degree or pursue a fellowship. After flirting for a short time with the idea of entering the bar, he was appointed the first Professor of Mathematics at UCL in 1828 at age 22. Although De Morgan had resigned this post in 1831, having taken a principled stand against the dismissal of a teaching colleague, he took up the post again in 1836 following the death of his successor in a drowning

---

[18] P. W. Hammond and Harold Egan, *Weighed in the Balance: A History of the Laboratory of the Government Chemist* (London: HMSO, 1992), 13.

[19] *University College London Calendar for 1853–54*, 34–36.

[20] "Notebook for Chemistry Session 1849–50," HC3419, Roscoe Collection, Royal Society of Chemistry.

accident.[21] Harry acknowledged De Morgan's influence for not only his mathematics but also the way he thought:

> De Morgan was certainly facile princeps among teachers of mathematics of his day, and he inspired the greatest enthusiasm for the subject in the minds of his pupils ... he was also one of the profoundest and subtlest thinkers of the nineteenth century. His was a most original mind, and his method on instruction was quite different from that of the ordinary schoolmaster. The trouble he took with his students was extraordinary.[22]

However, students were not always respectful to De Morgan (as with many other teachers at the time), as Harry illustrates in an amusing story about sparrows released during a lecture:

> On one occasion a number of sparrows were let loose in the lecture-room and flew about, perching on the blackboard, much to the amusement of the audience, who expected every moment that one would alight on the professor's bald head. After some time his attention was drawn to their presence, and he remarked, resting his nose on the pointer, as was his wont, and surveying the class with his only eye, from behind a very large white choker: "I see nothing to laugh at if a sparrow does come into the room, and I daresay there are many here who have not got the brains of a sparrow." After which the lecture proceeded without interruption.[23]

With UCL attracting such outstanding teachers and with its more open access for students, it is not surprising that Harry came in contact with several fellow students who later would make outstanding contributions in their chosen fields. Among those with whom Harry formed lifelong friendships were William Russell, William Kenrick, Alfred Fletcher, and James Bell. William Russell entered UCL the year before Harry, and having graduated in 1851, acted as assistant to Edward Frankland, Professor of Chemistry at Owens College, Manchester, before spending two years with Robert Bunsen at Heidelberg University where he was awarded a PhD. Later, having returned to England, he was appointed Professor of Natural Philosophy at Bedford College for Women and then lecturer in chemistry at St Bartholomew's Hospital medical college. He was elected Fellow of the Royal Society in 1872 and played a leading role in the British Association for the Advancement of Science.[24] Russell and Harry shared an interest in chemical education for medical students and the scientific education of women. William Kenrick was the son of Archibald Kenrick of the hardware manufacturing family in Birmingham and brother-in-law of Joseph Chamberlain, the Birmingham industrialist and politician. As a Unitarian he became a leading figure in the National Education League and the National Liberal Association, and was

[21] For William De Morgan, see *ODNB*. De Morgan resigned his post again in 1866, taking a principled stand regarding the appointment of James Martineau to the Chair of Moral Philosophy.
[22] Roscoe, *Life and Experiences*, 25.
[23] Roscoe, *Life and Experiences*, 26.
[24] For Russell, see *ODNB*. See also Roscoe, *Life and Experiences*, 34.

elected MP for North Birmingham in 1885.[25] Between 1866 and 1877 he was a council member of the Midland Institute, founded in 1854 to advance scientific and technical education, an interest shared with Harry.

Alfred Fletcher (1827–1920) was a Congregationalist who started at UCL the same year as Harry, having acted as an assistant surveyor on Brunel's South Wales Railway. Fletcher's personal diaries reveal a camaraderie between the two students, sharing many social occasions (often organized by Williamson) as well as participating together in physical activities such as walking and rowing on the Thames. At the conclusion of his course in 1851, Fletcher was awarded the gold medal for chemistry (much to Harry's chagrin) and the Fletcher family always added that Harry (who later became such a distinguished chemist in Manchester) was in second place.[26] Fletcher became a successful chemical manufacturer in London before his appointment as a sub-inspector of the Alkali Inspectorate in 1864, working under Chief Inspector Robert Angus Smith. Following Smith's death in-service in 1884, Fletcher was appointed his successor.[27] It was during the period 1864 to 1884 that Fletcher often approached Harry (both were conveniently based in Manchester) to act as an arbitrator in the apportionment of financial damages arising from pollution cases, or to act in a consultancy role for chemical companies and advise them on how best to modify their plant or production operations to comply with the legal limit for noxious vapor emissions.[28]

James Bell (1825–1908) joined the Laboratory of the Inland Revenue Department at Somerset House in 1846 and was chosen to attend UCL for training in chemistry and practical chemistry (among several other subjects), where he was a chemistry prize winner.[29] In 1867 Bell was appointed Deputy Principal of the Laboratory and then Principal in 1874, before retiring in 1894.[30] The Laboratory became part of the new Government Laboratory in 1894.[31] He published two outstanding books: *The Analysis and Adulteration of Foods* (1881–1883) and *The Chemistry of Tobacco* (1887), and was elected a Fellow of the Royal Society in 1884.[32]

Harry's move to London to pursue his studies at UCL gave him time to socialize with the Roscoe, Fletcher, and Jevons family members, while also drawing him into wider social networks formed principally, but not exclusively, of fellow Unitarians. Initially Harry stayed with his uncle Richard Roscoe, who was a doctor. It was this uncle who had tried to persuade Harry's mother not to allow her son to study medicine because "it was an extremely arduous profession which would probably kill me," even though Harry always felt he was well suited to being a doctor.[33] Harry found his uncle an amusing companion and always ready to share a humorous story. His uncle told a story relating to his doctor's rounds in an out-of-town practice:

---

[25] For William Kenrick, see *ODNB*. See also Roscoe, *Life and Experiences*, 34.

[26] Roscoe, *Life and Experiences*, 34.

[27] Peter Reed, *Acid Rain and the Rise of the Environmental Chemist in Nineteenth-Century Britain* (Farnham, Surrey: Ashgate, 2014), 111.

[28] Diaries of Alfred Evans Fletcher (Private Collection of the Fletcher family).

[29] The training was an arrangement between the Board of Excise and UCL (and funded by the Treasury).

[30] Thomas Edward Thorpe, "Dr. James Bell, C.B., F.R.S.," *Nature* 77 (1908): 539–540, on 539.

[31] Hammond and Egan, *Weighed in the Balance*, 123.

[32] James Bell, *The Analysis and Adulteration of Foods*, 2 vols. (London: Chapman and Hall, 1881–1883) and James Bell, *The Chemistry of Tobacco* (London: n.p., 1887).

[33] Roscoe, *Life and Experiences*, 12.

He saw the dead body of an old villager, who had met with an accident, being carried into his cottage. He went in to see if he could be of any use. The old wife Betty threw up her hands and said "Thank God it's no worse!" "Why, Betty," said the doctor, "how could it be worse? Here's John brought home dead on a stretcher." "Why," said she, "he might have been taken ill i' th' autumn and died i' th' spring."[34]

Harry always had a penchant for humorous stories, as his autobiography often shows.

Soon after commencing his studies at UCL, Harry, his mother, and sister established a home in Torrington Place (London) where two cousins also lived. The cousins, Frank and William Caldwell Roscoe, were sons of William Stanley Roscoe (son of William Roscoe Jr.) and Hannah Eliza Caldwell; William was a poet and essayist. Soon after, Harry, together with his mother and his sister, moved to a house in Oval Road, Camden Town, where two cousins, Henry Roscoe and Stanley Jevons, joined them.[35] Henry Roscoe had been head of the legal firm Field and Roscoe in Lincoln Inn Fields.[36] Stanley Jevons, who was to play a large part in Harry's life, both in London and later in Manchester, was the son of Thomas Jevons, a Liverpool iron merchant and inventor and his wife Mary Anne Jevons, daughter of William Roscoe.[37] He was educated at University College School, and then between 1851 and 1853 was a fellow student of Harry's at UCL. A letter to Stanley from his father, dated 18 March 1851, indicates that Stanley and the Roscoes attended the Unitarian Chapel in Little Portland Street (near Regent Street), London, where they were drawn to the ministry of Edward Tagart, who was general secretary of the British and Foreign Unitarian Association between 1842 and 1858. Charles Dickens had been a congregant there during the early 1840s.

Like Harry, Stanley studied mathematics with De Morgan and chemistry with Williamson and Graham; in 1852 he was awarded the silver medal in chemistry and the following year won the gold medal.[38] In 1853 Graham approached Harry about becoming assayer to the Royal Mint in Sydney but Harry declined it, whereupon Graham asked Jevons to take the post and, although initially a little diffident, Jevons finally accepted after encouragement from his father.[39] Before departing for Australia, Jevons was preoccupied with essential preparations, including training in assay work in London and Paris, and purchasing necessary equipment and chemicals that would have to make the long journey to Sydney. Jevons also had to plan his assay office in Sydney and appoint an assistant. Jevons first spent time at the Royal Mint in London with Thomas Graham and William Allen Miller (Professor of Chemistry at King's College, London, and non-resident assayer at the Royal Mint) learning to carry out assay analyses while also compiling a list of laboratory equipment and chemicals that needed to be purchased for the Mint in Sydney.[40] Graham then arranged for Jevons to

---

[34] Ibid., 31.

[35] Ibid., 31.

[36] Ibid., 31.

[37] R.D. Collison Black, "William Stanley Jevons (1835–1882)," *ODNB*.

[38] *The University College London, Calendar for the Session 1853–54* (London: Walton and Maberly, 1853), 175.

[39] Harriet Jevons, *Letters and Journal of W. Stanley Jevons* (London: Macmillan, 1886), 38.

[40] Black, *Papers and Correspondence of William Stanley Jevons*, volume II, Letter 27 (WSJ to Roscoe, 28 August 1853).

study with Eugène Péligot (Professor of Applied Chemistry in the Conservatoire des Arts et Métiers and assayer to the Paris Mint). Although Jevons didn't get on well with Péligot and the Paris Mint seemed to be over-staffed and functioning at a low level, Jevons passed the diploma examinations that confirmed his competence in assay analyses.

On 29 June 1854, Jevons set out from Liverpool on the 1,230-ton clipper *Oliver Lang*, bound for Sydney, with his personal possessions and two assay balances that he kept in his cabin; the 18 packing cases (including 15 cases of acid) had already sailed from London.[41] When the ship arrived in Sydney on 6 October 1854, Jevons found that the new Mint building would not be ready until June 1855, leaving time to explore several aspects of Australian life, with meteorology being a favorite interest. But even after only three months in Sydney, Jevons had decided to return to Britain within 5 to 10 years.[42] When Jevons resigned from the Sydney Mint in March 1859, his interests had turned to philosophy and political economy. Returning to Britain in September 1859, Jevons re-enrolled at UCL to complete his BA (having regretted not completing the degree earlier) and then gained an MA in mental philosophy and political economy. These qualifications, together with further research and the publication of two books (including *The Coal Question*), laid the basis for Jevons joining Harry on the staff of Owens College, Manchester, in 1863, as we will discuss later.[43]

Another close family connection for Harry was the Crompton family. Charles Crompton was a Justice of the Queen's Bench (who was knighted in 1852) and a close friend of Harry's father.[44] In 1832 he married Caroline, daughter of Thomas Fletcher, a Liverpool merchant. Sir Charles became a second father to Harry after the premature death of his father. The Cromptons lived for a period in Hyde Park Square, London, and it was in the drawing room of their house that Harry first met his future wife, Lucy Potter, daughter of Edmund Potter, a Unitarian, a calico printer and MP for Carlisle. Sir Charles's son Charles became a close friend of Harry's and studied at Trinity College, Cambridge.[45] It was on a visit to Charles at Cambridge that Harry first met the Scottish physicist James Clerk Maxwell, who later went on to develop his field theory and the electromagnetic theory of light, as Harry relates in his autobiography:

This young man [Maxwell] addressed me in a very broad Scotch as follows: "Come and see my devil; I've got a devil." So I went. The floor of his room was covered with sheets of white paper; upon these were drawn a most complicated series of curves; from the ceiling was hung a doubly suspended pendulum, the "bob" of which was a heavy weight ending in a point. On placing the point on one of these curves and releasing the weight, the point followed exactly these singular curves running all over the floor in a grotesque manner. This was my first introduction to Clerk Maxwell and his "devil."[46]

---

[41] Harriet Jevons, *Letters and Journal*, 39; and Black, *Papers and Correspondence of William Stanley Jevons*, volume II, Letter 41 (WSJ to Roscoe, 25 June 1854).

[42] Black, "William Stanley Jevons (1835–1882)."

[43] Roscoe, *Life and Experiences*, 39.

[44] For Sir Charles John Crompton, see *ODNB*.

[45] For Henry Crompton, see *ODNB*.

[46] Roscoe, *Life and Experiences*, 32.

While attending UCL, Harry was befriended by Henry Holland and Benjamin Brodie. Henry Holland was physician extraordinary to Queen Victoria and physician ordinary to Prince Albert, and President of the Royal Institution.[47] He was related to Josiah Wedgwood and Elizabeth Gaskell, and to Harry through the Caldwell family. For many years at the center of London social life, he attended soirées and organized breakfast parties, and at one of the former, Harry was unknowingly vetted as a possible tutor or companion to one of Victoria and Albert's sons, but the Court official at the party was obviously not sufficiently impressed to pursue the matter further.[48] Benjamin Brodie studied the classics, mathematics, and the law before deciding to study chemistry with Justus Liebig at Giessen and was awarded a PhD for his research on beeswax.[49] Harry met Brodie in London soon after starting at UCL and the two became close friends; initially probably because Brodie had adapted part of his house as a laboratory, where the two were able to carry out experiments. Later Brodie was elected to the Aldrichian Chair of Chemistry at Oxford (later renamed the Waynlete Chair) during a period when Harry was Professor of Chemistry at Owens College; besides chemistry, their shared interest in university and scientific education brought a lifelong friendship.[50]

Having secured a BA degree with honors in chemistry in 1852,[51] Harry had already decided to further his studies in chemistry by studying with Robert Bunsen, Professor of Chemistry at the University of Heidelberg, who had moved from the University of Breslau in late 1852. As Brock has suggested, it was probably Thomas Graham who influenced Harry's decision to study with Bunsen. Graham had met Bunsen in 1840 during his visit to Britain to present a paper on cacodyl compounds (i.e., organic compounds containing arsenic) and gas analysis at the Glasgow meeting of the British Association for the Advancement of Science, and recognized the importance of Bunsen's work in advancing analytical chemistry.[52]

His time in London and studying at UCL were important for Harry. He benefited from the teaching of two outstanding chemists of the time in Graham and Williamson. Williamson played a two-part role: furthering Harry's interest and skills in practical analytical chemistry; and through his encouragement of social discourse between teacher and students at parties and opportunities to expand social and academic networks when parties were organized for eminent chemists visiting London. Such networks were largely separate from Harry's extended family networks in London, but together these networks would enable Harry to move more easily among influential metropolitan circles at different times in his professional and political life. Having good London connections would prove essential to Harry's accomplishments during his long tenure at Owens College, Manchester.

[47] For Sir Henry Holland, see *ODNB*.
[48] Roscoe, *Life and Experiences*, 40–41.
[49] For Sir Benjamin Collins Brodie, see *ODNB*.
[50] Roscoe, *Life and Experiences*, 41–43.
[51] *University College London Calendar for 1854–55*, 233.
[52] William H. Brock, "Bunsen's British Students," *Ambix* 60 (2013): 203–233, on 207.

4

Seeing the Light

Why Did Roscoe Go to Heidelberg?

After Harry Roscoe had completed his degree at University College London (UCL), the position of the assayer to the new mint in Sydney came up and Thomas Graham recommended him. Anxious to go to Heidelberg, Roscoe proposed his cousin Stanley Jevons instead, probably because of his family's precarious financial position as a result of his father's bankruptcy in 1848 (for Jevons, see Chapter 3).[1] The attraction of Heidelberg was the new Professor of Chemistry there, Robert Bunsen (1811–1899).[2] Roscoe's desire to work with Bunsen was probably because of his high regard for Graham, who also carried out research on the border between chemistry and physics. He later remarked, "There is certainly much to be done in the land lying between these two Sciences—but whether I can manage to do anything remains to be seen."[3] Furthermore, if an aspiring chemist wanted to carry out research before taking up an academic position in Britain, it was more or less essential to go to Germany or France. By the 1850s, Germany was beginning to be the more attractive of the two countries, although France had a strong chemical-physical tradition (for example, the work of Marcellin Berthelot) which might have appealed to Roscoe.[4]

The founding of the University of Heidelberg in 1386 was a consequence of the papal schism of 1378 and the subsequent support of the Pope in Rome by the Count Palatine (along with the other German rulers).[5] As a consequence, students from the Palatinate had to leave the University of Paris, which supported the French-backed pope in Avignon. The Count Palatine, Ruprecht I, set up a new university with the help of the Roman Curia. Following the Reformation, Heidelberg and its university were racked by several wars and changes of religion, namely the Thirty Years' War (1618–1648), the Nine Years' War (1688–1697), during which Heidelberg was almost completely destroyed, and finally the French Revolutionary Wars. By 1800, four centuries

---

[1] Margaret Schbas, *A World Ruled by Number: William Stanley Jevons and the Rise of Mathematical Economics* (Princeton, NJ: Princeton University Press, 1990), 14; also see R.D. Collison Black, ed., *Papers and Correspondence of William Stanley Jevons*, vol. II, *Correspondence, 1850–1862* (Clifton, NJ: Augustus M. Kelley, 1973), Letter 21 (Thomas Jevons to WSJ, 9 February 1853) and Letter 22 (Thomas Jevons to WSJ, 18 February 1853).

[2] For the fame of Bunsen in Britain at the time, see W. H. Brock "Bunsen's British Students," *Ambix* 60 (2013): 203–233, on 207–210.

[3] Black, *Papers and Correspondence of William Stanley Jevons*, vol. II, Letter 87 (Roscoe to WSJ, 10 May 1856).

[4] For Berthelot, see Harold Baily Dixon, "Berthelot Memorial Lecture," *Journal of the Chemical Society, Transactions* 99 (1911): 2353–2371; Carl Graebe, "Marcelin Berthelot," *Berichte der Deutschen Chemischen Gesellschaft* 41 (1908): 4805–4872.

[5] Andreas Cser, *Kleine Geschichte der Stadt Heidelberg und ihrer Universität* (Heidelberg: Kleine Buch Verlag, 2007); Arleen Marcia Tuchman, *Science, Medicine and the State in Germany; The Case of Baden, 1815–1871* (New York: Oxford University Press, 1993), 16.

Henry Enfield Roscoe. Peter J.T. Morris and Peter Reed, Oxford University Press. © Oxford University Press 2024.
DOI: 10.1093/oso/9780190844257.003.0004

after it was founded, the university was at its lowest ebb. However, the territory of the Palatinate on the eastern bank of Rhine now became part of the Grand Duchy of Baden and the university was re-founded as a state institution in 1803 by Grand Duke Karl Friedrich, who gave the university the name of Ruprecht-Karls Universität, linking the name of its original founder with that of its re-founder. The Grand Dukes made themselves rectors of the University, but they were represented on a day-to-day basis by an academic pro-rector, a position held by the law professor Robert von Mohl (1799–1875) when Roscoe arrived in 1853.[6]

There was a revolution in Germany in 1848 and a national parliament was briefly convoked, of which several professors at Heidelberg were members (including von Mohl) before the revolution was put down by Austria and Prussia. It was thus still a time of turmoil in Heidelberg; a revolutionary army held the town until it was recaptured by the Prussian army, which left in 1851.[7] Once things had settled down, the university grew rapidly and gained an international reputation as a liberal institution, although women were not admitted until 1900. However, it was the 1850s and 1860s that were the "glory years" for the university, as Heidelberg quickly lost ground to the new German capital and its university after the unification of Germany in 1871.[8] Heinrich Helmholtz, Gustav Kirchhoff, and Heinrich Treitschke all left for Berlin— a move that parallels the departure of German scientists such as August Wilhelm Hofmann from England somewhat earlier. Roscoe was there at just the right time.

Chemistry in the form of pharmacy had been part of the medical faculty since 1652, and the first chemical laboratory was set up in 1686. However, chemistry as a scientific discipline first arrived at the University of Heidelberg when Georg Adolf Suckow (1751–1813) was appointed Professor of Natural History, Chemistry, and Botany in 1778.[9] Suckow's Chair was divided into pure and applied chemistry in 1803, and Karl Wilhelm Gottlob Kastner (1783–1857) was appointed to the Chair of Pure Chemistry in 1805.[10] Kastner went to the University of Halle in 1812. Leopold Gmelin (1788–1853) was appointed Extraordinary Professor of Chemistry in the medical faculty at his insistence in 1814 and became a full professor three years later. He remained at Heidelberg for over three decades until he retired following a stroke in 1850.[11] After 1818, he carried out his chemical teaching and research in a former Dominican friary. Chemistry had two rooms in the cloisters next to the Anatomy department. In 1850, after the Anatomy department left the friary for a new building, chemistry was given a third room. The lecture theater was in the former chapel. His successor Robert

---

[6] For von Mohl, see *DB*.

[7] Tuchman, *Science, Medicine and the State*, 103.

[8] David Cahan, *Helmholtz: A Life in Science* (Chicago: University of Chicago Press, 2018), 407–408.

[9] For the history of chemistry at the University of Heidelberg before Bunsen, see Rudolf Schmitz, *Die deutschen pharmazeutisch-chemischen Hochschulinstitute* (Ingelheim am Rhein: C. H. Boehringer Sohn, 1969), 177–189; August Bernthsen, "Die Heidelberger chemischen Laboratorien für den Universitätsunterricht in den letzten hundert Jahren," *Zeitschrift für Angewandte Chemie* 42 (1929): 382–384; Theodor Curtius and Johannes Rissom, *Aus der Geschichte des Chemischen Universitätslaboratoriums zu Heidelberg seit der Gründung durch Bunsen* (Heidelberg: Verlag Rochow, 1908), 1–2.

[10] For Kastner, see *DB*.

[11] For Gmelin, see *DB* and Bernd Wöbke, "Das Portrait: Leopold Gmelin (1788–1853)," *Chemie in Unserer Zeit* 22 (1988): 208–216; Petra Stumm, *Leopold Gmelin (1788–1853): Leben und Werk eines Heidelberger Chemikers* (Herbolzheim, Baden-Württemberg: Centaurus, 2012).

Bunsen was in the philosophy faculty (the standard faculty for the natural sciences at the time).[12] Wilhelm Delffs (1812–1894) was given Gmelin's Chair in the medical faculty; he had been an extraordinary professor in the philosophy faculty since 1843.[13]

Robert Bunsen

Robert Wilhelm Eberhard Bunsen was born in March 1811 in Göttingen and went to the University of Göttingen, where his father was the Head Librarian, to study the natural sciences, including chemistry under the leading teacher of chemistry, Friedrich Stromeyer.[14] However, he took his doctorate on hygrometers, which would now be considered physics, and his interests lay on the border between physics and chemistry. Göttingen was in the Kingdom of Hanover and when Roscoe was born in 1833, they shared a common monarch, William IV; Bunsen's father had served in the army of George III. Bunsen taught chemistry at Kassel Polytechnic School between 1836 and 1839. While he was at Kassel, Bunsen began to study the evil-smelling and poisonous fluid, Cadet's liquid, made by distilling potassium acetate and arsenic(III) oxide. Bunsen thought he had prepared free cacodyl as the organometallic equivalent of a metallic element. He had actually prepared the dimer of the cacodyl radical (an arsenic atom with two methyl groups), but his research spurred his student Edward Frankland (1825–1899) to seek the free ethyl radical—a search which ultimately led to the modern theory of valency and the field of organometallic chemistry.[15] Bunsen then went to the University of Marburg where he was promoted to a full professorship in 1841. During this period Hermann Kolbe (1818–1884) worked as his assistant.[16] While he was at Marburg, Bunsen visited Iceland in 1846 and became interested in the chemistry of igneous rocks.[17] Bunsen went briefly to the University of Breslau (now Wrocław in Poland) in 1851 before succeeding Gmelin at Heidelberg a year later. It

---

[12] *Addressbuch: Winter-halbjahr 1852–1853* (Heidelberg: Karl Winter, 1852), 5. The university calendar in Heidelberg was called the *Addressbuch* and there was one for each semester (*halbjahr*). They are not perfect, as they do not record every student, but they are very useful. They can be downloaded in five-year batches at Adreßbücher der Universität Heidelberg 1818–1922, https://www.ub.uni-heidelberg.de/helios/digi/unihdadressbuch.html.

[13] *Addressbuch: Sommer-halbjahr 1853* (Heidelberg: Karl Winter, 1853), 4; and *Addressbuch: Sommer-halbjahr 1843* (Heidelberg: Karl Winter, 1843), 5. For Delffs, see *DB*.

[14] For the life of Bunsen, see Georg Lockemann, *Robert Wilhelm Bunsen: Lebensbild eines deutschen Naturforschers* (Stuttgart: Wissenschaftliche Verlagsgesellschaft, 1949); Henry Enfield Roscoe, "Bunsen Memorial Lecture," *Journal of the Chemical Society, Transactions* 77 (1900): 513–554; and the excellent introduction to Christine Stock, ed., *Robert Wilhelm Bunsens Korrespondenz vor dem Antritt der Heidelberger Professur (1852): Kritische Edition* (Stuttgart: Wissenschaftliche Verlagsgesellschaft, 2007), section 1, "Einführung," XV–XXX; section 2, "Zu den Briefen," XXXI–XLVIII; section 3, "Reisen: ein wichtiges Kapitel in den Briefen," XLIX–LXXIV. The exact date of Bunsen's birth is disputed. For Stromeyer, see *DB* and Ernst Homburg, "The Rise of Analytical Chemistry and Its Consequences for the Development of the German Chemical Profession (1780–1860)," *Ambix* 46 (1999): 1–32.

[15] For Frankland, see *ODNB* and Colin A. Russell, *Edward Frankland: Chemistry, Controversy, and Conspiracy in Victorian England* (Cambridge: Cambridge University Press, 1996).

[16] For Kolbe, see *DB* and Alan J. Rocke, *The Quiet Revolution: Hermann Kolbe and the Science of Organic Chemistry* (Berkeley: University of California Press, 1993).

[17] Curt Wentrup, "Bunsen the Geochemist: Icelandic Volcanism, Geyser Theory, and Gas, Rock and Mineral Analyses," *Angewandte Chemie International Edition* 60 (2021): 1066–1081.

was in Breslau that Bunsen first met the physicist Gustav Kirchhoff (1824–1887), who followed Bunsen to Heidelberg in 1854.[18]

Like many chemists of the nineteenth century (including Roscoe), Bunsen was particularly interested in the application of chemistry to industry. In particular, he studied the chemistry of the blast furnace (see Chapter 8). He developed new methods of gas analysis for this purpose, employing a straight-tube eudiometer and balls of solid absorbents hanging on platinum wires. Bunsen was a prolific developer of chemical instrumentation, and the Bunsen burner will be discussed later in this chapter. Bunsen also developed the Bunsen cell, a zinc-carbon cell which was the forerunner of the modern battery, in 1841, the Bunsen "grease spot" photometer in 1843, the "constant level" hot-water bath whereby the water lost by evaporation was continually replenished in 1850, a bizarre-looking thermostat for the measurement of the specific heat of gases in 1867, the water aspirator in 1868, and the Bunsen ice calorimeter in 1870.

When Bunsen died in August 1899, his obituary in *The Times*, written by Roscoe, noted:

> Passionately devoted to his laboratory and to his students, he formed a school of chemistry second only—and about that there may be a question—to that of Liebig at Giessen. His lectures were original and most instructive, but it was in his patient teaching in the laboratory that his genius showed itself strongest, and the pupils from all parts of the world who were fortunate to work under him ... all conceived for him the deepest feelings of respect and affection, his precept and example having influenced their whole lives.[19]

Of no one is this truer than Roscoe himself.

Roscoe's Arrival in Heidelberg

At noon on a Sunday in August 1853, the *PS Baron Osy* of the Antwerp Steam Navigation Company set off from St. Katherine's Wharf, in the shadow of the Tower of London, on its way to Antwerp.[20] Roscoe was on board the ship, accompanied by his mother, sister, and several friends. His route from Antwerp the next day is not entirely clear, but assuming he was using the latest edition of *Bradshaw*, he would have taken a train to Cologne and then "journeyed by boat slowly up the Rhine, stopping at the places which were still old-fashioned"[21] to Mainz (then known by the British as

---

[18] For Kirchhoff, see *DB*.

[19] *The Times*, August 17, 1899, 4. For Roscoe's authorship of the obituary, see Henry Enfield Roscoe, *The Life and Experiences of Sir Henry Enfield Roscoe DCL, LLD, FRS, Written by Himself* (London: Macmillan, 1906), 79–80.

[20] The account of Roscoe's voyage to Heidelberg is taken from Roscoe, *Life and Experiences*, 44–46. The departure of the *PS Baron Osy* at noon every Sunday is given by "Antwerp, Brussels, Cologne, Hamburg, etc.," *The Times*, July 25, 1853, 1. For the history of the ship, see C.L.D. Duckworth and G. E. Langmuir, *Railway and Other Steamers* (Glasgow: Shipping Histories, 1948), 136.

[21] Roscoe, *Life and Experiences*, 45.

Mayence).[22] Here he probably took the train to Heidelberg (or continued by river to Mannheim and then took a short train journey to Heidelberg).

Having arrived in the university town, Roscoe presented his letters of introduction to von Mohl.[23] Roscoe and his family then found lodgings with a Frau Frisch close to the Karlstor gate in central Heidelberg, under the shadow of the ruined castle. The young Roscoe then set about learning German before the academic year got underway in mid-October[24] with the help of a retired teacher and the housemaid Gretchen.[25]

Eventually Bunsen returned to Heidelberg from his hometown of Göttingen, where his beloved mother had died on 12 July.[26] Roscoe later recalled his first meeting with Bunsen:

> Professor von Mohl kindly introduced me to the great chemist. I shall never forget my first sight of him—the man who afterwards became one of my dearest friends, and to whom I owe more than I can tell. He ... was then at the height of his mental and physical powers. He stood fully six feet high, his manner was simple yet dignified and his expression one of rare intelligence and great kindness.... His modesty was natural and in no degree assumed. In his old age, and looking back on his life-work, he writes me that he "feels as keenly as ever how modest and contemptibly small is the amount which I have added to the building of Science." And yet the contributions of this man have been equalled by few.[27]

Early Work in Heidelberg

Roscoe was initially listed in the university calendar in the philosophy faculty.[28] The philosophical faculty was the usual category in the calendar for science students. However, Roscoe was not registered for the chemistry laboratory (called the chemistry institute in German universities), presumably because of Bunsen's absence; evidently, he could not register for the laboratory in the director's absence.[29] When he entered the laboratory, Roscoe shared a bench with Lothar Meyer (1830–1895), the later pioneer of the periodic table, and August Pauli of Munich. Pauli later left Heidelberg to work with John Bennet Lawes (1814–1900) at Rothamsted on the study of fertilizers

---

[22] *Bradshaw's General Railway Directory, Shareholders' Guide, Manual, and Almanack etc.* (London; Manchester: W. J. Adams, Bradshaw and Blacklock, 1852), 13.

[23] See Mohl, see *DB*.

[24] Announcement in the *Agromische Zeitung*, 8:38 (1853): 607. The winter semester always started on 15 October at this time, regardless of the days of the week.

[25] The Canadian physician John Corson had remarked on the linguistic abilities of the women of Heidelberg in five years earlier. J. Corson, *Loiterings in Europe* (New York: Harper & Bros. 1848), 91.

[26] Lockemann, *Bunsen*, 187. For the death date of Auguste Friederike Bunsen, see the *CERL Thesaurus*, https://thesaurus.cerl.org/record/cnp01158784.

[27] Roscoe, *Life and Experiences*, 47.

[28] *Addressbuch: Winter-halbjahr 1853–1854* (Heidelberg: Karl Winter, 1853), 19.

[29] For the social structure of the chemistry institute at Heidelberg, see Peter J.T. Morris, "Aspects of the Social Organization of the Chemical Laboratory in Heidelberg and Imperial College, London," in *The Laboratory Revolution and the Creation of the Modern University, 1830–1940*, ed. Klaas van Berkel and Ernst Homburg (Amsterdam: Amsterdam University Press, 2023), 225–239.

but died a short time afterward of phthisis (i.e., tuberculosis).[30] Nearby places in the laboratory were occupied by his British confreres William Russell (1830–1909) and Edmund Atkinson (1831–1900), to whom we will return, and two German students. Heinrich Meidinger (1831–1905) became a physicist at Karlsruhe Polytechnic, while Moritz Hermann became involved with his family's chemical works "Hermania" at Schönebeck on the Elbe.[31]

The chemical laboratories at Heidelberg when Bunsen arrived in October 1852 were still located in the former Dominican friary used by Gmelin. In his autobiography, Roscoe recalled his experience of working in the cloisters five decades earlier:

It was roomy enough; the old refectory was the main laboratory, the chapel was divided in two, one half became the lecture-room and the other a storehouse and museum. Soon the number of students increased and further extensions were needed, so the cloisters were enclosed by windows and working benches placed below them. Beneath the stone floor at our feet slept the dead monks [sic], and on their tombstones we threw our waste precipitates! There was no gas in Heidelberg in those days; nor any town's water supply. We worked with Berzelius's spirit lamps, made our combustions with charcoal ... and went for water to the pump in the yard. Nevertheless, with all these so-called drawbacks, we were able to work easily and accurately.[32]

As Baden was keen to attract a leading chemist, Bunsen was offered a new laboratory building if he accepted the Chair at Heidelberg.[33] The construction of a new chemical institute began in May 1854, and it was opened a year later in time for the summer semester.[34] The architect was Heinrich Lang (1824–1893), who had just designed the

[30] For Lothar Meyer, see DB; Phillips Bedson, "Lothar Meyer Memorial Lecture," Journal of the Chemical Society, Transactions 69 (1896): 1403–1439; and Karl Seubert, "Lothar Meyer," Berichte der Deutschen Chemischen Gesellschaft 28 (1918): 1109–1146. For August Pauli (not be to be confused with the younger Heidelberg student Philipp Pauli [1836–1920] who was also a friend of Roscoe), see Roscoe, Life and Experiences, 48 and 58, and John Bennet Lawes, Reply to Baron Liebig's Principles of Agricultural Chemistry (London: W. Clowes, 1855), 78. For Lawes, see ODNB.

[31] For Meidinger see DB. For the Hermania works see German Wikipedia.

[32] Roscoe, "Bunsen Memorial Lecture," 545. The Berzelius spirit lamp was an Argand lamp which was modified to use alcohol and the glass chimney was replaced by a metal one; see David A. Wells, Wells's Principles and Applications of Chemistry: For the Use of Academies, High-Schools and Colleges (Chicago: Ivison and Phinney, 1859), 322.

[33] Frank James has argued that Baden's keenness to support chemistry was a matter of supporting culture rather than supporting industrialization (not least because it was not clear in the early 1850s that chemistry would aid industrialization); F.A.J.L. James, "Science as a Cultural Ornament: Bunsen, Kirchhoff and Helmholtz in Mid-Nineteenth-Century Baden," Ambix 42 (1995): 1–9. For the contrary view that the Baden government sought to improve agriculture by prompting the teaching of chemistry, see Peter Borscheid, Naturwissenschaften, Staat und Industrie in Baden, 1848–1914 (Stuttgart: Klett, 1976). Tuchman, in Science, Medicine and the State, sees the promotion of science generally as being about the education of the future citizen, a view which would have met with Roscoe's approval.

[34] This account is based on Peter J.T. Morris, The Matter Factory: A History of the Chemistry Laboratory (London: Reaktion, 2015), chapter 5, "Modern Conveniences: Robert Bunsen and Heidelberg, 1850s," 119–145; also see Christine Nawa, "A Refuge for Inorganic Chemistry: Bunsen's Heidelberg Laboratory," Ambix 61 (2014): 115–140; and the excellent booklet produced for the dedication of Bunsen's laboratory as a chemical landmark, Christine Nawa, Robert Wilhelm Bunsen und sein Heidelberger Laboratorium, Heidelberg, 12. Oktober 2011 (Frankfurt am Main: Gesellschaft Deutscher Chemiker, [2011]). For the date of the summer semester, see Black, Papers and Correspondence of William Stanley Jevons, vol. II, Letter 55 (Roscoe to WSJ, November 1856).

chemistry buildings at Karlsruhe Polytechnic. Lang was closely associated with the polytechnic, both as a student and a teacher.[35] Bunsen drew on his experience of designing the new laboratories at Breslau and made suggestions. As Roscoe noted, "the new building soon became inconveniently crowded, and many applications [by students] for working benches had to be refused."[36]

In addition to a preparation room and a balance room, the new building also had special rooms for gas analysis and electrochemical work. There was an element of doubling up—the preparation room was also the chemical stores and the balance room was also the library. There were two large laboratories with 28 and 22 workplaces for teaching; the smaller one was for beginners. Bunsen had his private laboratory and there was a small laboratory; not for research, as one might assume, but for the carrying out of operations that were not allowed in the large laboratories. As Christine Nawa has pointed out, this was the aptly named *Stinkzimmer*, a room for smelly and dangerous experiments—possibly a result of Bunsen's early work on cacodyl.[37] There was also a small and unassuming lecture theater, and the "handsome" living quarters of the director alongside the laboratory in the southern wing, reached by climbing marble stairs.[38] Edward Festing (1839–1912), the future Director of the Science Museum in London, later remarked: "Professor Bunsen is of the opinion that a laboratory should not be too 'elegant', and his certainly is not so; but it looks thoroughly workmanlike."[39] The laboratories had running water and gas (piped gas had reached Heidelberg in 1853), but as in Justus Liebig's laboratory at Giessen, the sinks drained into conical wooden barrels. The reagents were kept in glass-fronted cupboards; small ones above the bench and larger ones against the walls. The fume cupboards were against the inner walls and large pieces of apparatus were placed on long benches running under the windows.

The presence of piped gas in the new laboratory made possible Bunsen's most famous invention, the Bunsen burner (Figure 4.1). In the early nineteenth century, there were only two types of laboratory heating devices: wick-based lamps fed by alcohol, and the laboratory furnace.[40] The Swiss scientist Ami Argand (1750–1803) developed an oil lamp in 1784 which mixed oil vapor with air, thereby increasing the temperature of the flame.[41] The introduction of illuminating gas obtained from the destructive distillation of coal around 1820 presented an opportunity for a new type of burner. Michael Faraday sketched an early gas burner in his *Chemical Manipulations*, with a jet of gas mixing with air and burning above an inverted, stemless metal

[35] For Lang, see *DB*.

[36] Roscoe, "Bunsen Memorial Lecture," 547.

[37] Nawa, "A Refuge for Inorganic Chemistry," 126; Catherine M. Jackson, "Analysis and Synthesis in Nineteenth-Century Organic Chemistry," PhD diss., University College London, 2009, 250–253.

[38] Described as "a handsome suite" by J. J. Tayler in "Mr. Tayler on Religion in Germany," *The Christian Reformer; or, Unitarian Magazine and Review*, new series, 12 (November 1856): 651–660, on 654. The staircase is mentioned in R. E. Oesper, *The Human Side of Scientists* (Cincinnati, OH: University of Cincinnati, 1975), 28. I am grateful to Robert Baptista for his help with this matter.

[39] E. R. Festing, *Report of Visits to Chemical Laboratories at Bonn, Berlin, Leipzig, etc.* (London: HMSO, 1871), 20. For Festing, see *ODNB* (forthcoming).

[40] One of the authors (PJTM) had a chemistry set in 1971 which contained a spirit lamp with a wick.

[41] Michael Schrøder, *The Argand Burner: Its Origin and Development in France and England, 1780–1800* (Odense, Denmark: Odense University Press, 1969).

NH₃

" The peculiar pungent smell of this
compound is noticed if we heat a bit
of CHEESE in a test-tube."
                    Roscoe and Lunt

HBP De 99

Figure 4.1 "A Dream of Toasted Cheese" by Beatrix Potter, 1899. The mouse is sitting on a Bunsen burner and reading H. E. Roscoe and J. Lunt, *Inorganic Chemistry for Beginners* (1893), from which the quotation is taken.

funnel.[42] Bunsen created a reliable gas burner, based on these principles, in 1855, after Heidelberg was supplied with coal gas. He worked with Roscoe and the "mechanicus" of the Modell-Kabinett (the physics laboratory) Peter Desaga (1812–after 1879). Roscoe takes up the story:

> Returning from my Easter vacation in London, I brought back with me an Argand burner with copper chimney and wire gauze top, which was the form commonly used in English laboratories at that time for working with a smokeless flame. This arrangement did not please Bunsen … the flame was flickering, it was too large, and the gas was so much diluted with air that the flame-temperature was greatly depressed. He would make a burner in which the mixture of gas and air would burn at the top of the tube without any gauze whatsoever, giving a steady, small and hot non-luminous flame under conditions such that it not only would burn without striking down when the gas supply was turned on full, but also then the supply was diminished until only a minute flame was left. This was a difficult … problem to solve, but after many fruitless attempts, and many tedious trials, he succeeded….[43]

Roscoe reported to Bunsen in October 1855, while he was working in London, that

> [y]our little lamps [sic, presumably so-called because they were based on Argand lamps] made by Desaga are admired by everyone, & we find them very useful. A man here has tried to make them but has not succeeded.[44]

In Bunsen's original burner, the proportions of air and gas were controlled by closing air-holes in the base by small stoppers. The familiar rotating collar for adjusting the air supply was introduced by the British instrument maker John J. Griffin (1802–1877) a few years later.

Bunsen's chemistry teaching, in common with many other chemistry professors, was based on chemical analysis, but in his case, he gave new students a sample of igneous rock to analyze, an approach which stemmed from his interest in the geology of Iceland. Roscoe's first chemical paper, coauthored with his fellow student Franz Schönfeld (1834–1911), was submitted on 24 July 1854 to the *Annalen*.[45] It was an analysis of a sample of mica schist from Brixen in the South Tirol (despite the paper being about the composition of gneisses), a sample of gneiss from Brazil and two samples from Norberg in Sweden, and a sample of rock from the north face of Mont Blanc.

---

[42] Michael Faraday, *Chemical Manipulation* (London: W. Phillips, 1827), 107. W. B. Jensen, "The Origin of the Bunsen Burner," *Journal of Chemical Education* 82 (2005): 518.

[43] Roscoe, "Bunsen Memorial Lecture," 547. Georg Lockemann, "The Centenary of the Bunsen Burner," *Journal of Chemical Education* 33 (January 1956): 20–22.

[44] Letter from Roscoe to Bunsen, dated 30 October 1858, quote on page 3, letter 1, Heid. Hs. 2741 III A-41, University of Heidelberg library, reproduced by courtesy of the library. "We" in the letter presumably refers to the chemistry department of UCL, perhaps making it the first department to use them in Britain. There is no mention of the name of the man who tried to make them, but it may have been John J. Griffin.

[45] Franz Schönfeld and H. E. Roscoe "Notiz über die Zusammensetzung einiger Gneise," *Justus Liebigs Annalen der Chemie* 91 (1854): 302–306.

Schönfeld later founded an artist's pigments manufacturing firm in his home city of Düsseldorf which still exists.[46]

Roscoe had already taken his doctoral examinations, which were then entirely oral, and did not require any kind of research. According to his autobiography, Roscoe was required to translate a passage from Virgil's *Aeneid* which he was allowed, as a concession to a foreigner, to translate into English. Alongside his friend August Pauli, Roscoe was orally examined in general and analytical chemistry, stoichiometry, and eudiometry.[47] He was awarded a doctorate in philosophy "cum laude" and also the degree of Master of Arts. Bunsen then taught Roscoe gas analysis, probably alongside Meyer and August Pauli, before Roscoe moved into original research. Around this time, he prepared platinum salts and also studied thermochemistry on the basis of the papers by William Thomson (later Lord Kelvin) (1824–1907), the Danish chemist Julius Thomsen (1826–1909), and James Joule (1818–1889), but he found the mathematics in Thomson's papers difficult to follow.[48] Roscoe made light of his early research in his autobiography and he seemed to have been embarrassed by it in retrospect. His second paper was submitted to the *Annalen* on 4 April 1855 and was on the absorption of chlorine in water, specifically whether it obeyed Henry's Law which states that the absorption of a gas is proportional to pressure.[49] As Roscoe refers to the work of Schönfeld, it is likely that this paper was a direct result of his training by Bunsen in the handling of gases, as well as Bunsen's desire to try out his new absorptiometer which measured the absorption coefficients of gases in water or other solvents (e.g., alcohol). Roscoe concluded that chlorine and water do not obey Henry's Law and that the departure from the law decreases above the temperature at which the postulated chlorine hydrate was formed. Bunsen then entrusted Roscoe with the translation of his forthcoming book on eudiometry, which duly appeared as *Gasometry* in 1857, the same year as the German version, *Gasometrische Methoden*.[50]

In the spring of 1854, Roscoe, accompanied by his mother, his sister Harriet, and her fiancé Edward Enfield, traveled through Germany, visiting Giessen where he met Heinrich Will (1812–1890) and Marburg (where he met Hermann Kolbe) before making more touristic visits to Leipzig and Dresden.[51] The high point of the tour

---

[46] For Schönfeld, see German *Wikipedia*.

[47] The documents relating to his doctoral examination and his printed doctoral certificate are in file H-IV-102/50, University of Heidelberg archives. However, there is no mention in this file of any examination in Latin.

[48] Black, *Papers and Correspondence of William Stanley Jevons*, vol. II, Letter 35 (Roscoe to WSJ, 21 February 1854).

[49] A. E. Roscoe [sic], "Ueber das Verhalten des Chlors bei der Absorption in Wasser," *Justus Liebigs Annalen der Chemie* 95 (1855): 357–372.

[50] Letter from Roscoe to Bunsen, dated 2 March 1855, accepting Bunsen's offer, letter 3, Heid. Hs. 2741 III A-41, University of Heidelberg library; Black, *Papers and Correspondence of William Stanley Jevons*, vol. II, Letter 39 (Roscoe to WSJ, 25 May 1854). Robert Bunsen, *Gasometry, Comprising the Leading Physical and Chemical Properties of Gases*, trans. Henry E. Roscoe (London: Walton and Maberly, 1857); and Robert Bunsen, *Gasometrische Methoden* (Brunswick: F. Vieweg, 1857).

[51] Roscoe, *Life and Experiences*, 58–59; Black, *Papers and Correspondence of William Stanley Jevons*, vol. II, Letter 38 (Roscoe to WSJ, 20 April 1854), quote about Berlin on 72. For Will, see A. W. von Hofmann, "Heinrich Will: Ein Gedenkblatt," *Berichte der Deutschen Chemischen Gesellschaft* 23 (1890): 852–899.

was the visit to Berlin, which he called "a splendid city—fine streets—beats ours in London—fine shops—& magnificent public buildings" at the time, although he later deprecated it in his autobiography, saying it had been "an overgrown village, with detestable pavements, shabby vehicles, and medieval sanitary arrangements." He formed a good impression of its army in contrast to that of Hesse. In Berlin he met Heinrich Rose (1795–1864) and Carl Rammelsberg (1813–1899) (who later succeeded Rose in Berlin).[52] Finally the party went to Munich, where unfortunately Liebig was away, but Roscoe met Liebig's son (presumably Hermann, as Georg was in India at the time) and Max Pettenkofer (1818–1901).[53]

Harriet married Edward Enfield at the British Embassy in Bern on 19 August 1854.[54] This must have been a late change of plan, since on 25 May the marriage was going to take place in England in September.[55] Born in Nottingham in 1811, Enfield was the grandson of the Unitarian minister William Enfield and hence a cousin of the Roscoes.[56] He was educated at Manchester College while it was in York. After being a moneyer at the Royal Mint, he retired with a pension in 1851 when the Mint was reorganized and became a philanthropist with a focus on education. He was involved with University College and University College Hospital, and became President of Manchester New College after it had moved to Gordon Square in London (it is now Harris Manchester College in Oxford). He died in 1880, but Harriet lived until 1919. His first wife Honora, the daughter of the mining engineer and cofounder of UCL John Taylor FRS, had died three days after giving birth to her son on 25 November in 1849, only a year after they were married.[57] His son Ernest William Enfield became a cable engineer, laying underwater telegraph cables, and then a banker with Hart, Fellows & Company, which was taken over by Lloyds in 1882.[58] He died in 1925; his son Ralph Roscoe Enfield was the father of the journalist Edward Enfield and the grandfather of the comedian Harry Enfield.

Friends and Fellow Students

It is worthwhile to look at the wider social circles in which Roscoe moved while he was in Heidelberg. In December 1852 Heidelberg had a population of 14,564, about the current size of Ripon in North Yorkshire.[59] In the following year, Eugène Guinot, a French journalist and playwright, remarked:

---

[52] Both Heinrich Rose and Rammelsberg are in *DB*. Also see H. A. Miers, "Rammelsberg Memorial Lecture," *Journal of the Chemical Society, Transactions* 79 (1901): 1–43.

[53] For Pettenkofer, see *DB*.

[54] *The Gentleman's Magazine, and Historical Review* 196 (November 1854): 504.

[55] Black, *Papers and Correspondence of William Stanley Jevons*, vol. II, Letter 39.

[56] For Enfield, see *ODNB*.

[57] https://www.wikitree.com/wiki/Taylor-30326; for John Taylor, see *ODNB*.

[58] https://atlantic-cable.com/CableStories/Enfield/index.htm.

[59] Census figure from the entry on Heidelberg in *Wikipedia*.

The population of Heidelberg is composed of about fifteen thousand *Philistines* and seven hundred students. *Philistines* is the name given to the citizens of the town by the students, but why or wherefore it is not easy to ascertain.[60]

In actual fact, the university had 719 students in the winter semester of 1853–1854, the term when Roscoe arrived.[61] This is roughly the same size as a large Oxbridge college today. So clearly there was every opportunity for Roscoe to know his fellow students well. The chemistry cohort when Roscoe arrived was a rather unusual one. There was only one student from the previous semester and there were proportionally a large number of foreign students. There were 13 students all told, of which four were from outside Germany and three of whom were from Britain. Roscoe was good friends with Edmund Atkinson and William Russell. He sailed the skiff "Lady Margaret" with Atkinson on the River Neckar to the bemusement of the locals.[62] Atkinson was the son of a pharmacist in Lancaster and was later Professor of Experimental Science at the Royal Military Academy, Sandhurst. Russell was from Gloucester and studied chemistry at University College before he went to Heidelberg. He applied for the Chair in Chemistry at Owens College but was beaten by Roscoe, despite being Frankland's assistant there. He later taught at St. Mary's Hospital and then St. Bartholomew's Hospital, in both cases in succession to August Matthiessen (1831–1870). Matthiessen himself arrived at Heidelberg in the summer semester of 1854 and quickly became part of Roscoe's circle. He had studied under Liebig in Giessen before coming to Heidelberg. He suffered from a muscular twitch caused by a paralytic seizure during his infancy. Sadly, after carrying out useful work on alkaloid chemistry and the electrical properties of materials, he committed suicide, by swallowing hydrocyanic acid, in 1870.[63] Despite his closeness to these students, in his autobiography Roscoe did not mention Henry Eastlake (1835–1869), who arrived with Matthiessen in the summer semester of 1854, nor Walter Leaf (1835–?) and Thomas Nicholas who arrived in the winter semester of 1854–1855, although he must have surely known them.[64] Leaf then went to Owens College, where he studied medicine, and he then became a general practitioner and a surgeon. However, Roscoe was close to the other foreign student in his cohort, namely Friedrich Konrad Beilstein (1838–1906) from Russia, who became a leading chemist in his home city of St. Petersburg.[65] He is best known today for his handbook of organic chemistry. In his autobiography, Roscoe recalls that there were American students, but admitted he had forgotten their

[60] Eugène Guinot, *A Summer in Baden-Baden* (London: J. Mitchell, n.d.). The date for this work is unknown as it does not appear in the book; I have selected 1853, the year given several times in WorldCat, which fits with the actual figures for the population and the students.

[61] *Addressbuch: Winter-halbjahr 1853–1854* (Heidelberg: Karl Winter, 1853), 26.

[62] For Atkinson, see Brock "Bunsen's British Students," 219; and *Journal of the Chemical Society, Transactions* 79 (1901): 888–889; for Russell, see *ODNB* and *Journal of the Chemical Society, Transactions* 113 (1918): 339–350; for the skiff, see Roscoe, *Life and Experiences*, 63.

[63] For Matthiessen, see *ODNB* and *Journal of the Chemical Society* 24 (1871): 615–617.

[64] Brock, "Bunsen's British Students," 223 (Eastlake) and 227 (Leaf and Nicholas).

[65] For Beilstein, see Otto N. Witt, "Friedrich Konrad Beilstein," *Journal of the Chemical Society, Transactions* 99 (1911): 1646–1649; and Elena Roussanova, *Friedrich Konrad Beilstein: Chemiker zweier Nationen. Sein Leben und Werk sowie einige Aspekte der deutsch-russischen Beziehungen in der Chemie in der zweiten Hälfte des 19. Jahrhunderts im Spiegel seines brieflichen Nachlasses* (Hamburg: Norderstedt, 2007).

names.[66] The most prominent of these students was Edwin Reakirt (1832–1888), the son of Joseph Reakirt of the pharmaceutical firm of J & J Reakirt in Philadelphia and an alumnus of the University of Pennsylvania. He became an operator of coal mines and gas-works in later life.[67]

Curiously, Roscoe does not mention many of the German students in his cohort or in the following years. Few of these students became leading academic chemists and therefore we know nothing about the subsequent careers of most of the German chemists in his cohort. As we have already seen, Heinrich Meidinger became a physicist at Karlsruhe Polytechnic. Hubert Grouven (1831–1884) became an agricultural chemist and promoted the use of Peruvian guano as fertilizer.[68] Other students moved into industry, most notably Eugen de Haën (1835–1911), who joined in the winter semester of 1854 and later founded the firm of de Haën.[69] Perhaps the socially most prominent student in Roscoe's time was Ernst vom Rath (1836–1920), who came from a well-connected family of industrialists in Duisberg. Ernst vom Rath himself founded a relatively unimportant and short-lived banking firm (Köster, vom Rath) in 1868, but his brother Adolph, who had previously worked for the Deichmann Bank, cofounded the Deutsche Bank in 1870.[70] His nephew Walther vom Rath was a senior member of the Aufsichtsrat of IG Farben, having married the daughter of Wilhelm Meister of Hoechst.[71]

The students who arrived in the winter semester of 1855–1856 were an interesting group. The total number of students was 32, a large increase from 24 in the previous semester and presumably a result of the new laboratory now being fully open.[72] However, the number of foreign students remained at 12 (of whom 7 were British) and the new intake was largely from Baden (5) and its neighboring states (4), in contrast to no students from Baden in the summer semester of 1855. Jacob Volhard (1834–1910), who was the leading member of this cohort, eventually became Professor of Chemistry at Halle and is best known today for the Hell–Volhard–Zelinsky halogenation.[73] He trained under Liebig in Giessen and worked with him in Munich. Theodor Simmler (1833–1873) from Switzerland went on to work with Bunsen's successor in Breslau, Carl Löwig (1803–1890), and then held various academic positions in his homeland.[74] He cofounded the Swiss Alpine Club in 1863. Ludwig Mautner (1835–1918) from Bohemia, later Ritter von Markhof, seems to have been sent to Heidelberg

[66] Roscoe, *Life and Experiences*, 55.

[67] *University of Pennsylvania: Biographical Catalogue of the Matriculates of the College … 1749–1893* (Philadelphia: Society of the Alumni, 1894), 176. and history of J. & J. Reakirt at https://www.wdl.org/en/item/9330/.

[68] For Grouven, see German *Wikipedia*.

[69] For de Haen, see *DB*.

[70] http://histmath-heidelberg.de/heidelberg/Personen/Koester.htm; Werner Plumpe, Alexander Nutzenadel, and Catherine Schenk, *Deutsche Bank: The Global Hausbank, 1870–2020* (London: Bloomsbury Business, 2020), 12, 150, and 201.

[71] Jens Ulrich Heine, *Verstand und Schicksal* (Weinheim: VCH, 1990), 233–236.

[72] *Addressbuch: Winter-halbjahr 1855–1856* (Heidelberg: Julius Groos, 1855).

[73] Daniel Vorländer, "Jacob Volhard," *Berichte der deutschen chemischen Gesellschaft* 46 (1912): 1855–1902.

[74] For Simmler, see *Historisches Lexicon der Schweiz* at https://hls-dhs-dss.ch/de/articles/044281/2012-12-19/. For Löwig, see Hans Heinrich Landolt, "Carl Löwig," *Berichte der deutschen chemischen Gesellschaft* 23 (1890): 905–909 and *DB*.

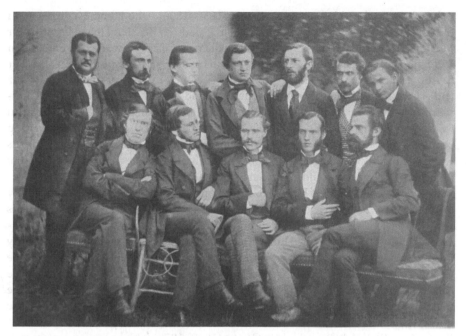

Figure 4.2   Photograph of Kekulé's circle, Heidelberg, September 1857. Front row, left to right: J. Bahr; H. Landolt; G. Carius; A. Kekulé; L. von Pebal; Back row: E. Gaupillat; A. Frapolli; A. Wagner; H. Roscoe; Lothar Meyer; A. Pavesi; F. Beilstein.
From R. Anschütz, *August Kekulé*, volume 1, *Leben und Wirken* (Berlin: Verlag Chemie, 1929), facing 64.

by his aristocratic family for business training, but afterward he financially ruined three companies in a row.[75]

Several of the people Roscoe recalled in his autobiography appear in a remarkable souvenir photograph (Figure 4.2) taken professionally for Reinhold Hoffmann (1831–1919), August Kekulé's first student and friend, who went into industry, and later became a partner in Kalle and a cofounder of Griesheim-Elektron.[76] It was one of the earliest photographs taken in Heidelberg and it had to be taken in the open air because the photographer could not fit such a large group in his studio. The date is uncertain, but it was probably taken in late September or early October 1856, soon after Hoffmann and Kekulé had returned from London, and just before Roscoe returned to London. Roscoe himself is in the middle of the back row, leaning on the shoulder of Lothar Meyer. He seems rather despondent; perhaps his departure from Heidelberg was imminent. In his autobiography, he mentions Agostino Frapolli (1824–1903), later a professor in Milan, in an institution which later became Milan

[75] "Mautner Markhof," https://www.dynastiemautnermarkhof.com/en/adolf-ignaz-ritter-mautner-von-markhof/ludwig/
[76] Richard Anschütz, *August Kekule*, vol. 1: *Leben und Wirken* (Berlin: Verlag Chemie, 1929), 64–65. I wish to thank Ernst Homburg and Christoph Meinel for their help with this photograph. For Hoffmann, see German *Wikipedia* and *DB* (outline only).

Polytechnic; Lothar Meyer; Angelo Pavesi (1830–1896), who became a professor in the Advanced School of Agriculture in Milan; Friedrich Beilstein; Johan Fredrick (or Johann Friedrich) Bahr (1805–1875), a Swedish chemist who worked in Stockholm and Uppsala; Hans Landolt (1831–1910), who became a professor in Bonn and later Berlin; Leopold von Pebal (1826–1887), who became Professor of Chemistry at Graz and was murdered by his laboratory technician; and August Kekulé (1829–1896) himself.[77] Ironically, Hoffmann was absent as the original photograph had to be retaken and he was ill when the second photograph was taken. It is clear that this was a group centered on Kekulé rather than Bunsen. One wonders if Roscoe recalled these friends while writing his autobiography by looking at his copy of the photograph, as all the members of the group would have been given a print.

Roscoe's recollections of his Italian friends are good examples of the unreliability of his autobiography.[78] He describes Frapolli as the well-known senator, but he was neither well-known (even in his lifetime), nor was he a senator.[79] One can only conclude that Roscoe has confused him with his elder brother Lodovico, who was a fairly prominent Deputy (not a Senator) in the Italian parliament and Grand Master of the Grand Orient of Italy lodge of freemasons between 1867 and 1870.[80] Similarly, he calls Pavesi a "professor at Pavia," which indeed he was for 10 years, but he spent the majority of his career before his death in 1896 in Milan.[81] For whatever reason, he does not mention two other members of the lunch group; namely, the French chemist Ernest Gaupillat and Georg Ludwig Carius (1829–1875) who became a professor in Marburg.[82]

Other friends and colleagues that Roscoe mentions in his autobiography included Adolf Baeyer (1835–1917), a student of Kekulé who was later a leading organic chemist and ennobled as Adolf von Baeyer, and Agostinho Vicente Lourenço (1822–1893), a Portuguese chemist from Goa who pioneered condensation polymerization while working with Adolphe Wurtz in Paris before becoming a professor in Lisbon (it appears that he was only briefly in Heidelberg).[83] Roscoe also mentions Leon Nikolaevich Shishkov (or Schiskoff) (1830–1909), an aristocratic graduate of the artillery school in St. Petersburg who worked with Bunsen on the chemistry of explosives, then went back to Russia to teach at the Mikhailovskaya Artillery Academy

---

[77] Roscoe, *Life and Experiences*, 54–55 and 415. For Bahr, see https://sok.riksarkivet.se/sbl/Presentation.aspx?id=19011; for Landolt (he is most famous today for inventing the iodine clock reaction), see Richard Pribram, "Hans Heinrich Landolt," *Berichte der deutschen chemischen Gesellschaft* 44 (1911): 3337–3394; for Pebal, see Lothar Meyer, "Leopold von Pebal," *Berichte der deutschen chemischen Gesellschaft* 20 (1887): 997–1015; for Kekulé, see *DB*; Francis R. Japp, "Kekulé Memorial Lecture," *Journal of the Chemical Society, Transactions* 73 (1898): 97–138; and Alan J. Rocke, *Image and Reality: Kekulé, Kopp, and the Scientific Imagination* (Chicago: University of Chicago Press, 2010).

[78] Roscoe, *Life and Experiences*, 54.

[79] Personal communications from Franco Calascibetta, both dated 19 May 2021. I wish to thank Franco for his assistance with the biographies of Italian chemists.

[80] *Wikipedia* entry on the Grand Orient of Italy.

[81] For Pavesi, see https://www.treccani.it/enciclopedia/angelo-pavesi_(Dizionario-Biografico).

[82] For Carius, see *DB*.

[83] For Baeyer, see *DB*; and William Henry Perkin, "Baeyer Memorial Lecture" *Journal of the Chemical Society, Transactions* 123 (1923): 1520–1546.

before retiring to his estates in 1865 after a series of family tragedies, where he set up a private laboratory.[84]

Roscoe's close relationship with two important physicists is particularly noteworthy. Hermann von Helmholtz (1821–1894) did not come to the University of Heidelberg until 1858 where he was Professor of Physiology; he was called to the Chair of Physics in Berlin in 1871.[85] However, he married Anna von Mohl (1834–1899), the daughter of Robert von Mohl, as his second wife in 1861. It was through Anna von Mohl that Roscoe came to know him well, and they remained close friends.[86] Helmholtz visited Roscoe in Manchester and met Joule on two occasions; in March 1864 and almost exactly 20 years later, in April 1884. The second occasion was a disappointment, as Joule was now suffering from what was probably dementia and he was unable to discuss scientific matters with his German visitor.[87] Roscoe also arranged for Helmholtz to be the first physicist to be invited to give the Chemical Society's Faraday Lecture in the spring of 1881 and organized his visit to Britain. Almost inevitably this involved a visit to Roscoe's home in Manchester.[88] Georg Quincke (1834–1924) came to Heidelberg to study physics in the winter of 1854, and after holding an extraordinary professorship in Berlin and a professorship in Würzburg, he returned to Heidelberg as the Professor of Physics in 1875.[89] He took part, with Bunsen and Roscoe, in the Karlsruhe conference of 1860 (see below); when he died in 1924, he may have been the last surviving participant of that conference. Another friend, who was also a friend of Bunsen and Helmholtz, was the Jewish mathematician and historian of science Leo Königsberger (1837–1921), who was a professor at Heidelberg between 1869 and 1875, and again between 1884 and 1920; he wrote a biography of Helmholtz.[90] Roscoe knew the chemist and historian of chemistry Hermann Kopp (1817–1892), whom he described as the "the ideal type of the old German professor," through Bunsen, as he did not arrive in Heidelberg from Giessen until 1864.[91] Finally, although Roscoe was mostly drawn to scientists, he was a friend of Adolph von Vangerow (1808–1870), a member of a noble Pomeranian family, who was Professor of Roman Law at Heidelberg from 1840 to 1870; Roscoe probably knew him through Professor von Mohl.[92] He also met the literary critic Leslie Stephens (1832–1904) by chance while the latter was punting

---

[84] Personal communications from David Lewis, both dated 21 May 2021. I wish to thank David for this assistance with the biography of Shishkov, for whom see http://calendar.lib48.ru/all-dates/shishkov-leon-nikolaevich-1830-1909 and http://person.lib48.ru/shishkov-leon-nikolaevich.

[85] For Helmholtz, see DB; George Francis Fitzgerald, "Helmholtz Memorial Lecture," Journal of the Chemical Society, Transactions 69 (1896): 885–912; and Cahan, Helmholtz.

[86] Roscoe, Life and Experiences, 89–92 and 94–95. Roscoe perhaps overstates the importance of his relationship with Helmholtz; from Cahan, Helmholtz, it appears that Helmholtz was much closer to Thomson (Baron Kelvin) in the circle of his British friends.

[87] Roscoe, Life and Experiences, 120; Cahan, Helmholtz, 296 and 593. Roscoe incorrectly says Helmholtz's visit in 1884 was his only meeting with Joule. Donald S.L. Cardwell, James Joule: A Biography (Manchester: Manchester University Press, 1989), says nothing about either visit, despite devoting an appendix of the book to Joule and Helmholtz.

[88] Roscoe, Life and Experiences, 89–92; Cahan, Helmholtz, 567–576.

[89] For Quincke, see DB.

[90] Roscoe, Life and Experiences, 94–95. For Leo Königsberger, see DB.

[91] Roscoe, Life and Experiences, 96–97, quote on 96. For Kopp, see DB; T. E. Thorpe, "Kopp Memorial Lecture," Journal of the Chemical Society, Transactions 63 (1893): 775–815; and Rocke, Image and Reality.

[92] Roscoe, Life and Experiences, 89. For Vangerow, see DB.

on the river in Heidelberg (he was in the town to learn German) and they remained lifelong friends.[93]

What influence did Roscoe's social circle have on his subsequent career, and what does it tell us about Roscoe's outlook? We can with certainty say that Roscoe's social circle at Heidelberg did not affect his future career; there is no one here who prompted Roscoe's career or worked alongside him (Carl Schorlemmer is absent from the social circle recorded in his autobiography). The one time these social connections came to the fore was during the British Association for the Advancement of Science meeting in Manchester in 1887, when several chemists listed here were photographed with Roscoe and Mendeleev (namely, Lothar Meyer, Quincke, Russell, and Atkinson). What does this social circle tell us about Roscoe? He was clearly more interested in scientists (but physicists as well as chemists) than non-scientists who spoke English, which suggests that he soon became fluent in German. Roscoe was clearly attracted to elite scientists such as Baeyer, Kekulé, and Landolt, rather than the run-of-the-mill students in his own cohort. This was perhaps because he was introduced to scientists who worked closely with Bunsen, such as Bahr and Shishkov. However, we can surmise that he was not anti-Semitic (as he was a friend of Köngisberger) or strongly racist (as he was friendly with Lourenço, who was of Indian origin).[94] It is possible that he was particularly attracted to people with progressive views (as in the case of Stephens), but this is not certain. Finally, Roscoe seems to have had a rather ambivalent relationship with the Americans, even though they were an intellectually strong group among the English-speaking students, since he could not remember the names of the American chemists and does not seem to have been friends with any of the American non-scientists despite sharing their language. Indeed, this seems to have also been true of the British students, apart from his closest friends.

Roscoe's Photochemical Research

In the spring of 1854, soon after taking his doctorate, Roscoe began his long-running research with Bunsen on photochemistry. Bunsen set Roscoe the task of determining the chemical activity of light using chlorine water, and solutions of bromine and iodine, in the presence of organic compounds. However, Roscoe did not find any specific reactions or any relationship between the illumination period and chemical action.[95] To Roscoe's horror, Wilhelm Constantin Wittwer (1822–1908), a Privatdozent in

---

[93] Roscoe, *Life and Experiences*, 63–64. For Stephens, see *ODNB*.

[94] On one occasion, writing to Jevons who was in Australia, Roscoe said "Don't you marry a native my boy, & bring back some black piccaninies [*sic*] I would not come to see you then." While this was clearly a joshing remark between two young men who knew each other well, Roscoe's remark about not visiting Jevons if he married a native woman shows that he was not completely immune from the racial prejudices of his period; Black, *Papers and Correspondence of William Stanley Jevons*, vol. II, letter 87 (10 May 1856), quote on 230. We have not found any other evidence of racism in Roscoe's behavior or writings.

[95] R. W. Bunsen and Henry E. Roscoe, "Photochemical Researches with Reference to the Law of the Chemical Action of Light," in *Report of the Twenty-fifth Meeting of the British Association for the Advancement of Science; Held at Glasgow in September 1855* (London: John Murray, 1856), 48–49. Reference is made herein to a paper on these experiments in the *Quarterly Journal of the Chemical Society* in October 1855, but no such paper seems to exist, although there is one in 1856 (see next footnote).

Munich, then published similar work on chlorine water; although Bunsen and Roscoe believed his conclusions were incorrect.[96] Bunsen then directed Roscoe's attention to the well-known photochemical reaction between chlorine gas and hydrogen gas, first announced by the French chemists Joseph-Louis Gay-Lussac (1778–1850) and Louis-Jacques Thénard (1777–1857) in 1809.[97] This was seemingly a convenient reaction to study as the electrolysis of hydrochloric acid using the Bunsen cell produced an equal mixture of chlorine and hydrogen. On reaction, the color of the chlorine disappears and the product hydrogen chloride is easily estimated. Although Bunsen and Roscoe's papers were published over a number of years, the core of the experimental work appears to have been carried out by Roscoe in the spring and summer of 1855, so there is no need to consider the chronological development of their ideas or the rather complex details of the experimental work. Bunsen and Roscoe did not really understand photochemistry and unsurprisingly had no inkling that the photochemical reaction between chlorine and hydrogen is a chain reaction initiated by the photochemical disassociation of the chlorine molecule. As a result, their work, although pioneering at the time, had practically no influence on the development of modern photochemistry.

Roscoe was again discomfited at the beginning of 1856 to discover that the Anglo-American chemist John William Draper (1811–1882) had already investigated this reaction in 1843 and had invented his "tithonometer" (a type of actinometer) to measure the chemical activity of light.[98] He discovered that there was a peak in the blue-violet region of the visible spectrum of chlorine which faded toward each end of the spectrum. Although he had not been aware of Draper's findings, Bunsen reassured Roscoe that "I think Draper has left us plenty to do."[99] With hindsight, the

[96] Black, *Papers and Correspondence of William Stanley Jevons*, vol. II, letter 39. and letter 75 (Roscoe to WSJ, 15 October 1855). W. C. Wittwer, "Ueber die Einwirkung des Lichts auf Chlorwasser," *Annalen der Physik* 170 (1855): 597–612. Bunsen and Roscoe published their criticisms of Wittwer's paper in "II. Photochemische Untersuchungen," *Annalen der Physik* 172 (1855): 373–394, and "X. Photochemical Researches," *Quarterly Journal of the Chemical Society* 8 (1856): 193–211. Wittwer repeated his experiments and found the same result as before in "XI. Ueber die Einwirkung des Lichtes auf Chlorwasser," *Annalen der Physik* 173 (1856): 304–310. He continued to work on this reaction; see Wittwer, "Ueber die Einwirkung des Lichtes auf Chlorwasser," *Annalen der Physik* 182 (1859): 266–289.This episode is nicely covered by J. R. Partington in his *History of Chemistry*, vol. 4 (London: Macmillan, 1964), 717; Partington clearly had some sympathy for Wittwer and cites Wilhelm Ostwald in his support. There are two oddities about Bunsen and Roscoe's response to Wittwer. They state that their work took almost two years to complete, which would imply that it started around October 1853 as their paper is dated 16 August 1855, and it is stated to be the second paper in the series "Photochemische Untersuchungen," but there is no trace of the first paper in the literature.

[97] Joseph-Louis Gay-Lussac and Louis-Jacques Thénard, "Des Mémoires lus à l'Institut national, depuis le 7 mars 1808 jusqu'au 27 février 1809," *Mémoires de physique et de chimie, de la Société d'Arcueil* 2 (1809): 295–358 on 349–350. Available online at https://gallica.bnf.fr/ark:/12148/bpt6k1050356z/f517. item. For Gay-Lussac, see Maurice Crosland, *Gay-Lussac, Scientist and Bourgeois* (Cambridge: Cambridge University Press, 1978).

[98] Letter from Roscoe to Bunsen, dated 2 December 1855, Heidelberg University Library, file 2741. John W. Draper, "XXX. On Some Analogies between the Phenomena of the Chemical Rays, and Those of Radiant Heat," *Philosophical Magazine*, series 3, 19 (1841): 195–210; Draper, "XLIX. Description of the Tithonometer, an Instrument for Measuring the Chemical Force of the Indigo-Tithonic Rays," *Philosophical Magazine*, series 3, 23 (1843): 401–415. Both volumes are available at the Internet Archive. For Draper, see Klaus Hentschel, "Why Not One More Imponderable?: John William Draper and His Tithonic Rays," *Foundations of Chemistry* 4 (2002): 5–59.

[99] Letter from Bunsen to Roscoe dated 13 January 1856, reproduced in Roscoe, *Life and Experiences*, 60–61. The original letter is in the Deutsches Museum archives, HS 908–1033, letter 00910.

chlorine-hydrogen reaction was an unfortunate choice upon which to base the new field of photochemistry as it is a chain reaction and the chain-length for the formation of hydrogen chloride is so variable, and dependent on so many factors not really within the chemist's control; namely, the pressures, surfaces, impurities, and even the intensity of the light source itself.[100] Initially, Roscoe struggled with his research. In a letter to Jevons in March 1855, he said:

> You must know that I have been working nearly a year upon it—& up to within 6 weeks of the present time I had not got any regular actions, & I was as you may imagine in something like despair. I act upon a mixture of equal vols: of dyCl [*sic*, presumably dry chlorine gas] & H with light & I used to determine the amount of decomposition by estimating the quantity of free chlorine w^h remained undecomposed. This was a very tiresome & troublesome process & although I made a great number of expts I could get no regular action—Now however the apparatus is altered & instead of measuring the action by an analysis, I measure it by the rising of water in a tube connected with the Insolated apparatus, & this rise shows exactly the amount of HCl formed in any given time—water is of course always present in the part of the apparatus w^h is exposed to light—In this way I can make a great number of observations every day—& I have proved
> 1. That the action is proportional to the time of exposure.
> 2. That it is proportional to the amount of light.[101]

These two laws became known as the Bunsen-Roscoe Law, for which see below. Roscoe later recalled it was also uncomfortable work:

> I remember that much of my work was done in the loft of the laboratory, a portion of which was boarded off for me—as I had to work in the dark—and the heat during that summer was abnormal. I used to work with very little clothing on, but stuck at it hard, week after week, meeting with all sorts of discouragements, but at last succeeding in obtaining the wished-for results.[102]

In some respects, however, the two chemists welcomed these difficulties, as they enabled them to make the point that the chemicals had to be as pure as possible and the experimental conditions had to be rigorously controlled. It was fortunate that Bunsen was a master of making apparatus for gas reactions and was an outstanding expert at gas analysis.

Bunsen and Roscoe established that photochemical reactions were a result of the absorption of light by the reactants. They went on to state the Bunsen-Roscoe Law, namely:

---

[100] Personal communication from Richard Wayne, 6 June 2021.
[101] Black, *Papers and Correspondence of William Stanley Jevons*, vol. II, letter 55 (Roscoe to WSJ, 12 March 1855); also see letter 75 (Roscoe to WSJ, 15 October 1855).
[102] Roscoe, *Life and Experiences*, 59.

1. The amount of chemical effected by a constant source of light is directly proportional to the time of exposure.
2. The amount of chemical action effected by the light in equal times is directly proportional to the amount of light.

It is now usually stated as: the effect of radiation is always the same regardless of the variation in the intensity or in the time of radiation as long as the product of the two is kept constant. The law was announced publicly by Roscoe at the British Association for the Advancement of Science meeting at Cheltenham in August 1856.[103] In propounding the Bunsen-Roscoe Law, they were doing little more than restating the earlier Grotthuss-Draper Law. This was first proposed in 1819 by Theodor, Freiherr von Grotthuss (1785–1822), but then largely forgotten until it was independently revived by Draper in 1842.[104]

To use their actinometer with sunlight, Bunsen and Roscoe had to mitigate the strength of the sunlight in some way, as direct sunlight would cause an explosion. They either used a small amount of diffuse sunlight from the zenith or a tiny pencil of sunlight using a heliostat. Bunsen and Roscoe created a new unit to quantify the chemical action of sunlight, the unfortunately named light-meter, which was the height of the column of hydrogen chloride that would be produced by that light passing through an unlimited atmosphere of chlorine and hydrogen. They then determined the chemical action of a given wavelength of light by passing sunlight shining through a slit through two quartz prisms, then placing the actinometer at specific points across the solar spectrum. In this way, Bunsen and Roscoe determined that the reaction was principally caused by the absorption of light around 400 nm and 380 nm, that is to say, in the blue-violet and extreme violet regions of the visible spectrum.[105] We now know that this particular reaction is best carried out with ultraviolet light on the edges of a major absorption band of chlorine at 330 nm, so roughly at 350 nm and 310 nm.[106] Bunsen and Roscoe realized that this so-called action spectrum was specific to the chlorine-hydrogen reaction and that other photochemical reactions, for example the decomposition of silver salts, would have different maxima.

---

[103] Robert W. Bunsen and Henry E. Roscoe, "Photo-chemical Researches," in *Report of the Twenty-Sixth Meeting of the British Association for the Advancement of Science; Held at Cheltenham in August 1856* (London: John Murray, 1857), 62–68, on 67. It is now mainly applied to biological systems rather than photochemistry. See the bibliography of Roscoe's publications for all their papers on photochemistry. An excellent summary of their work was produced by Roscoe in "Light, Chemical Action of," in *Dictionary of Chemistry*, ed. Henry Watts, vol. 3 (London: Longman, Green, 1865), 678–695. Also see Christine Nawa and Christoph Meinel, eds., *Von der Forschung gezeichnete Instrumente und Apparaturen in Heidelberger Laboratorien skizziert von Friedrich Veith (1817–1907)* (Heidelberg: heiBOOKS, Universitätsbibliothek Heidelberg, 2020), https://books.ub.uni-heidelberg.de/heibooks/catalog/book/793.

[104] Angelo Albini, "Some Remarks on the First Law of Photochemistry," *Photochemical and Photobiological Sciences* 15 (2016): 319–324. For Grotthuss, see *DB*.

[105] Robert Wilhelm Bunsen and Henry Enfield Roscoe, "Photo-chemical Researches, Part IV," *Philosophical Transactions of the Royal Society* 149 (1859): 879–926.

[106] D. Marie, J. Burrows, R. Meller, and G. K. Moortgat, "A Study of the UV-Visible Absorption Spectrum of Molecular Chlorine," *Journal of Photochemistry and Photobiology A: Chemistry* 70 (1993): 205–214; Yutaka Matsumi, Masahiro Kawasaki, Tetsuya Sato, Tohoru Kinugawa, and Tatsuo Arikawa, "Photodissociation of Chlorine Molecule in the UV Region," *Chemical Physics Letters* 155 (1989): 486–490.

Bunsen and Roscoe also discovered the inhibition of the reaction.[107] Although they were aware of the inhibiting effect of oxygen, they misinterpreted this delay in the reaction as an "induction period" produced by the resistance of the molecules to taking part in the reaction, which has to be overcome by an initial absorption of photochemical energy. This could be considered to be a forerunner of the concept of activation energy, but (as Jacobus van't Hoff later pointed out) they were fundamentally in error here (just as Draper had been with his idea of light producing an allotropic form of chlorine).[108] The later research of David Leonard Chapman (1869–1958) in Manchester (Chapman moved to Oxford soon afterward) and the even later research of Ronald Norrish (1897–1978) in Cambridge showed that this inhibition was caused by trace impurities which can be very difficult to remove beforehand.[109] These impurities break the chain reaction between chlorine and hydrogen by providing an alternative pathway, and the main chain reaction cannot proceed until all of the impurity has been consumed. Norrish and his student John Griffiths (who later worked at the Laboratory of the Government Chemist) focused on the role of ammonia and nitrogen trichloride (produced by the reaction between ammonia and chlorine). However, any impurity will produce this effect (for example, oxygen, as Chapman pointed out), so it is still not clear which impurity was the cause in Bunsen and Roscoe's research. The contention of Bunsen and Roscoe that excess chlorine or hydrogen produced inhibition was disproved by Chapman and Patrick Sarfield MacMahon.[110]

The long delay in publishing much of this work arose because the two chemists were only able to spend part of the summer together carrying out this research after 1856. The writing up of their papers was also delayed by their heavy workload; rescuing Owens College in the case of Roscoe, and the development of spectroscopy by Bunsen after 1859.[111]

In conclusion, Bunsen and Roscoe's decision to use the unreliable chlorine-hydrogen reaction as the basis for a chemical actinometer to measure the intensity of light was unfortunate, owing to the many unpredictable effects on the reaction; not least because of the explosive nature of the chlorine-hydrogen reaction at high levels of light intensity. The first reliable chemical actinometer was developed by British Admiralty Materials Laboratory chemists C. G. Hatchard and Cecil Parker as late as

[107]   Robert Bunsen and Henry Enfield Roscoe, "XVIII. Photo-chemical Researches, Part II. Phenomena of Photo-Chemical Induction," *Philosophical Transactions of the Royal Society of London* 147 (1857): 381–402.

[108]   J. H. van't Hoff, *Etudes de Dynamique Chimique* (Amsterdam: Frederik Muller, 1884), 74–75 and 82.

[109]   Charles Hutchens Burgess and David Leonard Chapman, "CXXXVIII. The Interaction of Chlorine and Hydrogen," *Journal of the Chemical Society, Transactions* 89 (1906): 1399–1434. John G.A. Griffiths and Ronald G.W. Norrish, "The Photosensitised Decomposition of Nitrogen Trichloride and the Induction of the Hydrogen-Chlorine Reaction," *Transactions of the Faraday Society* 27 (1931): 451–458. Griffiths and Norrish, "The Induction Period of the Photochemical Reaction between Hydrogen and Chlorine," *Proceedings of the Royal Society of London, A. Mathematical and Physical Sciences* 147 (1934): 140–151. For Chapman, see *ODNB* and Edmund John Bowen, "David Leonard Chapman," *Biographical Memoirs of Fellows of the Royal Society* 4 (1958): 34–44; for Norrish, see *ODNB* and Frederick Sydney Dainton and Brian Arthur Thrush, "Ronald George Wreyford Norrish," *Biographical Memoirs of Fellows of the Royal Society* 27 (1981): 379–424.

[110]   David Leonard Chapman and Patrick Sarsfield MacMahon, "XIX. The Interaction of Hydrogen and Chlorine," *Journal of the Chemical Society, Transactions* 95 (1909): 135–138.

[111]   In his letter to Roscoe, dated 11 October 1857, Bunsen counsels him to put his photochemical work aside and concentrate on his teaching; Deutsches Museum Archives letter 00919.

1956, using the photoreduction of potassium ferrioxalate in solution.[112] Bunsen and Roscoe's interest in the photochemical action of daylight is understandable given the importance of sunlight in biological processes, but sadly their research did not have any lasting impact.[113]

London Interlude

On 8 July 1855, when Roscoe was preoccupied with his photochemical research, his old teacher Alexander Williamson wrote to him, asking if he would be willing to become his assistant for the winter semester of 1855–1856. As this letter shows Williamson's regard for Roscoe and his teaching methods, it is worth quoting at length:

The Council appointed me yesterday afternoon Professor of Chemistry, including Practical Chemistry, and my first act in that capacity is to ask you to give the College the benefit of your services as assistant to the general Chemistry Class. The duties of that office will be laborious enough, especially during the first session, & require a man of scientific acquirements & skill. In fact, they will occupy in all probability the full business time of every day in the week for six months. They will consist of preparing the experiments for the lectures & often assisting me to devise them, in recording the experiments in a journal, and last, though not least, in assisting me in the instruction of the class in the following manner:

I intend giving the students exercises in the most important points taught in the lectures, and shall be anxious to get the whole class to do them. These will of course have to be corrected; and in many cases it will be desirable to explain to individual students whatever they may have failed to understand in the lectures. Now, the correcting of these exercises & conferring with individual students will be one of the most important duties of the assistant, and one which I imagine will be far from disagreeable or unprofitable. I may state that I intend mentioning the assistant's name in the prospectus if, as I trust, no objection is made to my doing so by the Senate. So the engagement I propose to you is for six months, commencing from the first October, and as honorarium I would propose the sum of sixty guineas (£63), or it will be better to say from the 15th September to the end of March at the above rate, as we shall have a good deal to prepare before beginning.

I need hardly assure you how pleasant it will be to me personally, & satisfactory officially, to have you with me, and that I should do my best to make the arrangement comfortable to you & conducive to your advancement. But I ought to say at once that

[112] C. G. Hatchard and C. A. Parker, "A New Sensitive Chemical Actinometer: II. Potassium Ferrioxalate as a Standard Chemical Actinometer," *Proceedings of the Royal Society of London, A. Mathematical and Physical Sciences* 235 (1956): 518–536. For a recent discussion of this actinometer, see B. Wriedt and D. Ziegenbalg, "Application Limits of the Ferrioxalate Actinometer," *ChemRxiv* (Cambridge: Cambridge Open Engage, 2021) https://chemrxiv.org/engage/chemrxiv/article-details/60c7597af96a00617128901c.
[113] Bunsen and Roscoe, "Photo-Chemical Researches, Part IV"; Bunsen and Roscoe, "Photo-chemical Researches, Part V. On the Direct Measurement of the Chemical Action of Sunlight," *Philosophical Transactions* 153 (1863): 139–160.

I don't think you would under this arrangement have time during the day for private work, so that the only kind of work which you could carry on simultaneously with it would be evening work, such as translation.[114]

Roscoe later remarked to Jevons that "as nothing better seems likely to turn up & as the £60 which he offered me for the winter although not much still was better than 'a poke in the eye with a burnt stick.'"[115] Roscoe went to Paris in the middle of August having "worked away like a trooper all spring & summer" on his photochemical research. He saw all the sights in Paris and went to the international exhibition with several German friends and then arrived in England with one German friend (presumably William Dittmar).[116] He then acted as Williamson's assistant on the introductory chemistry course between mid-September 1855 and the end of March 1856, returning to Heidelberg for the summer semester at the end of April.[117] It was not altogether an easy relationship, as Williamson was vexed that Roscoe had not followed his line of research and did not regard him as being of his school.[118] It was also hard work which left no time for research. According to Roscoe, "We had many new & striking experiments, & in the arrangement of a new Museum & making apparatus & preparations I can assure you that I had enough to do."[119] This was probably a reference to the new geological museum at University College, but it may have been a chemical museum.[120] In the late spring and summer of 1856, Roscoe returned to Heidelberg to resume his research on photochemistry.[121] Bunsen had planned a trip to Italy and Sicily to visit the volcanic areas such as Etna, Vesuvius, and the Liparic islands, as he had done in 1843, but it was delayed until the autumn of 1857.[122]

Roscoe left Heidelberg in the autumn of 1856, soon after the group photograph with Kekulé's circle and probably around the end of September or early October. His training under Bunsen had been completed and he now sought a proper career as a

---

[114] Letter from A. Williamson to H. E, Roscoe, dated 8 July 1855, LT0247, Roscoe Collection, Royal Society of Chemistry, reproduced in Roscoe, *Life and Experiences*, 37–38, with slight differences, mainly the replacement of "&" with "and." However, in the original, the passage "in assisting ... manner:" is written in the original as "in the following manner, assisting me in the instruction of the class" and then amended. The underlinings are in the original document. Reproduced with permission of Royal Society of Chemistry Library. £63 in 1855 would be worth £5,400 in 2023. Also see J. Harris and W. H. Brock, "From Giessen to Gower Street: Towards a Biography of Alexander William Williamson (1824–1904)," *Annals of Science* 31 (1974): 95–130.

[115] Quote in Black, *Papers and Correspondence of William Stanley Jevons*, vol. II, Letter 75 (Roscoe to WSJ, 15 October 1855), on 191. It is not clear why Roscoe refers to £60, not 60 guineas, perhaps he confused guineas and pounds in his mind despite Williamson helpfully giving the conversion.

[116] Ibid., quote on 190.

[117] Roscoe's letter to Bunsen dated 2 March 1856 letter 3, Heid. Hs. 2741 III A-41, University of Heidelberg library.

[118] Black, *Papers and Correspondence of William Stanley Jevons*, vol. II, Letter 87.

[119] Ibid., Letter 87, quote on page 228.

[120] Note 4 on page 228 of letter 87 says it was a geological museum; Morris, *The Matter Factory*, chapter 8, "Chemical Museums," 198–231. The noted historian of UCL's chemistry department, Alwyn Davies, is not aware of a chemical museum from this period; personal communication from Alwyn Davies, 6 June 2021.

[121] Black, *Papers and Correspondence of William Stanley Jevons*, vol. II, Letter 87. Also see Roscoe's letter to Bunsen, dated 2 March 1856, Heidelberg University Library, file 2741.

[122] Black, *Papers and Correspondence of William Stanley Jevons*, vol. II, Letters 87; letter from Bunsen to Roscoe, dated 11 October 1857, Deutsches Museum Archives, letter 00919, in which he states that he received Roscoe's letter, which does not seem to be extant, in Naples where he saw Vesuvius erupting.

chemist, as we will see in the next chapter. Roscoe believed that he would have been a stronger candidate than Charles Bloxam (1831–1887) for the Chair of Chemistry vacated by the early death of John Eddowes Bowman (1820–1856) at King's College, London, but he was barred by the fact he was not an Anglican, exclaiming, "What humbug that is!"[123] Clearly, he would have to find a position where there was no religious test. In the meantime, he would have to subsist on private teaching.

Heidelberg Summers

After 1856, Roscoe returned to Heidelberg in the summer vacations to work with Bunsen on their joint photochemical researches until he was married in 1863, after which he had more conventional summer holidays. Nonetheless, even in this period, he was able take some time traveling around:

> I was usually able to get away from Manchester at the beginning of July and worked steadily in Heidelberg until the middle or end of August, in the years from 1857 to 1862, when Bunsen and I, accompanied sometimes by Kirchhoff and sometimes by Häuser, the well-known historian, or by both of these intimate friends, made excursions into the Bavarian Highlands, the Tyrol, and Switzerland. I have often regretted that I never made notes of the incidents which occurred on these excursions, and of the humour and wit of my companions, especially of Häuser, which was a never failing source of amusement.[124]

Spectroscopy

By the 1850s, chemists were aware of the value of colored flames to detect small amounts of certain metals. There were, however, three problems with this method. A metal with a bright flame color (notably the yellow flame of sodium salts) would mask other metals (in particular potassium).[125] The existing burners were not very effective at producing

---

[123] Black, *Papers and Correspondence of William Stanley Jevons*, vol. II, Letter 87, quote on 229. For Bloxam, see *ODNB* and D. I. Davies, "Charles Loudon Bloxam," *Analytical Proceedings* 18 (1981): 327–331; for Bowman, who died of influenza aged 35, see *ODNB* and *Quarterly Journal of the Chemical Society* 9 (1857): 159–160.

[124] Roscoe, *Life and Experiences*, 61–62. Ludwig Häusser (1818–1867), to give the more usual spelling of his name, was born in Alsace. He became a Professor of History in Heidelberg and a leading liberal politician in Baden, being a cofounder of the forerunner of the *Süddeutschen Zeitung*. He is best-known for his four-volume *Deutsche Geschichte vom Tode Friedrichs des Grossen bis zur Gründung des deutschen Bundes* (1854–1857). He is in the *DB* and is mentioned in passing in Tuchman, *Science, Medicine and the State*, with Vangerow.

[125] There is a huge literature on the origins of spectroscopy. M. A. Sutton, "Sir John Herschel and the Development of Spectroscopy in Britain," *British Journal for the History of Science* 7 (1974): 42–60; M. A. Sutton, "Spectroscopy and the Chemists: A Neglected Opportunity," *Ambix* 23 (1976): 16–26; F.A.J.L. James, "The Creation of a Victorian Myth: The Historiography of Spectroscopy," *History of Science* 13 (1985): 1–24; M. A. Sutton, "Spectroscopy, Historiography and Myth: The Victorians Vindicated," *History of Science* 24 (1986): 425–432; John B. Hearnshaw, *The Analysis of Starlight: Two Centuries of Astronomical Spectroscopy* (Cambridge: Cambridge University Press, 2014); John C.D. Brand, *Lines of Light: The Sources of Dispersive Spectroscopy, 1800–1930* (Luxemburg City: Gordon and Breach, 1995); William McGucken,

these colors. Furthermore, not all metals produce colored flames. The first of these problems was solved by another British chemist working in Bunsen's laboratory. In 1858 Rowlandson Cartmell (1824–1888) suggested the use of an indigo solution in a hollow glass wedge, or the rather more portable cobalt glass to filter out the yellow sodium flame.[126] Bunsen now started working with a range of filters to analyze materials in the new Bunsen burner with its hot gas flame, which solved the second problem.

His friend and collaborator Kirchhoff then suggested it would be easier to use a prism to distinguish the different metals, rather than filters. This step also solved the problem of detecting metals which did not produce a colored flame. Bunsen and Kirchhoff converted what had just been a prism in the hands of other scientists— literally so in the case of Wollaston, who remarkably discovered the dark lines while holding a prism in his hand—into an analytical instrument. Thanks to Bunsen's exquisite experimental skills and his care to obtain high purity in his reagents—thus eliminating the D line of sodium previously found in nearly all spectra—these two German scientists converted spectroscopy into a practical technique.[127]

Kirchhoff made a major breakthrough when he realized that the bright lines produced in the Bunsen burner flame corresponded to the dark lines discovered in the solar spectrum by William Hyde Wollaston (1766–1828) in 1802 and independently by Joseph Fraunhofer (1787–1826) in 1814.[128] At first, Bunsen and Kirchhoff thought that passing sunlight through a flame colored by sodium, for example, would convert the dark solar line into a bright line, but in fact this process actually enhanced the dark line if the sunlight was bright enough. Kirchhoff then realized that the dark line was created by the absorption of sunlight by the sodium in the flame. If that was the case, they could reproduce the dark lines in the laboratory and thus easily work out which element corresponded to these dark lines—at least in theory.[129] But how did they first get the idea of passing sunlight through their apparatus? It is claimed that it was a result of their detection of barium and strontium in the blaze of a fire in Mannheim, 12 miles away, which made Bunsen wonder if they could repeat this process with the even more distant Sun.[130] Leaving aside whether they could have seen

Nineteenth-Century Spectroscopy: Development of the Understanding of Spectra, 1802–1897 (Baltimore, MD: Johns Hopkins Press, 1969).

[126] For Cartmell see Brock, "Bunsen's British Students," 221, and James, "The Creation of a Victorian Myth."

[127] F.A.J.L. James, "The Establishment of Spectro-Chemical Analysis as a Practical Method of Qualitative Analysis, 1854–1861," *Ambix* 30 (1983): 30–53. Also see Jochen Hennig, *Der Spektralapparat Kirchhoffs und Bunsens* (Berlin: Verlag für Geschichte der Naturwissenschaften und Technik, 2003) and Hennig, "Die spektroskopischen Arbeiten von Gustav Kirchhoff und Robert Bunsen," in *Kanonische Experimente der Physik: Fachliche Grundlagen und historischer Kontext*, ed. Peter Heering (Berlin: Springer, 2022), 153–168.

[128] For the discovery of the dark lines, see Melvyn C. Usselman, *Pure Intelligence: The Life of William Hyde Wollaston* (Chicago: University of Chicago Press, 2015), 72–75. For Fraunhofer, see *DB* and Myles W.W. Jackson, *Spectrum of Belief: Joseph von Fraunhofer and the Craft of Precision Optics* (Cambridge, MA: MIT Press, 2000).

[129] Owen Gingerich, "The Nineteenth Century Birth of Astrophysics," in *Physics of Solar and Stellar Coronae: GS Vaiana Memorial Symposium*, ed. Jeffrey F. [*sic*, should be L.] Linsky and Salvatore Serio (Dordrecht: Kluwer, 1993), 47–58, on 47–48.

[130] "Some Scientific Centres: VI. The Heidelberg Physical Laboratory," *Nature* 65 (24 April 1902): 587–590. It is unfortunate that this report is anonymous, so we cannot gauge the credibility of its author. Could it have been Roscoe?

the fire from their laboratory and whether their spectroscope would have worked at such a distance, there was a much simpler reason why Bunsen would have thought of examining the Sun with the spectroscope. With Roscoe he was examining the action of sunlight on their hydrogen-chlorine actinometer, and they were specifically concerned with the colors of the solar spectrum and the dark lines in that spectrum as a way of determining the approximate wavelengths of their action spectrum.[131] It would have been a natural extension to examine these dark lines in terms of the new spectral lines revealed by the spectroscope; not least because they were already using the heliostat, which was central to the analysis of the solar spectral lines by Bunsen and Kirchhoff.

Kirchhoff gave a lecture on this new discovery to the Berlin Academy of Sciences on 27 October 1859, and Bunsen wrote to Roscoe on 15 November telling him that he and Kirchhoff could not sleep because of their excitement about their wonderful breakthrough, which enabled them to determine the chemical composition of the Sun and stars with the same certainty as they could analyze chemicals in the laboratory.[132] This was presumably when Roscoe first heard about Bunsen and Kirchhoff's work on atomic spectra.

Bunsen and Kirchhoff were soon casting around to find new samples to subject to their relatively simple spectroscope, which forced Bunsen to put his photochemical work with Roscoe to one side.[133] In the spring of 1860, they tried the waters from Bad Dürkheim, a spa town best known for its good wines and its giant wine vat, in the neighboring Bavarian Palatinate, 22.5 miles (as the crow flies) from Heidelberg. At first, Bunsen and Kirchhoff only observed the lines of known alkali and alkaline earth metals. However, when the alkaline earth metals and most of the lithium had been removed, they saw two remarkable blue lines close together for the first time. Bunsen called this new element "cæsium," after *cæsius*, the Latin for the light blue of the upper sky, according to Bunsen.[134] A few months later, in early 1861, Bunsen and Kirchhoff examined lepidolite, a mica-like fluorosilicate mineral found in Saxony. After removing the potassium present in the mineral, they observed two violet lines in the spectrum, which they recognized as a new element. Bunsen called it "rubidium" because of a beautiful pair of deep red lines in its spectrum, after the Latin *rubidus* for deep red.[135] It turned out that rubidium was quite widespread, but only in low concentrations. In contrast to cæsium, which was only isolated as the metal in 1882, Bunsen prepared rubidium metal in 1863.[136]

---

[131] For example, see the letters from Bunsen to Roscoe dated 8 May 1857 (Deutsches Museum 00916); 22 May 1857 (Deutsches Museum 00917); and 7 June 1857 (Deutsches Museum 00917).

[132] Letter from Bunsen to Roscoe dated 15 November 1859 (Deutsches Museum 00930). Roscoe provides two rather different translations, one in the Bunsen Memorial Lecture and one in *Life and Experiences* (81–82), which is also incorrectly dated as being 13 November 1869, another example of the inaccuracy of his autobiography. Gingerich produced his own translation in his paper.

[133] Letter from Bunsen to Roscoe dated 11 April 1860 (Deutsches Museum 00932).

[134] Bunsen's intention was doubtlessly poetic, but he was in error. According to Lewis and Short's *Latin Dictionary*, cæsius was rare and was only used for gray eyes.

[135] Bunsen had again chosen a very rare Latin word, probably derived from the more common *rubia* for madder.

[136] Mary Elvira Weeks, *Discovery of the Elements*, 6th ed., enlarged and revised (Easton, PA: Journal of Chemical Education, 1960), 619–634.

Roscoe was shown the new technique when he arrived in Heidelberg for the summer vacation of 1860, between the discovery of cæsium and rubidium. He later recalled:

> I shall never forget the impression made upon me by looking through Kirchhoff's magnificent spectroscope, arranged in one of the back rooms of the old building in the Hauptstrasse, which then served for the Physical Institute, as I saw the coincidence of the bright lines in the iron spectrum with the dark Frauenhofer's [sic] lines in the solar spectrum. The evidence that iron, such as we know it on this earth, is contained in the solar atmosphere, struck one instantly as conclusive.[137]

Roscoe became an ardent advocate for the technique in Britain, translating Bunsen and Kirchhoff's German papers and giving a public lecture about their work at the Royal Institution in March 1861, which he described as being one of the most successful he ever gave there.[138] He argued for the priority of Bunsen and Kirchhoff's work and the use of Kirchhoff's scale of the solar spectrum, but he would have regarded these points as self-evident as one of their collaborators.[139]

Karlsruhe Conference

It was also during this vacation that Roscoe persuaded Bunsen to attend the famous Karlsruhe conference on the standardization of atomic weights and other related issues, held between 3 and 6 September 1860.[140] It was Kekulé who first had the idea of holding such a conference, and he initially broached the idea with his French friend and ally Adolphe Wurtz (1817–1884) in October 1859.[141] They settled on Karlsruhe, on the border between Germany and France, as the best site. They then won the backing of Karl Weltzien (1813–1870), the professor at Karlsruhe who had worked under Eilhard Mitscherlich (1794–1863) in Berlin; his laboratory at the Polytechnic was one of the models for Bunsen's new laboratory.[142] Kekulé called a meeting in Paris to organize the conference in late March 1860. In addition to the obvious participants, Kekulé, Wurtz, and Weltzein, this meeting was also attended by Roscoe and Baeyer. As far as we can tell, Roscoe did not play an important role in the organization of the conference, and it is perhaps significant that Kekulé, rather than Roscoe, was given the task of contacting British chemists about the event.[143] Despite the closeness of

---

[137] Roscoe, *Life and Experiences*, 69.

[138] Ibid., 69–70.

[139] For Roscoe's support for the Kirchhoff scale, see Klaus Hentschel, *Mapping the Spectrum: Techniques of Visual Representation in Research and Teaching* (Oxford: Oxford University Press, 2002), 55. I wish to thank Frank James for drawing my attention to this book.

[140] For an overview of the Karlsruhe conference, see Alan J. Rocke, *Nationalizing Science: Adolphe Wurtz and the Battle for French Chemistry* (Cambridge, MA: MIT Press, 2001), 226–233.

[141] For Wurtz, see A. W. von Hofmann, *Erinnerung an vorangegangene Freunde*, 3 (Brunswick: Vieweg, 1888), 171–432 and Rocke, *Nationalizing Science*.

[142] For Weltzien, see *DB*; for Mitscherlich, see *DB* and Hans-Werner Schütt, *Eilhard Mitscherlich: Baumeister am Fundament der Chemie* (Munich: Oldenbourg, 1992).

[143] Alan J. Rocke, *Chemical Atomism in the Nineteenth Century: From Dalton to Cannizzaro* (Columbus: Ohio State University Press, 1984), 292.

Karlsruhe, the capital of Baden, to Heidelberg, Bunsen was reluctant to attend any conference to discuss what he considered to be a purely theoretical issue. He was probably also preoccupied by the development of atomic spectroscopy. Roscoe and Bunsen stayed with Weltzien. However, Bunsen and Roscoe had little influence on the discussions during the conference, and Roscoe did not take part in the correspondence between Kekulé, Weltzein, Wurtz, and Roscoe's close friend Lothar Meyer after the conference.

Bunsen and Kirchhoff Come to England

Roscoe persuaded Bunsen and Kirchhoff to visit England in the summer of 1862 to see the International Exhibition in South Kensington. They visited "some of the usual sights of the metropolis" when they were in London. At a garden party held by John Peter Gassiot (1797–1877), the amateur scientist and port shipper (his firm Martinez Gassiot is now part of the Symington group) at his home on Clapham Common, they met the aged Michael Faraday.[144] Bunsen, Kirchhoff, and Roscoe also visited Sir Charles Wheatstone, who was among other things a pioneer of spectroscopy, at his home in Regent's Park.[145] Roscoe relates in his autobiography that the two German scientists visited Charles Arnold, an assistant master at Rugby School, whose wife Susanna was from Heidelberg. Despite Bunsen and Kirchhoff being reluctant, the Arnolds forced them to attend the communion service in the school chapel by changing the time of the service so they could take part before the group's departure to Manchester, where they met Joule.[146] The three men had their photograph taken during their stay in Manchester (Figure 4.3). Bunsen and Kirchhoff visited the new Mersey Bank works in Warrington of the dyestuff firm Roberts, Dale with Roscoe. They met Roberts, Dale's chemist, and the German colorist Heinrich Caro. This visit was propitious for Caro as he was later to enter Bunsen's laboratory on his return to Germany in 1866. Two years later, he then became the research leader at the new firm of Badische Anilin-und-Soda Fabrik, now better known as BASF, in Ludwigshafen-am-Rhein in the Bavarian Palatinate.

Roscoe's close collaboration with Bunsen was now drawing to a close. The photochemical work had now been completed, and Roscoe was increasingly involved with the expansion of Owens College and the running of the chemistry department. Furthermore, after his marriage in July 1863, his family took precedence over his relationship with Bunsen during the summer. The two men continued to correspond up to Bunsen's death in August 1899, although the last extant letter from Bunsen is from 1892. From Roscoe's letters to Bunsen it is clear that he was anxious to see his old friend again. They came close on several occasions, mostly notably in September 1889, when Roscoe was keen to see Bunsen's new house and attend a conference.[147]

---

[144] For Gassiot, see ODNB.

[145] For Wheatstone, see ODNB and Brian Bowers, Sir Charles Wheatstone FRS, 1802–1875 (London: HMSO, 1975).

[146] For Joule, see ODNB and Cardwell, James Joule.

[147] Letter from Roscoe to Bunsen, dated 29 August 1889, letter 28, Heid. Hs. 2741 III A-41, University of Heidelberg library; also see letters 29 (dated 7 September 1889) and 30 (8 September 1889).

Figure 4.3 Gustav Kirchhoff, Robert Bunsen (sitting), and Harry Roscoe. Taken during the visit of Bunsen and Kirchhoff to Manchester in 1862.

He got as far as Dover, where he had to turn back because of his gout.[148] In March 1899, Roscoe told Bunsen that he was going through the letters from him.[149] After Bunsen's death, he had these letters bound and then donated them to the Bunsen Gesellschaft for safekeeping in November 1904.[150] The society in turn gave them to the newly founded Deutsches Museum in Munich by 1914.[151]

Influence of Bunsen and Heidelberg on Roscoe

The years that Roscoe spent in Heidelberg as student and researcher shaped the remainder of his long life. Under Bunsen, he received a thorough training in experimental skills and the manipulation of apparatus. His apprenticeship with Bunsen

---

[148] Letter from Roscoe to Bunsen, dated 17 September 1889, letter 31, Heid. Hs. 2741 III A-41, University of Heidelberg library; also see letter 33 (dated 21 September 1889).

[149] Letter from Roscoe to Bunsen dated 31 March 1899, letter 40, Heid. Hs. 2741 III A-41, University of Heidelberg library.

[150] Note at the beginning of the bound volume, HS 908–1033, Deutsches Museum archives, and Roscoe, *Life and Experiences*, 80.

[151] Personal communication from Matthias Röschner, head of the archives of the Deutsches Museum, 14 April 2023.

determined the path of his chemical career which, like his master's own research, was to lie between chemistry and physics; especially in photochemistry, the physical properties of gases and solutions, and in spectroscopy (including solar spectroscopy). Like his friend Meidinger, Roscoe could have even become a physicist under different circumstances. His experiments with Bunsen on photochemistry were probably his most important research, and the Bunsen-Roscoe Law is still cited. His later interest in vanadium stemmed from Bunsen's work on minerals. It was thanks to Bunsen (and perhaps Thomas Graham) that Roscoe never entered the increasingly important field of organic chemistry and in particular synthetic dye chemistry in which he could have been very influential, given Manchester's key role as a world center of the textile industry. However, Bunsen's work on blast furnaces also showed Roscoe how chemists could contribute to the modernization of industry, leading to his close involvement with the chemical industry, as we will see in Chapter 8.

Even more importantly, his time in Germany molded his views on chemical education. His own teaching and his laboratories were closely modeled on the Heidelberg model. We will examine this in more detail in Chapter 7. More generally, Roscoe quickly came to wholeheartedly embrace the German system of tertiary education and believed that Britain should emulate this model, as we will explore in Chapter 9. This was a system based on the active engagement of professors in teaching students and in the importance of experimental research in the case of the sciences, in contrast to the tutorial system of Oxford and Cambridge with its emphasis on book learning and cramming for examinations. Furthermore, it was an approach which favored university education rather than the culture of "on-the-job" training with its day release teaching and night schools which remained important in Britain up to the 1960s. The celebrated maxim of "Lehrnfreiheit and Lehrfreiheit" of German education stressed the value of the maximum freedom to choose what to learn and how to teach, rather than the rule-bound systems of British education with its frequent examinations. Roscoe had encountered this system in the liberal and progressive environment of Heidelberg and by being taught by a first-rate professor, which inevitably meant that he saw it through rose-tinted spectacles. Even Roscoe deplored some aspects of German universities, especially the student corps with their duels and dueling scars.[152]

His experience at Heidelberg also reinforced Roscoe's belief in liberal progressive politics. The Grand Dukes of Baden introduced a liberal constitution in 1818 (partly to gain popular support for the succession to the throne of the morganatic line of the ruling dynasty) and generally ruled in a relatively liberal manner, although their policy became more conservative after the revolution of 1848–1849.[153] Heidelberg was effectively a liberal enclave in the reactionary German Confederation. In 1859–1863, while Roscoe was still traveling regularly to Heidelberg, an attempt to introduce clerical control of Catholic schools (although the dynasty itself was Protestant, most of the country was Catholic) was resolved in favor of secular control, which would have presumably met with Roscoe's approval. Hence the rather anomalous situation in

---

[152] Roscoe, *Life and Experiences*, 62–63.
[153] Tuchman, *Science, Medicine, and the State*, chapter 2, "Political Changes and Educational Reforms in Baden, 1815–1848," 34–53, and chapter 5, "Revolution, Reaction and the Politics of Education," 91–112.

Baden and particularly in Heidelberg gave Roscoe a rather favorable view of Germany as a whole, and he fell in love with the country and its people:

> My knowledge of the Germans and Germany has led me to love the Fatherland, and, I venture to think, to understand as well as to respect and admire the nation. As to any feelings antagonistic to England and the English existing in the minds of the many Germans with whom I became intimate I never found a trace, for Treitschke I did not know.[154]

We will see in Chapter 11 how this belief in a fundamentally liberal and progressive German people (as opposed to the autocratic regimes that ruled most of the German states) was badly shaken in 1914 near the end of Roscoe's life.

---

[154] Roscoe, *Life and Experiences*, 97–98. Heinrich Gotthard von Treitschke (1834–1896) was Professor of History at Heidelberg between 1867 and 1873, and then in Berlin; he was also a National Liberal member of the Reichstag. An ultra-nationalist, he called for the annexation of the other kingdoms in Germany by Prussia and opposed the involvement of Catholics, Poles, and Jews in German political life. He strongly disliked England. For Treitschke, see Andreas Dorpalen, *Heinrich von Treitschke* (New Haven, CT: Yale University Press, 1957).

5

From Instability to Stability

When Roscoe arrived in Manchester in October 1857 to take up the vacant Chair in Chemistry, Owens College had been open for six and half years. The College's foundation had been the culmination of much debate and rumination among leading Mancunians going back into the previous century and was welcomed finally as a further accolade to Manchester's position as Britain's principal industrial and commercial center.

Roscoe's arrival in Manchester coincided with the College facing several challenges hampering its future. Low student numbers were undermining the College's financial viability, and Roscoe quickly sought to mount a campaign (with other staff) to convince leaders in local industry and commerce of the benefits of studying at the College. When the U.S. Civil War (1861–1865) caused mass unemployment in Manchester due to the cessation of cotton imports, Roscoe took a leading role organizing a series of public lectures that advanced the Working Men's College movement and forged a close relationship between Owens College and the Manchester Working Men's Colleges. Seeing the success of this lecture series, Roscoe later organized a more extensive program of science lectures that for most of the working population was their first connection with science. These ventures, alongside his lectures at prestigious national institutions, elevated his standing and that of Owens College. Later, Owens College would go on to become the first civic university in England and would act as a model for other civic universities in England through the second half of the nineteenth century.

The Growth of Manchester as a Major Industrial and Commercial Centre

By the early decades of the nineteenth century, Manchester had become Britain's principal industrial and commercial center, driven by the mass industrialization (often referred to as the Industrial Revolution) that had transformed Britain since the 1750s. The preeminent commodity was cotton, and Manchester became known as "Cottonopolis": importing raw cotton from India and the United States; spinning and weaving cotton fabrics; dyeing and printing; and exporting fabrics for making up as clothing. Gradually the cotton industry metamorphosed from a domestic trade into industrial-scale production in large factory buildings accommodating innovative machinery (the spinning jenny, the spinning frame, and the spinning mule), all driven by steam power, and with a dramatic increase in output.[1] As mentioned in

---

[1] The spinning jenny was invented by James Hargreaves in 1764, the spinning frame by Richard Arkwright in 1768, and the spinning mule by Samuel Crompton in 1799.

Henry Enfield Roscoe. Peter J.T. Morris and Peter Reed, Oxford University Press. © Oxford University Press 2024.
DOI: 10.1093/oso/9780190844257.003.0005

Chapter 1, the import of raw cotton in the 1800s was influenced by periods of war with France. By 1815 the Manchester area was the main producer of cotton textiles in Britain.

The spur in industrial and commercial enterprise was accompanied by a rapid rise in population between 1801 and 1841. In 1801 the population was 70,409; by 1821 it had risen to 108,016; and by 1841 it was 242,983.[2] Over 60% of the population during this period was younger than age 24.[3] Later, the population rose further: in the mid-1840s due to those fleeing the famine in Ireland caused by the potato blight; and from about 1848 by the many émigrés escaping persecution in continental Europe.[4] Manchester's limited housing stock forced these destitute migrants into squalid cellars and other overcrowded forms of housing that worsened public health and contributed to the spread of diseases.[5]

The factories of Manchester relied on coal from the surrounding south Lancashire coalfields as a source of steam with which to drive their machinery. It was estimated that 2 million tons of coal were burned in Manchester during the 1860s. The amount of coal consumed (and smoke produced) became an indicator of industrial activity, while the number of chimneys was taken as a measure of economic output and of the town's prosperity.[6] Research on the chemical destruction of building stone in Manchester conducted by Robert Angus Smith, the Manchester-based analytical chemist and later Chief Inspector of the Alkali Inspectorate, identified the role of sulfur in the coal and led to Smith using the term "acid rain" for the first time in 1859.[7] Smoke added to the squalid living conditions in Manchester, about which Friedrich Engels (in *The Conditions of the Working Class in England*) and Charles Dickens (in *Hard Times*) wrote so passionately.[8]

Politically, Manchester benefited from the reforms initiated during the Whig government's retention of power from 1831 to 1840 (except for the period of Peel's first administration, 1834–1835) and believing "they had a mandate to try to identify and address the major economic, social, religious and political problems by which the whole of the United Kingdom seemed beset."[9] The Reform Act of 1832 expanded the voting mandate, the Factory Act of 1833 focused on the welfare of children working in factories (bringing to prominence the social reformer Edwin Chadwick), and the Municipal Corporations Act of 1835 established municipal boroughs, required annual elections to town councils, the appointment of a town clerk and a treasurer, and adoption of regular and legitimate accounting procedures. It is interesting to note that

---

[2] Manchester's population historically (download at manchester.gov.uk).

[3] David Cannadine, *The Victorious Century: The United Kingdom, 1800–1906* (London: Viking, 2017), 60.

[4] Peter Reed, *Acid Rain and the Rise of the Environmental Chemist in Nineteenth-Century Britain: The Life and Work of Robert Angus Smith* (Farnham, Surrey: Ashgate, 2014), 47.

[5] Ibid., 49–50.

[6] Stephen Mosley, *The Chimney of the World: A History of Smoke Pollution in Victorian and Edwardian Manchester* (Cambridge: White Horse, 2001), 17–18.

[7] The acid rain was due to the air containing sulfur acids, formed when coal with a high sulfur content was burned. The first use of the term was in Robert Angus Smith, "On the Air of Towns," *Journal of the Chemical Society* 9 (1859): 196–235, on 232.

[8] Friedrich Engels, *The Conditions of the Working Class in England in 1844*, trans. F. K. Wischnewetzky (London: Allen and Unwin, 1892) and Charles Dickens, *Hard Times* (London: Bradbury & Evans, 1854).

[9] Cannadine, *Victorious Century*, 166.

Roscoe's father was a member of the 1833 Royal Commission that laid the foundation of the Act. Manchester as an incorporated town was outside the terms of the Act but was allowed to seek corporation status, which it did in 1838, though city status was not attained until 1853. The fight for corporation status was led by Richard Cobden, the Manchester calico-printer and radical politician. It was Cobden who in 1839 became leader of the Anti-Corn Law League that led the fight against the 1815 Corn Laws, passed during the Napoleonic Wars to restrict the importation of all cereal grains, thereby increasing the price of bread, a staple food for a large section of the population, especially the poor.[10]

Foundation of Owens College and the Early Years to 1857

By the early 1830s London had two universities, University College and King's College, while Manchester as Britain's major industrial and commercial hub lacked a university or a college for the liberal arts or for the sciences, though flirting with the possibility of such institutions. Manchester Academy (later New College), founded in 1789 and a successor to the Warrington Academy, had offered the prospect of such an institution, especially with John Dalton, the eminent chemist, physicist, and meteorologist teaching mathematics, natural philosophy, and chemistry, but it did not survive in Manchester, and after moving to York and then back to Manchester, it finally relocated to London in 1853.[11] For many prominent figures in Manchester, a university institution remained an important priority. At the inaugural meeting of the Manchester Athenaeum in January 1836, several members including its President, James Heywood, spoke again in support of a university since "[n]ow such a college is not only wanted, but actually called for in Manchester...."[12] Publication of *A Plan of a University for Manchester* which had been read at a meeting of the Statistical Society garnered further interest. The *Plan*, while acknowledging the grievances of the religious tests of English universities that barred dissenters, highlighted the intellectual benefits that such an institution could bring to Manchester:

> Not one [William] Roscoe but many are wanted; not one Dalton standing like a patriarch of science at the head of the scientific part of the nation but a body of men who may appreciate and follow up his experiments or make similar advances in other branches of natural philosophy.[13]

Many of the town's influential figures met at the York Hotel on 10 November 1836, and while there were expressions of general support for a university and various plans were even discussed, enthusiasm quickly dissipated even though a committee had been formed. It was not until 10 years later and with the announcement of a generous

---

[10] Sir Robert Peel repealed the Corn Laws in 1846.

[11] Robert H. Kargon, *Science in Victorian Manchester* (Manchester: Manchester University Press, 1977), 11.

[12] *Manchester Guardian*, 16 January 1836, 3.

[13] Quoted in E. Fiddes, *Chapters in the History of Owens College and of Manchester University 1851–1914* (Manchester: Manchester University Press, 1937), 21.

bequest by John Owens, a wealthy Manchester businessman, that the aspiration for a university was rekindled and momentum was reborn.

John Owens was born in Manchester in 1790 and at age 18 became a partner in his father's hat-making business.[14] The business was to later expand into export markets, initially North and South America, but later into the Middle East, India, and China. By 1840 Owens had withdrawn from the family business to become a financial speculator in corn, cotton, and railway shares. In religious outlook, Owens conformed to the established church, though as a low churchman who contributed to nondenominational schools.[15] He spent little, abhorred borrowing, worked long hours, and took few holidays but, nevertheless, kept abreast of current affairs and made donations to local medical and educational charities. He died a very wealthy businessman in 1846. Although he appears not to have openly expressed support for a university in Manchester, or to having been influenced during his lifetime by his close friends George Faulkner and Samuel Fletcher who wanted a university, Owens left a bequest of £96,954 for the establishment of an institution "in such branches of learning and science as now and may be hereafter taught in the English universities."[16] The bequest would subsequently lay the foundation for what would become Owens College and later, Victoria University and the University of Manchester.

When the bequest was drawn up in 1844 (soon after the death of his father, Owen Owens), Owens attached specific conditions and stipulated a trustee constitution. The two conditions were: "improving and instructing the male sex (not less than 14 years)"; "all involved not required to make any declaration about religious opinions"; and "no instruction in theological or religious subjects."[17] The age stipulation was to ensure that teaching was focused at the university level, rather than school level. As if anticipating a likely struggle ahead in fulfilling the conditions, the bequest named specific trustees from across Manchester's diverse cultural and political spectrum, including: George Faulkner (industrialist and business partner of Owens); Samuel Fletcher; James Heywood (politician and philanthropist); William Neild (industrialist); Richard Cobden (manufacturer and politician); the mayor of Manchester; the dean of Manchester; and the local members of Parliament.[18] Several of those named were members of the committee linked to the unsuccessful 1836 initiative.

The trustees, under the chairmanship of George Faulkner, adopted the model of the Scottish universities and the university colleges in London. It was Faulkner who secured accommodations for the new college (to be named Owens College) by purchasing Richard Cobden's former house in Quay Street and leasing it back to the trustees for £200 per annum. The trustees agreed to make four full-time professorial appointments: Alexander (Sandy) Scott (Chair of Logic, Mental and Moral Philosophy, and English Language); Joseph Greenwood (Chairs of Classics and Ancient and Modern History); Archibald Sandeman (Chair of Mathematics); Scott

---

[14] For John Owens, see *ODNB*.

[15] B. W. Clapp, *John Owens, Manchester Merchant* (Manchester: Manchester University Press, 1965), 172.

[16] Details of Owens's will quoted in ibid.,173; £96,954 in 1846 was worth nearly £9 million in 2023.

[17] Ibid., 173.

[18] Kargon, *Science in Victorian Manchester*, 155.

was also appointed Principal of the College. Salaries were set at £350 plus two-thirds of the class fees, with Scott having an additional salary of £200 as Principal.[19] In addition, the trustees made four half-time professorial appointments that included chemistry, all with a salary of £150 plus fees.

Competition for the Chemistry Chair was especially fierce, with two good local candidates. Robert Angus Smith had worked as Lyon Playfair's assistant at the Royal Manchester Institution before becoming an analytical and consultant chemist. Frederick Crace Calvert had spent his formative period in France and, following his return to Manchester, investigated the disinfectant properties of carbolic acid (phenol) that led Joseph Lister to use carbolic acid in his antiseptic spray when treating postoperative infections. Other candidates included: John Stenhouse, who had studied with Thomas Thomson (Glasgow), Thomas Graham (Glasgow), and Liebig (Giessen); and Frederick Penny, who had studied with Liebig (Giessen) and was the current Professor of Chemistry at Anderson's University in Glasgow. However, Edward Frankland was the successful candidate, probably because his application was supported with several outstanding testimonials. He had worked with Lyon Playfair at the Office of Woods and Forests in London, gained his PhD at Marburg under Robert Bunsen, and then studied with Liebig at Giessen.[20]

Frankland was a leading proponent of laboratory-based learning that his mentors Liebig and Bunsen had adopted and encouraged. Either at his interview or soon after his appointment, Frankland must have insisted on a laboratory for practical chemistry classes. Finding space for a chemistry laboratory within the confined space of Cobden's former house was impossible, so the trustees raised a further £9,550 with which to purchase the old stable buildings adjoining the house and construct purpose-built facilities on four levels for the study of chemistry.[21] The facilities included a lecture room (to accommodate up to 150 students), a laboratory for practical classes (for 42 students), a private room and laboratory for Frankland, a rooftop area for experiments involving gases, a basement furnace room, a weighing room, and a store for chemicals and equipment.[22] The building was heated, ventilated, and well-lit. Although Frankland was not appointed until January 1851 and the College opened in March for a limited session, the chemistry facilities were ready for the full winter session starting the following October.

When Owens College opened in March 1851, it joined the ancient universities of Oxford and Cambridge, the two universities in London (University College and Kings College), Durham University, and four universities in Scotland providing higher education in Britain.[23] Among universities and those colleges aspiring to be universities, a major dichotomy was emerging: Was the pursuit of higher education to provide general intellectual development and acquisition of important skills, such

[19] Ibid., 156.

[20] For Frankland, see *ODNB*.

[21] Philip J. Hartog, ed., *The Owens College Manchester, Founded 1851: A Brief History of the College and Description of Its Various Departments* (Manchester: Cornish, 1900), 4.

[22] "Owens College—Chemical Lecture Room and Laboratory," *Manchester Guardian*, 13 August 1851, p. 5.

[23] The four Scottish universities were St. Andrews (1413), Glasgow (1451), Aberdeen (1495), and Edinburgh (1582).

as understanding, analysis, reasoning and synthesis, or was it to provide training for specific careers? Such questions were regularly debated during the rest of the nineteenth century whenever national educational policy was under review. It is a topic that is still debated today, and we will return to it regularly during the remainder of this book.

The trustees and the Principal Sandy Scott recognized from the beginning the immense struggle ahead for Owens College and "did not underestimate the challenge of bringing university education to 'the metropolis of the world's industrial activity.' "[24] Scott also recognized the importance of the College having a professor of chemistry—the value of chemistry as an academic discipline and the important connection between chemistry and the industrial and mercantile interests in Manchester and the wider regional area. Scott had been Professor of English Language and Literature at University College London from 1848 until his appointment at Owens College in October 1850, a period during which Thomas Graham and Alexander Williamson, two of the outstanding chemists of the nineteenth century, were creating a prestigious school of chemistry at UCL (as discussed in Chapter 3). Scott was deeply religious and one of the founders (alongside Charles Kingsley, Maurice Hare, and John Ludlow) in 1848 of Christian socialism, a philosophy that brought together Christianity and socialism.[25] As Principal, Scott set the highest academic standards and resisted pressure from many in Manchester for a more practically based education. Throughout the College's first six years the demands on Scott were burdensome and relentless, and with failing health he resigned as Principal in May 1857. Undeterred, he continued his teaching at the College, promoted education for the working classes, and, with others, founded the Manchester Working Men's College in 1858.

The future success and financial stability of the College depended on a regular intake of students. While the trustees must have anticipated some variation in student numbers over the first few years, the number did in fact vary quite markedly: 62 students in 1851–1852; 71 in 1853–1854; 58 in 1854–1855; and 33 in 1856–1857 (see Table 7.1). For the professors, this was concerning because their salary included a proportion of the student fees; in the 1856–1857 session, Frankland had 15 students working in the chemistry laboratory.[26] Rumors that Owens College might close were in constant circulation throughout Manchester. No doubt concerned over its future, Frankland resigned in July 1857 to take up a lectureship in chemistry at St. Bartholomew's Hospital in London. During his six years Frankland had shown, through his teaching, research, and consultancies, how the chemistry department could not only advance chemical science but also serve the interests of local industry and commerce.[27]

[24] Colin Lees and Alex Robertson, "Owens College: A J Scott and the Struggle against Prodigious Antagonistic Forces," *Bulletin of the John Rylands Library* 78 (1996): 155–172; 156.

[25] For Alexander John [Sandy] Scott (1805–1866), see *ODNB*.

[26] Colin A. Russell, *Edward Frankland: Chemistry, Controversy and Conspiracy in Victorian England* (Cambridge: Cambridge University Press, 1996), 190.

[27] Ibid., 160.

Roscoe, the Chemistry Chair, and Rising Student Numbers

When the newspaper advertisement for the Chair of Chemistry at Owens College appeared in August 1857, Roscoe was determined to apply.[28] Professorial vacancies in chemistry were quite rare given the relatively small number of universities, while the cadre of well-qualified chemists trained in Germany or graduating from the Royal College of Chemistry was increasing steadily, making each vacancy ever more competitive.

At the time of Frankland's resignation, Roscoe, aged 24, had completed his research with Robert Bunsen in Heidelberg and had been living in London for about a year. Roscoe had not been idle while reflecting on his future career options and securing an income; he was teaching at the East India Company Military College at Addiscombe in south London; and conducting research in his private laboratory with the assistance of Wilhelm Dittmar, a friend who had come to London from Bunsen's laboratory.[29] As time allowed, Roscoe acted as a consulting chemist and undertook a major investigation with Lyon Playfair, Secretary of Science in the Department of Science and Art, on the ventilation of dwelling houses.[30]

Roscoe's application to Owens College was supported by testimonials from Bunsen, Williamson (who had succeeded Graham as Chair of Theoretical Chemistry at University College London), Thomas Graham (Master of the Mint), and Justus Liebig (Giessen). Such a well-supported application was not a foregone conclusion because of two strong local candidates, Frederick Crace Calvert and Robert Angus Smith, both of whom were candidates when Frankland was appointed. Given the precarious state of Owens College, Roscoe's enthusiasm might have been misplaced, even though such appointments did not appear regularly, but he also wanted to make the acquaintance of James Prescott Joule. Roscoe received testimonials from Alexander Williamson and Robert Bunsen and would very likely have grasped the synergy between the College and Manchester's industrial and commercial enterprises and especially with the developing chemical manufacturing centers in nearby southwest Lancashire.

Roscoe's sound qualifications and outstanding research experience must have impressed the trustees. But they also likely sensed the important leadership role the young chemist could play working in tandem with the Principal Professor Greenwood (who had replaced Professor Scott in July 1857) to secure university status for the College. An announcement of Roscoe's appointment as Professor of Chemistry was confirmed in the *Manchester Guardian* about 10 days before the 1857–1858 session was due to start in October 1857.[31]

---

[28] Henry E. Roscoe, *The Life and Experiences of Sir Henry Enfield Roscoe DCL, LLD, FRS, Written by Himself* (London: Macmillan, 1906), 101.

[29] The East India Company Military College trained officers for the Company's private army in India. In 1858 the name changed to the Royal India Military College. See J. M. Bourne, "The East India Company's Military Seminary, Addiscombe, 1809–1858," *Journal of the Society of Army Historical Research* 57 (1979): 206–222.

[30] Henry E. Roscoe, "Some Chemical Facts Respecting the Atmosphere of Dwelling-Houses," *Quarterly Journal of the Chemical Society* 10 (1858): 251–269.

[31] "Owens College, Manchester," *Manchester Guardian*, 3 October 1857, 1.

While searching for a suitable house in Manchester, Roscoe had an unwelcome experience that must have caused him to reflect on the wisdom of taking the appointment. While Roscoe had satisfied himself that the college was not in imminent danger of demise, the public apparently held contrary opinions, as he relates in his autobiography:

> I was standing one evening, preparing myself for my lecture by smoking a cigar at the back gate of the building, when a tramp accosted me and asked me whether this was the Manchester Night Asylum. I replied that it was not, but that if he would call again in six months he might find lodgings there! That this opinion as to the future of the college was also generally prevalent is shown by the fact that the tenancy of a house in Dover Street was actually refused to me when the landlord learnt that I was a professor in that institution [Owens College].[32]

Such exchanges must have reminded Roscoe, if he had forgotten, of the issue of student numbers. The College attracted 34 students during the 1857–1858 session as Roscoe began to organize the chemistry department and deliver his lectures and laboratory exercises in his usual authoritative manner to the 15 students studying chemistry.[33] However, he must have been heartened to find an old friend, Frederick Guthrie, working as an assistant in the chemistry department.[34] Guthrie and Roscoe were fellow students at UCL and had studied chemistry with Alexander Williamson. They both had undertaken research with Robert Bunsen in Heidelberg, with Guthrie going on to undertake research with Hermann Kolbe in Marburg before his appointment as an assistant to Edward Frankland at Owens College in 1856.[35] The chemistry department staff also included Joseph Heywood as an assistant to the professor, while Wilhelm Dittmar continued to act as Roscoe's private research assistant as he had in London (see Chapter 7).

Attracting more students to ensure financial stability proved a conundrum for the Principal and staff of Owens College. Not only did the issue require the College to convince local industrialists and merchants of the benefits of studying at the higher education level (and even gaining a degree), but a parallel requirement was to attract well-qualified students who had the prerequisite knowledge to undertake the advanced-level courses.

Manchester would also have to adjust to having a college for advanced study. While recognized as a major industrial and commercial center, most of the existing businesses (especially those in the cotton industry) had relied on entrepreneurs sensing good business opportunities and exploiting new technical inventions (made by others). The origin of such ventures was far removed from academia, as represented by Owens College. Those associated with the College believed strongly in the benefits of studying at such an institution, but this was far from the case among local

---

[32] Roscoe, *Life and Experiences*, 102–103.
[33] Sir Henry E. Roscoe, *Record of Work Done in the Chemical Department of the Owens College, 1857–1887* (London: Macmillan, 1887), 1. See also Roscoe, *Life and Experiences*, 102.
[34] Roscoe, *Life and Experiences*, 104.
[35] For Frederick Guthrie, see *ODNB*.

industrialists, and merchants who had more likely relied on lengthy "work experience" and wealth would need to be convinced otherwise. Even when business leaders erred toward recognizing the benefits, they demanded that instruction should focus only on narrow areas of knowledge that would aid their business, rather than following a fuller course that would benefit businesses more broadly.[36] The College's aim was to instill expertise and knowledge to sustain businesses at the cutting edge. Early in his tenure, Roscoe determined that consultancy work could usefully demonstrate the benefits of the College and its chemistry department for Manchester's industrial and commercial enterprises, especially for the chemical manufacture center in southwest Lancashire. For a fuller account of Roscoe's consultancy work, see Chapter 8.

Attracting well-qualified students remained for many decades an insurmountable issue for Owens College and largely outside its control. Roscoe understood the advantage of the German system where the *Abitur* examination at the Gymnasium (or school) was "the prerequisite for studying at universities, for taking other higher-level examinations, or to gain entry to the mid-level of the civil service."[37] Unlike Germany, Britain was hampered by the absence of universal school education. While access to elementary schools was provided by some local authorities and churches, secondary education in the 1850s was only available in fee-paying schools. It was not until the 1870 Education Act that there was a government commitment to a national education system for England and Wales, and a system established of school boards to build and manage schools; a separate Act in 1872 extended this provision to Scotland.[38] A further Education Act in 1880 made school attendance compulsory for children aged 5–10 years; compulsory attendance was extended to age 11 by further legislation in 1893. Even with this limited provision, classical education prevailed and science education was very limited, in alignment with Oxford and Cambridge Universities. The full extent of the national deficiencies in science education would only emerge through evidence given to a succession of select committees and royal commissions during the 1860s and 1870s; Roscoe and Joseph Greenwood gave evidence to several of these bodies, making clear how limited access to good school education was holding back higher education (see Chapter 9). Roscoe remained a strong advocate through the rest of his life for schools to have well-qualified teachers and well-designed laboratories.[39]

Besides his demanding schedule at Owens, Roscoe was also undertaking courses of lectures at several of Britain's prestigious scientific institutions, including the Royal Institution, the London Institution, and the Society of Apothecaries. Most lectures focused on spectrum analysis and related phenomena for which Roscoe had become a recognized international expert following his work with Robert Bunsen and Gustav Kirchhoff in Heidelberg. These lectures not only reinforced Roscoe's standing within the scientific community and led to the publication of several journal articles, but also

[36] Roscoe, *Record of Work Done*, 9.

[37] Peter Reed, "Learning and Institutions. Emergence of Laboratory-Based Learning, Research Schools and Professionalization," in *A Cultural History of Chemistry in the Nineteenth Century*, ed. Peter Ramberg (London: Bloomsbury Academic, 2022), 191–215, on 195.

[38] The 1870 Education Act, https://www.parliament.uk/about/living-heritage/transformingsociety/livinglearning/school/overview/1870educationact/.

[39] Peter Reed, "Learning and Institutions," 196.

added to Owens College's growing reputation as a leading institution in Britain for studying the sciences. In June 1863, Roscoe was elected a Fellow of the Royal Society for his eminence as a chemist and for his scientific papers, including his papers on photochemistry with Robert Bunsen.[40] His proposers were headed by Thomas Graham, Michael Faraday, and August Wilhelm Hofmann. He was now part of the British scientific elite and yet he was still only 30 years old.[41]

The Potter Family and Home Life

The short period between his appointment at Owens College and the start of the 1857–1858 session gave little time for Harry Roscoe and his mother to find a house and establish a home in Manchester. They were forced to fall back on occupying the house of Roscoe's predecessor, Edward Frankland. No matter where Henry lived, there was always a warm welcome and a convivial atmosphere, whether visitors were family relations, friends, or colleagues from Owens College.

Harry regularly stayed with relatives when attending meetings in London. It was while staying with one of his uncles (and a close friend of his father), Sir Charles Crompton, that Harry met his future wife, Lucy Potter, in the drawing room at 22 Hyde Park Square in 1858.[42] In 1852 Crompton had been appointed a Judge of the Court of Queen's Bench (a senior court of Common Law with civil and criminal jurisdiction) and earlier had worked closely with Harry's father writing legal reports on decisions of the exchequer.[43] He took a strong paternal interest in Harry after his father's death and "would have sent him to Cambridge had he been disposed to go there," rather than UCL.[44] In 1829 Jessy, one of Crompton's daughters, married Edmund Potter, a calico printer who later became a leading figure in Manchester society and political life, one of the founders with John Bright and Richard Cobden of the Anti-Corn Law League and the Liberal Member of Parliament for Carlisle (1861–1874). Lucy was the youngest of his three daughters and was just 17 when Harry first met her.

One of Potter's granddaughters was Beatrix Potter, the artist, children's writer, sheep breeder, and farmer.[45] Beatrix was born in 1866, the eldest child of Rupert Potter (son of Edmund Potter) and Helen (née Leech). During a lonely childhood Beatrix became interested in drawing and painting, and a visit to her grandfather (Edmund Potter) at Camfield Place, Hatfield, began a lifelong interest in the natural world.[46] When her family visited the Lake District in 1882 an interest was sparked that would come to

---

[40] Certificate of election, EC/1863/12, Royal Society, https://catalogues.royalsociety.org/CalmView/Record.aspx?src=CalmView.Catalog&id=EC%2f1863%2f12.

[41] For a detailed examination of Roscoe's role creating the school of chemistry at Owens, see Chapter 7.

[42] Sir Edward Thorpe, *The Right Honourable Sir Henry Enfield Roscoe, PC, DCL, FRS: A Biographical Sketch* (London: Longman, Green, 1916), 20; see also *ODNB* entry for Edmund Potter.

[43] For Sir Charles John Crompton, see *ODNB*.

[44] Roscoe, *Life and Experiences*, 31.

[45] For Beatrix Potter, see *ODNB* and Linda Lear, *Beatrix Potter: A Life in Nature* (London: Allen Lane, 2007).

[46] Judy Taylor, *Beatrix Potter, 1866–1943: The Artist and Her World* (London: F. Warne and the National Trust, 1987), 11 and 12.

occupy a major part of her life and elevate her name in botanical circles; she painted and studied in fine detail the strange world of fungi and became an acknowledged mycologist.[47] It was probably at Camfield and during visits to the Lake District that Harry met Beatrix and took a keen interest in her scientific work. In 1897 he assisted Beatrix in submitting a paper, "On the germination of the spores of Agaricineae," to the Linnean Society of London; women were not able to attend meetings or read or publish papers, but the paper was read on her behalf on 1 April 1897.[48] Beatrix became famous for her children's books based on the lives of her pet animals. *The Tales of Peter Rabbit* was the first of 22 such books published by Frederick Warne & Company and translated into 13 languages. The royalties from the books allowed Beatrix to buy Hill Top, a working farm in the Lake District, where she became a revered breeder of Herdwick sheep and acquired land in cooperation with the National Trust to prevent building developments.[49]

Harry Roscoe and Lucy married in September 1863 at Marylebone, London, and subsequently had three children—Edmund (born 1864), Margaret (1866), and Lucy Theodora (1870). Their honeymoon on the Continent included a visit to Heidelberg so Lucy could meet Bunsen and Harry's other friends there. Upon their return, the Roscoes moved to a house in Deniston Road, Victoria Park, in the Manchester township of Rusholme. Victoria Park was a 70-acre private estate established in 1837 that during the second half of the nineteenth century attracted many leading professional people, including Charles Halle (founder of the Halle Orchestra), Robert Darbishire (lay Secretary of Manchester College and founder of Manchester High School for Girls), Ford Madox Brown (a member of the Pre-Raphaelite brotherhood of artists), and Richard Cobden; the estate also attracted many dissenters.[50] A notable family who lived in Victoria Park was the Pankhurst family. Richard Pankhurst, wife Emmeline, and daughters Sylvia and Christabel played a leading role in both the national and Manchester-based women's suffrage movement from 1867.[51]

Later (around 1870), as the Roscoe family grew, they moved to a larger house in the same area, Regent House in Lower Bank Road, specially built for them.[52] The architectural historian, Nicholas Pevsner described the house as:

Gothic, grey and yellow brick with stone dressings. Tricky details with much variety in the texture and wall surfaces. Splendid entrance hall with a Gothic arcade and a gorgeous inglenook with Arts and Craft detail, including stained glass, brass repoussé work, mosaic and carved wood panels and settles, part of a later refurbishment.[53]

---

[47] Taylor, *Beatrix Potter*, 71–94.
[48] Potter finally received recognition from the Linnean Society on 24 April 1997, when Professor Roy Watling of the Royal Botanic Garden, Edinburgh, gave a lecture, "Beatrix Potter as mycologist."
[49] Taylor, *Beatrix Potter*, 25, 31, 197–198.
[50] Victoria Park Conservation Area—History, https://secure.manchester.gov.uk/info/511/conservation_areas/932/victoria_park_conservation_area/2.
[51] Rachel Holmes, *Sylvia Pankhurst: Natural Born Rebel* (London: Bloomsbury, 2020).
[52] Thorpe, *Roscoe*, 195. See the 1871 England and Wales Census and the 1881 England and Wales Census on Ancestry.com.
[53] "The Western Half of the Park," https://manchesterhistory.net/LONGSIGHT/VICTORIA/west.html.

Today, the house is called Marylands and forms part of Xaverian College, a Catholic Sixth Form College.

The spacious house and grounds (with a servant staff of six) provided the relaxed conditions for Harry when exercised by matters concerning the future of Owens College or the day-to-day demands of the chemistry department. Harry and Lucy thoroughly enjoyed home life with their growing family, and visitors (even the unexpected) were always welcome to join in their activities. As his friend and colleague at Owens, Sir Edward Thorpe, captured in his biography of Roscoe:

> Nowhere did Roscoe appear to greater advantage than in his home.... He was fond of the society of his fellows, hospitably disposed, and of a warm, genial nature. Indeed he had a genius for friendship, and a boundless capacity for sympathy and kindness— instinctive, spontaneous, impulsive—the sort of sympathy where action follows hard upon the heels of inclination, and the kind of kindness which is doubled because it acts quickly. Innumerable instances of his little nameless, but not unremembered acts of kindness and of love might be culled from a correspondence which stretches over half a century.[54]

Alongside his trait of comradeship, Harry Roscoe had the ability to see humor in almost any incident or event, as his autobiography frequently bears out. Both traits were also found in his grandfather, but Harry's sense of humor was probably much more pronounced, as readily shown during gatherings with family and friends.

Lucy played an important social role alongside her husband. Again, as Thorpe highlights:

> Lady Roscoe was a strong and sincere character, of wide sympathies and generous impulses, with a rich fund of common sense, and a high standard of duty and performance. She had many intellectual interests and a cultivated taste; was well read, a good judge of literary work, an assiduous collector of old rare and beautiful prints.... She was an admirable host, and all who have had the privilege of partaking of her hospitality cherish an unfading memory of her kindly manner, her quiet dignity, and unfailing tact.[55]

In the early years of her married life Lucy took up photography and was happy to work with wet collodion or prepare her own dry plates.[56] Her photographs were well received for their technical and aesthetic qualities; 10 of her photographs (including a portrait of her brother Edmund Compton Potter and a landscape of Ullswater, where the family often spent their summers) were displayed in the 1877 exhibition of the Photographic Society of Great Britain in Pall Mall, London.[57] Her photographic portrait of James Joule (Figure 5.1) is famous.

---

[54] Thorpe, *Roscoe*, 191–192.
[55] Thorpe, *Roscoe*, 190–191.
[56] Thorpe, *Roscoe*, 190.
[57] "List of Exhibitors," *The Photographic Journal, Including the Transactions of the Photographic Society of Great Britain*, new series, 11 (1877–1878): 19. In 1894, the Photographic Society of Great Britain became

Figure 5.1 James Joule by Lucy Roscoe, ca. 1876.

The later larger house, Regent House, in Victoria Park had extensive grounds that included a good-sized garden and a lawn tennis court. Harry was a good competitive player, with "his reach and length of stride making him a formidable player."[58] Friends, colleagues, and senior students had an open invitation to play tennis during the summer months. Harry and Lucy also shared an interest in horses and they were often seen riding together. Holiday periods, especially when the children were young, were spent in the Lake District, in Scotland or at Camfield Place, the country house of Edmund Potter.[59] The extended holidays also gave Harry time for writing; many of his books were written during holiday visits, though his scientific books were mainly written in Manchester, given the need for good libraries.[60]

the Royal Photographic Society. Roscoe, in *Life and Experiences*, refers to Lucy winning gold and silver medals for her photography, for which see "Amateur Photographic Competition (First Notice)," *Amateur Photographer* 3 (16 April 1889): 189.

[58] Thorpe, *Roscoe*, 196.
[59] Camfield Place later became the home of Barbara Cartland, the author and step-grandmother of Princess Diana.
[60] Thorpe, *Roscoe*, 196.

Engagement with Manchester's Scientific and Cultural Life

With Roscoe's professional work focused on Owens College, Manchester became the center of his scientific and cultural life. Commitments in London to give lectures, to participate in parliamentary commissions and select committees, or to attend meetings of professional bodies, were for only short periods. After only a few months in Manchester, Roscoe was elected a member of the Manchester Literary and Philosophical Society (MLPS) (often referred to as the Lit and Phil) in January 1858. The Lit and Phil's program of events became (and remained during his working life in Manchester) the focus of his intellectual and cultural interests outside chemistry and brought Roscoe into regular contact with many of the outstanding and influential members of Manchester society.

The Lit and Phil had been founded in 1781 and over time evolved into Manchester's leading scientific institution.[61] Perhaps the most illustrious member was the chemist John Dalton, who from 1793 until his death in 1844 took a leading role in the affairs of the Society. He was Secretary between 1800 and 1809, Vice-President from 1809 to 1816, and finally President from 1816 until his death in 1844. His lectures to the Society number some 130, of which over 30 were published in the Society's *Memoirs*.[62]

At the time of Roscoe's election, prominent scientific members included James Prescott Joule, Edward Schunck, William Fairbairn, Robert Angus Smith, and Frederick Crace Calvert. James Prescott Joule, discoverer of the mechanical equivalent of heat and the law of conservation of energy in the 1840s, became a prominent figure in the Lit and Phil, which served as his substitute for an institutional affiliation, as he was a gentleman scientist.[63] He served as Librarian, Honorary Secretary, Vice-President, and President (1860 and 1868). Roscoe greatly admired him as an "original thinker" and they became close friends over very many years through the Lit and Phil.[64] In the 1850s Joule's financial situation became precarious due to his connection with the family brewery business at Stone (Staffordshire), and it was Roscoe (together with Joseph Hooker) who led a campaign to secure a civil list pension in recognition of Joule's outstanding service to science. Such pensions were awarded to those suffering financial hardship who had made an important contribution to the country. Securing such a pension was "hit and miss" even for the most eminent of scientists, and final approval (as well as the amount of the pension) largely depended on the personal view of the prime minister. Although supporting letters had to refer to educational background, family responsibilities, and financial circumstances, some pertinent details were often omitted; in one letter Roscoe failed to include that the Joule household had six house servants.[65] Nevertheless, a pension of £200 from 1878 was successfully secured by the campaign.[66] Both Dalton and Joule have memorials in

---

[61] Robert Angus Smith, "A Centenary of Science in Manchester, for the 100th Year of the Literary and Philosophical Society of Manchester," *Memoirs of the Manchester Literary and Philosophical Society*, series 3, 9 (1883): 1–487.

[62] For John Dalton, see *ODNB*.

[63] Donald S.L. Cardwell, *James Joule: A Biography* (Manchester: Manchester University Press, 1989), viii.

[64] Thorpe, *Roscoe*, 37.

[65] Cardwell, *James Joule*, 249.

[66] For Joule, see *ODNB*.

Manchester; a marble statue of Joule by John Gibson stands in Manchester Town Hall, opposite a statue of John Dalton, and was unveiled in 1893 by Lord Kelvin. Roscoe related an amusing story about his vote of thanks to Lord Kelvin on the occasion of the unveiling:

> That one inducement that drew him [Roscoe] to Manchester was that he might sit at the feet of Joule, whose name was as well known on the Continent as that of Newton, but he found that all the Manchester of that day knew of Joule was his Stone Ales. One of his lady auditors, in complimenting him upon his little speech, observed: "Of course I quite understood your remark about sitting at Dr. Joule's feet, but why make allusion to his *toe-nails!*"[67]

Roscoe was a very committed member of the Lit and Phil and he served as an officer on several occasions: joint Secretary with Edward Schunck (1860–1862), joint Secretary with Joseph Baxendell (1862–1873), Vice-President (1876), and President (1882).[68] Forty-one of his articles were published in the society's *Memoirs* and *Proceedings*. In 1900 Roscoe became the second recipient of the society's Dalton Medal for his research on the Dalton manuscripts and laboratory notebooks in the Lit and Phil's possession (for which see Chapter 7).

The Public Lecture Series and the Working Men's College Movement

The dramatic rise in unemployment in Manchester during the U.S. Civil War, due to reduced cotton supplies for local mills, led to the organizing of programs of public events and lectures to mitigate against any potential civil unrest. Later, these initiatives invigorated the local Working Men's College movement, as well as the evening classes at Owens College.

The first signs of economic trouble had occurred just before the U.S. Civil War in the boom years of 1859 and 1860 when overproduction of raw cotton and woven textiles led to a depression in world markets. The start of the U.S. Civil War in 1861 exacerbated this market adjustment; even though the shortage of cotton was caused by the Union's blockade, Confederate states used cotton supplies as leverage to gain Britain's support in their conflict with the Union. During the war, many thousands of factory workers were laid off, causing economic hardship and also sparking concerns that unrest might escalate into dangerous civil rebellion. To mitigate against this threat, a committee was formed in Manchester to provide a program of evening recreational events. Roscoe acted as one of the secretaries of the committee that included the Rev. William Gaskell (the Unitarian minister and husband of Elizabeth Gaskell, the novelist) and Arthur Ransome (physician and expert on tuberculosis).

---

[67] Thorpe, *Roscoe*, 38.
[68] Papers by Henry Roscoe, see *Accumulated Index to the Memoirs and Proceedings, 1780–1989* (Manchester: Manchester Literary and Philosophical Society, 1991), 288–289; and the bibliography of Roscoe's publications at the end of this volume.

The program over the winter months of 1862 included over a hundred events, attracting weekly audiences of about 4,000. Roscoe was proactive and persuaded friends and colleagues at Owens College to contribute: he gave a lecture titled "Chemistry of the Candle"; Frederick Crace Calvert, "A Few Words on Chemistry"; Arthur Ransome, "Air and Water, the Great Purifiers"; Professor Joseph Greenwood (Principal of Owens College), "A Tour in Switzerland" with a magic lantern show; and Professor Robert Clifton (Professor of Physics at Owens College), "Heat," illustrated with numerous experiments.[69]

The success of these lectures demonstrated to Roscoe the public's genuine interest in scientific matters when given the opportunity. He saw an opening for a series of "penny science lectures," as he referred to them. The first was "delivered in a large hall in a poor part of Manchester in the spring of 1866"; it was well attended and enthusiastically received, prompting a more extensive series the following autumn.[70] The lectures were published and sold for a penny with considerable success. The following winter the series was extended, with a charge of 2s. 6d. (12.5p, or about £12 today) for the 13 lectures.

In response to the initial enthusiasm and demand, these lecture series continued for 11 consecutive winters. Lectures were delivered to audiences of up to 4,000 in Hulme Town Hall or the Free Trade Hall, two major meeting venues in Manchester. Again, Roscoe drew on not only his friends and colleagues at Owens College, but also his wider community of scientific friends who shared his enthusiasm for engaging with such large public gatherings on scientific topics. Lectures were given across a range of disciplines: Thomas Huxley (biologist); William Carpenter (biologist); John Tyndall (physicist); William Huggins (astronomer); Frederick Abel (chemist); and T. E. Thorpe (chemist). There were in total 11 series of lectures between 1866 and 1880, with each series published as a separate volume.[71] The series in Manchester spawned similar series in other major cities, including Glasgow where Roscoe gave a lecture on the "Chemical Action of Light."[72]

Roscoe saw these science lectures as an extension of his work at Owens College and his advocacy for higher education. But they are also indicative of the fervent support of Roscoe and many of his colleagues at Owens for the education of Manchester's working men; the foundation of the Manchester Working Men's College (MWMC) occurred in 1858. From MWMC's foundation there were strong connections with Owens College; Greenwood was Secretary and a Lecturer in Classics and Religion, and was joined by Roscoe and other professors, including Richard Christie (history), Sandeman (mathematics), and Scott (logic and philosophy); Alfred Neild (a trustee of Owens College) was treasurer and taught political economy.[73] In fact, the link was so strong that MWMC acted unofficially as an extension of Owens College's evening courses. The evening courses for 1858–1859 were diversified as part of Owens

---

[69] Roscoe, *Life and Experiences*, 125–126.

[70] Roscoe, *Life and Experiences*, 126.

[71] *Science Lectures for the People: Science lectures delivered in Manchester, 1866–1880.* Series 1–11 (Manchester: John Heywood, 1866–1879).

[72] Roscoe, *Life and Experiences*, 128.

[73] C. Lees and A. B. Robertson, "Community Access to Owens College, Manchester: A Neglected Aspect of University History," *Bulletin of the John Rylands Library* 80 (1998): 125–152, on 133.

College's strategy to increase student numbers, and the number of evening students rose to 107. But to avoid creating competition between the two institutions, in 1861 MWMC merged with Owens College and the latter's trustees agreed to reduce the fees for evening courses and to appoint supplementary staff to teach the growing number of students. The following year saw 235 registrations for evening courses.[74] While the trustees and staff fully recognized how this rising number of evening students was an important source of income, they little realized the College would come to depend on this source of income for the next 40 years.[75] Roscoe had presented courses at both Owens College and MWMC. From 1861 the chemistry department was a strong recruiter of both day and evening students; between 1873 and 1883 the latter number was 1,530, though Roscoe made few references to them in his account of the chemistry department's work. It is interesting to note that two of his evening students were Thomas Edward Thorpe and Arthur Schuster. Thorpe later went on to gain a PhD at the University of Heidelberg (see Chapter 7 for his later career). Schuster also gained his PhD at the University of Heidelberg, and was appointed professor of applied mathematics at Owens College, and dean of the faculty of science at the University of Manchester.

Recognizing the urgent need to improve the educational level of students entering Owens College, Roscoe became a strong advocate for better training of science teachers. Strangely, there is no reference to this aim in his *Record of Work Done*, but Roscoe did make the point forcefully in his evidence to the Select Committee on Scientific Instruction (also known as the Samuelson Committee) in March 1868; Roscoe emphasized the great demand for science teachers in Manchester, a demand that would increase every year, and that the training of science teachers was one of the most important functions of Owens College.[76] To illustrate this point, Roscoe made reference to two of his chemistry students who went on to be teachers: one had worked in a cotton mill and had attended the Manchester Working Men's College before becoming a qualified teacher for the Government's Science and Art Department; the other was Dr. Marshall Watts, who was awarded a Dalton Scholarship to study chemistry at Owens College before studying at the University of Heidelberg and later becoming the first science master at Manchester Grammar School in 1868.[77] Unfortunately, the records of Owens College do not provide specific numbers for teachers trained.

Changes in Academic Disciplines While Pressure Grows on Accommodations

As with all academic institutions, Owens College continued to adjust its departmental structure and revise its academic disciplines as opportunities allowed. With

---

[74] Ibid., 134.
[75] Ibid., 130.
[76] "Henry Roscoe's Evidence," *Report of the Select Committee on Scientific Instruction 1867–68*, P.P. 1868 (432), 280. See also Robert F. Bud and Gerrylynn K. Roberts, *Science versus Practice: Chemistry in Victorian Britain* (Manchester: Manchester University Press, 1984), 123, 129, and 133–135.
[77] "Henry Roscoe's Evidence," 279–280.

its commitment to degrees awarded by the University of London, Owens had to respond to changes in curricula and degree regulations, as well as having adequate space for examinations. These exacerbated the already serious pressure on the accommodations at Quay Street caused by rising student numbers, prompting the Principal and other senior staff, including Roscoe, to begin a serious review of what accommodations Owens College would require as a university.

Securing university status began increasingly to underpin decisions about academic disciplines and the structure of departments, though the benefits of particular disciplines for the Manchester region (for example, engineering and physics) carried added weight. Such issues had occupied the attention of Greenwood and Roscoe for some time. Roscoe felt strongly that Owens should have a department of physics and was highly critical that the subject was not being served well by Professor Sandeman, who as Professor of Natural Philosophy was essentially "the purist of pure mathematicians, and had not the slightest idea of experimentation...."[78]

Further consideration to appointing a professor of natural philosophy (or physics, as understood today) was forced on the trustees in 1858 after the University of London established the new B.Sc. and D.Sc. degrees in science.[79] Candidates seeking these degrees had to show a competency in several branches of the physical sciences, leading the trustees to seek the views of the professors. At a college meeting, all the professors, with the exception of Sandeman, felt the college should appoint a professor of natural philosophy able to teach a course of experimental science complementary to the course in chemistry as preparation for the new science degrees of the University of London. Outside support for such an appointment came in letters from Augustus De Morgan (Professor of Mathematics at University College London) and George Stokes (Lucasian Professor of Mathematics at the University of Cambridge), who both expressed the value of students engaging in experimental natural philosophy (physics) while acknowledging that most often professors of mathematics did not have the training to conduct such experimental work.[80] In March 1860, the trustees finally agreed to appoint a professor of natural philosophy.

Over its first 10 years and during a period of major change for the College, the professorship unfortunately changed hands three times and undermined the standing of the new department. In July 1860 Robert Clifton, sixth wrangler in the 1859 Cambridge tripos, was appointed the first Professor of Natural Philosophy.[81] Although he made a promising start and enabled Roscoe to relinquish his teaching of the subject, his tenure was short lived because in 1865 Clifton was appointed to the new Chair of Experimental Philosophy at Oxford.[82] The following year, a close friend of Roscoe, William Jack (HM Inspector of Schools for Scotland) succeeded to the Chair, but he left in 1870 to become editor of the *Glasgow Herald*.[83] Jack's successor in 1870 was

[78] Roscoe, *Life and Experiences*, 109.
[79] Joseph Thompson, *Owens College: Its Foundation and Growth* (Manchester: J. E. Cornish, 1886), 212.
[80] Thompson, *Owens College*, 213–215.
[81] The sixth wrangler is the student who came sixth in the Mathematical Tripos Cambridge in a given year.
[82] Roscoe, *Life and Experiences*, 109.
[83] "Biography of William Jack," https://universitystory.gla.ac.uk/biography/?id=WH0246&type=P. Later he was a member of the Macmillan & Company publishing firm and from 1879 until 1909 Professor of Mathematics at Glasgow University.

Figure 5.2 The original premises of Owens College in Quay Street, Manchester.

Balfour Stewart, who had been Director of the Kew Observatory during a period when it became a national center for the standardization of meteorological instruments.[84] Soon after his appointment, Stewart was badly injured in a railway accident and was only able to return to his duties at Owens in 1871. He went on to achieve an early impact by establishing a new physics laboratory and becoming a prolific writer of textbooks, including *Conservation of Energy* (1873) and *Primer in Physics* (1872).[85] His students at Owens College included J. J. Thomson (discoverer of the electron in 1897), Arthur Schuster (Professor of Applied Mathematics at Owens College in 1881), and John H. Poynting (Professor of Physics at Mason College, later the University of Birmingham, in 1880). Stewart was an outstanding Chair of Physics, who worked in tandem with Roscoe to strengthen the scientific direction of Owens while also reinforcing the connection with Manchester's major businesses.

While these professorial chairs were mapping out the future scope of Owens College's academic credentials and setting a course to achieve university status, the college was facing a more serious and immediate challenge; to "escape from the stifling confinement and the increasing squalor of its home in Quay Street" (Figure 5.2).[86] To address this issue, on 25 January 1865 the trustees appointed a small committee (that

[84] Lee T. Macdonald, *Kew Observatory and the Evolution of Victorian Science, 1840–1910* (Pittsburgh: University of Pittsburgh University, 2018).

[85] For Balfour Stewart, see *ODNB*. Balfour Stewart, *Primer in Physics* (London: Macmillan, 1872) and *The Conservation of Energy: Being an Elementary Treatise on Energy and Its Laws* (London: Henry S. King, 1873).

[86] H. B. Charlton, *Portrait of a University: 1851–1951* (Manchester: Manchester University Press, 1951), 63.

included Greenwood and Roscoe) to report on the future accommodation requirements for the College.[87] The report-back of February 1865 highlighted the College's
success in raising the number of students; by 1864–1865 more students were being
taught at Owens College than in the main departments of University College London
in 1862–1863.[88] The report also drew attention to the limited laboratory space and the
lack of an examination hall. More serious and alarming were the reports by Roscoe
and Clifton on the poor air quality in their lecture rooms due to both the buildup of
carbon dioxide with so many students in the confined spaces and the release of carbon
monoxide and sulfur dioxide from the gas-lighting jets.[89] To avoid alarming students
and parents, this section of the report was not made public at the time.

The report marked the beginning of what was to be a long, uncertain, and intense
period for the management, staff, and supporters of Owens College in their endeavor
to relocate to an expanded campus and to secure university status.

---

[87] W. H. Chaloner, *The Movement for the Extension of Owens College, Manchester, 1863–73* (Manchester:
Manchester University Press, 1973), 3.
[88] Ibid., 3.
[89] Ibid., 4.

6

Creating a Civic University

By early 1865 the trustees, Principal, and staff of Owens College were focusing all their efforts on relocating to more extensive buildings and securing university status. While the first objective relied on two straightforward but demanding goals—finding a suitable location for the new buildings and mounting a successful fundraising campaign—the second objective had a much more uncertain outcome, because Owens College would become the first civic university in England.

What constitutes a university was a constant refrain during Owens College's endeavors to secure university status. Several factors would only emerge during the process:

- What faculties and departments would be appropriate?
- How important was having a medical school?
- Was the existing constitution appropriate and, if not, what constitution was, and would parliamentary legislation be necessary?
- Should women be admitted even though they were specifically excluded under the terms of the 1846 Owens bequest?
- Should the new university have the authority to set its own syllabuses and examinations, rather than relying on those of the University of London?

These factors (among others) were to be fiercely debated within Owens College and among its many supporters and detractors, some of whom could well influence the final outcome. If Owens College were to achieve university status, its travails would create a model for other aspiring civic universities to pursue for the remainder of the nineteenth century.

Bringing about both objectives over the period 1866–1880 required strong intellectual insight and tested the organizational and management skills of the trustees, Principal, and staff. Harry Roscoe was to play a leading role in realizing these objectives, exercising his outstanding networking ability, as this chapter will explore in detail.

Roscoe's Influence within the Owens College Leadership Group

By 1866 the acute lack of space in Quay Street forced Owens College to seek more commodious buildings in anticipation of securing university status. The new buildings would need to not only provide the additional space required by existing faculties, but also accommodate the new faculties deemed essential for a university. To achieve these goals would require both detailed planning and gathering support and

Henry Enfield Roscoe. Peter J.T. Morris and Peter Reed, Oxford University Press. © Oxford University Press 2024.
DOI: 10.1093/oso/9780190844257.003.0006

finance from far and wide. Under Joseph Greenwood's guiding hand as Principal, a small leadership group emerged, comprising Greenwood, Roscoe, and Adolphus Ward (Professor of History and English Language and Literature). How did this leadership group come about, and why were Roscoe and Ward chosen by Greenwood?

Soon after his appointment in 1857, Roscoe had supported Greenwood in his efforts to increase student numbers and stabilize the College's financial position. Greenwood for his part had come to appreciate (following his appointment as Principal) the importance of chemistry as an academic subject at Owens College because it underpinned many industrial sectors in the region around Manchester. The future prosperity of these businesses, the region, and the nation (given Manchester's position as one of Britain's largest cities) depended not only on a large number of trained chemistry students, but also on the research output of the chemistry school's staff and postgraduate students. Greenwood recognized the importance of the combination of training and research for chemistry, and under Roscoe's influence came to embrace this combination more broadly for the physical sciences and engineering. Henry Charlton noted how:

> [Greenwood gave] Roscoe full scope to re-orientate the pattern of academic objectives. The prosecution of the physical sciences became the instrument by which the universities renewed their vitality by restoring their contact with the pressing needs of the world of their own day.[1]

Joseph Greenwood had graduated from University College London with a BA in both classics and mathematics. In 1850 he was appointed to the Chair of Classics and History at Owens College, and then in 1857 replaced Scott as Principal following the latter's ill health. He also took a leading role in improving the educational opportunities for working men. In 1853 he supported opening classes for schoolmasters of primary schools; five years later he was appointed honorary secretary of the Manchester Working Men's College, and in 1861 brought about its amalgamation with Owens College.[2] Greenwood was an accomplished administrator and he enabled Owens College to "move by progressive and cautiously calculated stages towards its academic objectives."[3] He also had the uncanny ability, having chosen his objective, to identify and then work closely with his most appropriate academic colleagues. At Roscoe's interview in 1857, or soon afterward, Greenwood came to realize that together they could forge a bright future for Owens College. Roscoe (like his father and grandfather) had the ability to work closely with people from different backgrounds and with different perspectives to achieve shared objectives and goals. It was exactly this character trait that had aided Roscoe's grandfather, William Roscoe, in the cultural transformation of Liverpool and his successes in Parliament (as discussed in Chapter 2).

Adolphus Ward's appointment in 1866 as Professor of History and English Language and Literature brought another important ally into the leadership group alongside Greenwood and Roscoe, thus bringing the humanities and sciences

---

[1] H. B. Charlton, *Portrait of a University, 1851–1951* (Manchester: Manchester University Press, 1951), 60.
[2] For Joseph Greenwood, see *ODNB.*
[3] Charlton, *Portrait of a University*, 58.

together. Ward had studied in Germany as a boy while his father was consul-general in Leipzig and admired the German education system. When he was 16 he returned to Britain to continue his education at King Edward VI School, Bury St. Edmunds. He then studied at Peterhouse, Cambridge, where he graduated with a first in the Classical tripos and in 1862 was elected a Fellow of the college.[4] After briefly considering a legal career, Ward held a position at the University of Glasgow for a short time before his appointment in Manchester. Ward also accepted the value and importance of research in advancing knowledge and understanding in the humanities, although the process and the benefits often proved more difficult to articulate outside the sciences.[5] Hence all three of the leadership group recognized the crucial role of research in reshaping Owens College and moving progressively toward university status.

The death of Alexander Scott in 1866, following a long period of poor health, allowed the trustees and Principal to reorganize several of the academic posts. The new Chair of Logic, Mental and Moral Philosophy, and Political Economy was created, and Roscoe's cousin, Stanley Jevons, was appointed to the post. Jevons had taken the post of assayer at the branch of the Royal Mint in Sydney (Australia) between 1854 and 1859, before returning to University College London in 1859 to complete his degree and then pursue an MA degree in mental philosophy and political economy.[6] Having secured his MA with a gold medal, Jevons decided to undertake independent research in applied economics. Financial stringencies later forced him to seek a university appointment, but as a nonconformist (like his cousin) his opportunities were somewhat limited. Following a conversation with Roscoe at a family gathering, Jevons applied successfully in 1863 for a junior tutorship at Owens; there is no evidence that Roscoe canvassed support.

The year 1866 was an important one for Jevons. In addition to his appointment to his Chair at Owens, his book, *The Coal Question*, attracted the attention of such influential figures as John Stuart Mill, who had just become a member of Parliament for the City of Westminster, and William Gladstone, who was then Chancellor of the Exchequer (and later became Prime Minister). In due course, along with many other staff members at Owens, Jevons became frustrated by the College's ties to the examinations of the University of London that prevented the broadening of the courses and thereby developing his more able students. These views would become a clarion call for change to allow Owens College to set its own examinations and award its own degrees, but these changes were dependent on securing university status.

Roscoe had expressed concern over the lack of an engineering department, given Manchester's many outstanding engineers and engineering firms whose future would depend on a steady supply of well-qualified young engineers. In response, a group of leading Manchester engineers met in December 1866 and agreed to raise £10,000 for a department of civil engineering and mechanical engineering. The group included: a cofounder of the Institution of Mechanical Engineers (IME), the German-born Charles F. Beyer, who also cofounded the leading Manchester firm Beyer Peacock; two former presidents of the IME, Joseph Whitworth and William Fairbairn;

[4] For Sir Adolphus William Ward, see *ODNB*.
[5] Charlton, *Portrait of a University*, 55.
[6] For William Stanley Jevons, see *ODNB*.

and one future president, John Robinson.[7] Thanks to their personal intervention, the appeal had raised over £9,500 by the end of 1867, largely through local support. However, the setting up of an engineering department was held up by the space constraints at Quay Street.

Plans for the New College Buildings

With pressure mounting to resolve the accommodation constraints, plans were drawn up to launch a public appeal in early 1866, but unfortunately it had to be postponed because of instability in the banking system. In May 1866 Overend Gurney, the largest discount house in the City of London which bought and sold bills of transfer at a discounted rate, collapsed with huge debt after the Bank of England refused to provide financial support, raising concerns over the stability of the banking system.[8]

During the ensuing period Roscoe became impatient with the general lack of progress and decided further action could not wait. In 1866 Roscoe made an unannounced Sunday visit to the home of Thomas Ashton, a Unitarian Manchester cotton manufacturer and philanthropist at Ford Bank in Didsbury, to seek his support for the expansion of Owens College. Roscoe was aware of Ashton's business interests, his philanthropy, and his support for education. Ashton was well connected in Manchester cultural and political circles.[9] Just as important for Roscoe was Ashton's attendance at the University of Heidelberg, where he had studied chemistry and printing techniques before returning to Manchester and taking over the family-owned cotton mills and associated businesses. Having outlined the College's future plans, Roscoe was not put off by Ashton's initial unenthusiastic response. He had little sympathy for Owens College as the "Governors of the private trust were strong Churchmen and mainly Tories. The place was moribund, and he did not like to undertake the task of resuscitation."[10] As a Unitarian, Ashton's objection was not entirely unexpected and was true to character. Roscoe was not easily thwarted, holding out hope of convincing Ashton otherwise. Eventually Ashton agreed to join the steering committee for the College's extension and lead the public financial appeal. This involvement was crucial, and as a result of his enterprise and effort, Ashton effectively became the second founder of Owens College, after John Owens.

By late 1866–1867 the trustees decided to formally launch the public appeal, knowing that the majority of the funding would have to be raised in Manchester and the surrounding area. With considerable anticipation, but with a degree of anxiety because of the implications for the future of Owens College, the launch was made

---

[7] For Sir Joseph Whitworth and Sir William Fairbairn, see *ODNB*; see *Grace's Guide* for Charles Beyer, https://www.gracesguide.co.uk/Charles_Beyer; and John Robinson, https://www.gracesguide.co.uk/John_Robinson.

[8] The demise of Overend Gurney, https://www.bankofengland.co.uk/-/media/boe/files/quarterly-bulletin/2016/the-demise-of-overend-gurney. The debt was about £11 million, equivalent to around £1 billion in 2023.

[9] For Thomas Ashton, see *ODNB*.

[10] Henry Enfield Roscoe, *The Life and Experiences of Sir Henry Enfield Roscoe DCL, LLD, FRS, Written by Himself* (London: Macmillan, 1906), 111.

at a public meeting hosted by the mayor in Manchester Town Hall on 1 February 1867.[11] More than 90 representatives of Manchester's commercial, industrial, and cultural elite joined the College trustees and staff and its long-standing supporters. In an effort to guide the meeting to a positive outcome, two papers were circulated for discussion. The first appeared in the previous day's *Manchester Guardian* and was a progress report on the rise in student numbers, with its implications for the inadequate buildings at Quay Street. This paper was accompanied by a letter from Sir James Kay-Shuttleworth, the civil servant and well-known education reformer, expressing support for the College's expansion plans.[12] The second paper was based on a report by the staff (and endorsed by the trustees) summarizing the necessary expansion of academic disciplines if the College were to fulfill its role as a regional center for higher education and gain university status. The disciplines were configured in two broad groups—a department of general literature and science (as the liberal arts), and a department of theoretical and applied science. Most of the additional disciplines resided within the latter, and included civil and mechanical engineering (on which local engineers had already expressed support and were fundraising), astronomy and meteorology, and applied geology and mining. The proposals received general support, even though they added to the already rising demand for space in the new buildings and hence increased the funding target.

To create some forward momentum, two important resolutions were passed. The first, proposed by the local MP, Sir Thomas Bazley, and by Thomas Ashton (who had spoken enthusiastically about the College, in sharp contrast to his negative comments during his first meeting with Roscoe), was to establish a committee to raise a fund of £100,000 (and if possible, £150,000) for the proposed expansion. The second, proposed by William Callender Sr., a Manchester merchant, and John Platt, a local MP, was for an executive committee to oversee the different organizational parts of the extension, comprising Thomas Ashton (Chairman), Owen Heywood (Treasurer), Charles Beyer (engineer), Professors Greenwood, Richard Christie, and Roscoe, and six trustees. Both resolutions were passed unanimously, reflecting the enthusiasm and support openly expressed at the meeting. Though the hard work had only just begun, there was now a greater sense of momentum toward implementing the crucial improvements for Owens College.

Senior staff directly involved at different times carried heavy workloads. Roscoe's time and energy had to be split between his often-heavy workload for several committees and his responsibilities to the emerging school of chemistry, while also maintaining his lecturing to outside bodies and acting as joint-secretary of the Manchester Literary and Philosophical Society. Such demands might have exhausted a less indefatigable person, but Roscoe was strongly motivated by the benefits that would accrue to Owens College.

Two weeks after the town hall gathering, the Executive Committee at its first meeting (held at the Royal Institution in Moseley Road—now the Manchester City

[11] W. H. Chaloner, *The Movement for the Extension of Owens College, Manchester, 1863–73* (Manchester: Manchester University Press, 1973), 6–7.

[12] For Sir James Kay-Shuttleworth, see *ODNB*. "Owens College," *Manchester Guardian*, 1 February 1867, 3.

Art Gallery) agreed to set up four sub-committees to take responsibility for different strands of the extension plans.[13] These were: the Funding Sub-Committee for canvassing and raising finance; the Site Sub-Committee for selecting and purchasing a new campus site; the Building Sub-Committee for general planning of the new buildings; and the Constitutional Sub-Committee for all constitution-related matters. While the focus of the Executive Committee and the sub-committees was on the accommodation extension and the broadening of academic studies, they also had to grapple with the College's amalgamation with other educational institutions (that would form an integral part of the application for university status). The Executive Committee additionally took on the leadership and coordination role. The overall organizational arrangements became quite complex, with many issues changing daily, but the devolved responsibilities worked well.

The Site Sub-Committee was chaired by Murray Gladstone, a Manchester merchant with trading connections with India and a cousin of William Ewart Gladstone MP. The first task was to review possible sites, either those close to the city center or those in the immediate suburbs allowing convenient student access. Property speculation raised its head almost immediately, since no sooner were inquiries made about a particular site than the asking price rose dramatically. In March 1868 Gladstone secretly purchased a site of about 127 acres fronting Oxford Road, a major thoroughfare to the city center.[14] With approval of the Executive Committee, the title of the land was transferred from Gladstone to Thomas Ashton as Chairman of the Executive Committee.

As part of its due diligence and before appointing an architect, the Building Sub-Committee had begun a thorough review of newly designed and constructed buildings at a number of universities in Britain, including Glasgow, Oxford, Cambridge, and London. With his admiration for the German and Swiss education systems, Roscoe pressed the trustees and the sub-committee to take account of developments in Germany and Switzerland before the plans for Owens College were finalized. The trustees authorized Greenwood and Roscoe to undertake a tour in these countries during the summer vacation of 1868 and to gather information about the facilities and management of institutions in these countries, so practices and accommodation designs (in particular, the science laboratories) might benefit Owens College.[15]

The Tour of German and Swiss Universities and Polytechnics

The tour by Greenwood and Roscoe of German and Swiss universities and high schools was undertaken during July 1868. Their information-gathering was structured under eight headings:

1. the scale of science teaching in universities and schools;
2. the remuneration for staff and apparatus;

---

[13] Chaloner, *Movement for the Extension*, 7.
[14] Ibid., 8.
[15] Ibid., 8.

3. the buildings and their costs;
4. the profile of students;
5. the relationship between universities and polytechnics, their relationship with schools, and the curriculum balance between the sciences and the classics;
6. the relationship of academic bodies with state governments;
7. the role of seminaria (classes led by professors, and separate from lectures, to cater to outstanding students likely to go on to undertake original research or to teach);[16]
8. conclusions.

The tour was both intensive and extensive, taking in institutions in Bonn, Göttingen, Hanover, Berlin, Leipzig, Freiberg, Heidelberg, Karlsruhe, Munich, and Zurich.

The report by Greenwood and Roscoe provided detailed information on all the institutions they had visited. In summary, they noted that:

the number of teachers in the first and second grade, is considerably larger than in English Universities and Colleges, even after account has been taken of the larger number of students, and this discrepancy is probably greatest in the science departments.... The fees are much lower in Germany than with us.... It is, however, to be borne in mind, that one great expense in working a laboratory is the cost of apparatus and chemicals used by the students, and this is specially paid for in German Universities by the State.... In England, the greater part of this charge falls upon the professor. The scale of fees in the Chemical Department of Owens College is rather more than double that of Heidelberg.[17]

Laboratories in Germany were extensive and well-equipped for chemistry, biology, and physics, accommodating large numbers of students and funded by the state. The report was critical of the level of spending, and in his evidence to the Royal Commission in March 1871, Roscoe stated firmly that:

I think that more money has been spent upon several of those laboratories than was necessary. I do not mean that the accommodation which they give is more than is wanted, because they are now, many of them, full, but I think that the money has in some cases been somewhat lavishly spent. With regard to the assistants, and to the grants for working expenses and for the purchase of apparatus, I believe that there is not at all too much spent.[18]

A major concern at Owens College was the lack of educational achievement of students when entering their higher education studies. Although there were several types of schools in Germany, most students entering universities left their gymnasium

---

[16] A report of the tour, completed in July 1868, was presented to the Executive Committee in December 1868 and was given in evidence to the Royal Commission on Scientific Instruction and the Advancement of Science in March 1871. See *Report of the Royal Commission on Scientific Instruction and the Advancement of Science, Volume 1*, P.P. [C. 536], 501–508.

[17] Ibid., 501.

[18] Ibid., 501.

(school) at age 18 with a testimonial (or leaving certificate) accredited by an examination board, as well as with a review of the student's school career.[19] From 1871 the *Abitur* leaving certificate was required for admittance to all German universities.[20] The university courses usually extended over three years (six semesters), though some students remained for four years or even five years, given the extended range of studies.[21] In contrast, in England, formation of the Science and Art Department in 1853 and creation of examinations for science teachers and for students (with associated grants by result) from 1859 were slow to have an impact, given the very fragmented and non-universal secondary education provision.[22] Improvement on a large scale would only come gradually, in a series of stages through the remainder of the nineteenth century and the early twentieth century. Roscoe remained a passionate campaigner for universal secondary education linked to an examination, similar to the German provision.

While Greenwood and Roscoe did not view their findings as providing a template for Owens College, they believed that universities in Germany provided a successful model for higher education that had produced major industrial advances and economic benefits for their country. Their findings could certainly influence the direction of Owens College's pathway as a higher education institution. The implementation of specific parts of that plan would depend on financial, organizational, and operational opportunities; some were externally determined (by government, for example) and others, internally by the College governors. The report, nevertheless, aided the Executive Committee's planning and their assessment of likely costs.

Seeking Financial Support from the Government

With the projected extension plans likely to cost between £100,000 and £150,000 and the doubt as to whether such a fund could be raised privately in Manchester, the trustees remained uncertain about the viability of their plans and the timetable. However, in May 1867 Roscoe raised the possibility of a government grant following the government approval of an annual grant for a college of science in Dublin. For the financial year 1868–1869 the government had allocated £127,138 for higher education and learning across the United Kingdom, with £57,405 for buildings and £69,733 for other related purposes.[23] The fact that grants were being approved (albeit without a clear policy direction) gave the Executive Committee some optimism that if the government was pressed sufficiently, Owens College might receive a grant and thereby offset a significant part of the total cost of the extension. The Executive Committee thought it would be wise to seek advice on the best tactics to adopt when lobbying the government for financial support.

---

[19] Ibid., 506.
[20] Peter Reed, "Learning and Institutions: Emergence of Laboratory-Based Learning, Research Schools and Professionalization," in *A Cultural History of Chemistry in the Nineteenth-Century*, ed. Peter Ramberg (London: Bloomsbury Academic, 2022), 191–215, on 195.
[21] *Report of Royal Commission* [C.536], 506.
[22] D.S.L. Cardwell, *The Organisation of Science in England* (London: Heinemann, 1972), 89.
[23] Scotland received £39,384, Ireland £25,000, leaving England and Wales with £62,754.

In January 1868 the Executive Committee met with Sir Bernard Samuelson, who had just undertaken a detailed study of technical education provision in Europe and chaired the 1867–1868 parliamentary select committee on scientific instruction. Whether Roscoe was responsible for canvassing Samuelson's advice is not known, but later in 1868 Roscoe was to give evidence to the select committee based on his work at Owens College. Thereafter he and Samuelson, with their keen interest in scientific and technical education, became friends. Samuelson had managed an engineering firm in Manchester in the 1840s. He was thus well aware of the Manchester engineering scene and the benefits that Owens College could deliver for the Manchester region as a major center of theoretical and applied science.[24] Samuelson had previously raised the aspirations of Owens College with Lord Robert Montagu MP, Vice-President of the Committee of Council on Education, the committee of the Privy Council with responsibility for education and for distributing educational grants.[25] While Samuelson was unclear whether the government would approve a grant, he must have expressed optimism because soon after the meeting the Executive Committee initiated its new strategy. It circulated information to MPs and others who might be supportive, and a small committee comprising Thomas Ashton, William Houldsworth (Manchester industrialist and trustee), and Professor Greenwood was charged with arranging a meeting with Lord Derby's government. The mayors of local authorities in Lancashire, Cheshire, and the West Riding of Yorkshire were then invited to attend this meeting with the government.[26]

The first meeting took place with the 7th Duke of Marlborough (Lord President of the Council) and Montagu on 5 March 1868, but the timing was not helpful, unfortunately, because only a few days before, Disraeli had succeeded Lord Derby as prime minister. Nevertheless, the Duke of Marlborough, while admitting indirectly the lack of a clear policy on such grants, was generally supportive of grants for institutions such as Owens College.[27] A second meeting was arranged for 24 March, although there was no indication as to the form of the discussions or with whom.

The second meeting proved as unsatisfactory as the first, even though it was at 10 Downing Street with the new prime minister, Benjamin Disraeli, and George Ward Hunt, his Chancellor of the Exchequer. Disraeli expressed the view that should the government be unable to award a grant, he was certain that Lancastrians would come to the aid of Owens College.[28] The delegation was left with mixed reactions; some were hopeful, while others feared flat refusal, but in the end there was to be no positive outcome. Further interventions by Jacob Bright, MP for Manchester in Parliament, and by Sir Thomas Bazley MP and Hon. Algernon Egerton MP also proved fruitless. Later, evidence emerged that the treasury had drafted a minute allocating a grant to Owens College, but any possible approval was overtaken by a further change in government. In November 1868 the Disraeli Conservative government was replaced by a Liberal government headed by Gladstone. Further attempts to gather support

---

[24] For Bernard Samuelson, see *ODNB*.
[25] From 1859 a vice president was appointed to oversee policy.
[26] Chaloner, *Movement for the Extension*, 11.
[27] Ibid., 11.
[28] Ibid., 12.

were made in Parliament and during ministerial meetings, but by February 1869 the Executive Committee was refocusing its efforts on raising the finance from private sources in Manchester. This was at least in part to avoid any government interference in the college's affairs if a grant were approved.[29]

By early 1869 the Executive Committee was making steady progress on the extension plans. The Oxford Road site for the college extension was agreed, the fundraising in Manchester was proceeding well, and the architect, Alfred Waterhouse, had been appointed to design the extension buildings. Waterhouse was a well-respected architect in the gothic style who had won the competition to design the new Manchester Town Hall in 1868.[30] By this time the Executive Committee had been compelled to address important legal issues as a matter of some urgency since their resolution might require parliamentary legislation, a process likely to be protracted and subject to the vagaries of party politics or a change in government.

Progress toward the "Owens College, Manchester"

During the same period in 1868 when Samuelson was discussing a possible government grant, the Executive Committee was also seeking advice from James Bryce, Regius Professor of Civil Law at Oxford University, about the legal issues likely to arise with the extension plans. The issues facing the Executive Committee were: changing the legal status from a charitable foundation to a corporate body; amalgamating Owens College with the Manchester Royal School of Medicine; transferring the collections of both the Manchester Natural History Society and the Manchester Geological Society; and the admission of women.

John Owens's will of 1845 had closely defined the governing body for Owens College as a charitable trust, with about 25 named trustees representing a broad political spectrum of opinion, the town councils, education institutions, and other community groups serving the Manchester and Salford areas. When plans to reform Owens College were under consideration, the trustees felt the governance should be revised and a corporate body created to bring a closer partnership between the public and private sectors. Such a change would require careful legal review, as well the agreement of all the parties concerned, including the current trustees.

The amalgamation with the Manchester Royal School of Medicine was felt to be important academically and would enhance Owens College's application for university status. The Manchester Royal School of Medicine had been founded in 1824 by the Manchester surgeon Thomas Turner, and as early as 1836 proposals were put forward for the School to become the nucleus of an academic college. The opening of Owens College in 1851 had revived the idea of a higher education college with university status embracing both medicine and traditional academic disciplines. For many educationalists, a college without a medical school would not warrant university status.[31]

---

[29] Ibid., 14–15.
[30] For Alfred Waterhouse, see *ODNB*.
[31] Charlton, *Portrait of a University*, 62.

From early in his time at Manchester, Roscoe had come to understand the importance of Owens College having a medical faculty as it moved toward university status. The medical curriculum drew heavily on the physical sciences, and a medical faculty for Owens College would provide opportunities for collegiate teaching and purposeful research. The College's role in professional training would be enhanced, and the links between the College and the Manchester community it served further strengthened. A medical faculty would certainly enhance any university charter application made by Owens College.

In 1866 and probably at Roscoe's behest, Greenwood had started discussions on a possible merger, but these overtures were not always welcomed by all the staff at Owens College or by the Royal School of Medicine. William Williamson, Professor of Natural History, Anatomy and Physiology at the foundation of Owens College, was also a qualified physician and with his membership of the College's Senate exacted considerable influence on any merging of the College with the Royal School of Medicine. He raised the possible resistance of medical teachers, medical practitioners, and medical students to the proposed reorganization, but his main concern pointed to the likely negative effect of medical students on the social welfare of other students. His view held that:

> the inclusion within the college of medical studies would be a menace to the moral welfare of the normal arts and science students; for medical students, and especially medical students in the provinces—for the best went to London—were intellectually less sensitive and morally coarser than the average student, and so they would be a source of potential immorality.[32]

Even the Medical School's Senate shared this view since they suggested that "when medical students are admitted they might be segregated from all possibility of contact with Arts and Sciences students."[33] During 1870 and as part of work on the two parliamentary bills, the amalgamation with the Manchester Royal School of Medicine became a higher priority in Owens's plans. Two years later, with Waterhouse commissioned to design a new medical school, Roscoe and George Southam, Director of the Royal School of Medicine, visited London medical schools to assess their modern facilities.[34]

Manchester, like many other provincial towns in Britain in the late eighteenth century, had its avid collectors of natural history specimens. John Leigh Philips, whose family owned cotton and silk mills in Manchester, built up an eclectic collection of paintings, prints, etchings, books, and natural history specimens, particularly insects. He was a member of the Manchester Literary and Philosophical Society and was well connected with other cultural networks, including William Roscoe's circle in Liverpool. After Philips's death in 1814, his "cabinet" of natural history specimens was purchased by a group of interested individuals who in 1821 set up the Manchester

---

[32] Ibid., 61.
[33] Ibid., 61.
[34] Stella V.F. Butler, "A Transformation in Training: The Formation of University Medical Faculties in Manchester, Leeds, and Liverpool, 1870–84," *Medical History* 30 (1986): 115–132, on 120–121.

Natural History Society, with the objective of displaying Philips's collection and other private natural history collections. The Society steadily increased its holdings and had to open a new building on Peter Street in 1835 so more of the collection could be displayed. In 1850 the Society bolstered its collection further by acquiring the collection of the Manchester Geological Society. However, building maintenance costs and declining subscribers created severe financial difficulties by the 1860s, and the Society was forced to consider transferring the collections to Owens College. While accepting the synergy between the collections and its academic work, the College was not in a position to accept the collections at the time because of the very cramped conditions at Quay Street. Accommodating the collections would depend on suitable space in the new Oxford Road buildings.[35]

The Need for Parliamentary Acts

Having received legal advice from Bryce, the Constitutional Sub-Committee recommended that two parliamentary bills would be necessary, because of Owens College's charitable status. The first, to revise the constitution and create the Owens College Extension College, and the second, to amalgamate the Owens College Extension College with Owens College, to form the "Owens College, Manchester." This strategy and the draft bill were approved by the Executive Committee in February 1869 and by the trustees of Owens College in February 1870. To provide a degree of continuity, of the 21 governors of the Extension College, 10 were trustees of Owens College.

As predicted, the parliamentary path of the first bill proved to be both contentious and tortuous. The draft bill had initially to be approved by the Standing Orders Committee of the House of Lords and here the legal challenge occurred; whether a charitable foundation (as Owens College) could be enlarged by an outside body (as Owens Extension College) rather than by the charitable foundation itself (Owens College). This was referred to the attorney-general for a ruling, but after revising his initial ruling following a further legal challenge, he sent a clarifying note to the House of Lords where the bill was due to be debated. However, before the bill could move forward, a Manchester solicitor, Stephen Heelis, representing the interests of John Owens, raised several objections to the bill.[36] These concerned Owens College educating women students; and admitting students under 14 years of age and then teaching them subjects at the school level rather than university level.[37] To avoid undue delay at this crucial stage, Owens's Extension Committee felt compelled to remove the clause admitting women to the new college. Although other objections were raised, the government decided to proceed and the bill passed its second reading in the House of Lords by 33–6. However, before the bill got to the House of Commons, objectors continued their fight over the admission of women after the Manchester MP, Jacob Bright, had restored the earlier clause. As further negotiations continued, Bright weakened the contentious clause during the third reading of the bill to "such

---

[35] "History of Manchester Museum," http://www.museum.manchester.ac.uk/about/history/.
[36] Stephen Heelis was an executor of George Faulkner, who was an executor of John Owens's estate.
[37] Chaloner, *Movement for the Extension*, 16.

persons as the proper authorities of the college may from time to time direct."[38] The bill passed its third reading and received the Royal Assent on 4 July 1870.

Progress with the Oxford Road Buildings

With the parliamentary approval for the Owens Extension College as a corporate body, the Extension Committee had achieved another vital stage in reforming the college and now felt able to refocus its efforts on developing the Oxford Road site. A stone-laying ceremony by the President of the College, the Duke of Devonshire, was held on 23 September 1870. Ashton, as chair of the Funding Sub-Committee, told the distinguished audience that the total cost of the buildings (that would accommodate about 600 day students and many more evening students) and their fitting out would cost about £90,000, and that another £30,000 still needed to be raised. Following Ashton's speech, "a glass bottle containing current coins, copies of the three Manchester daily newspapers, the London Times and printed documents issued by the Extension Committee" was placed in a cavity beneath the foundation stone and covered with a lead plate, bearing the inscription, "The first stone of this building erected for the Owens College, Manchester, was laid by His Grace the Duke of Devonshire, KG, FRS, etc., the first president, September 23rd, 1870. Architect: Alfred Waterhouse, Esq."[39] This was another momentous occasion in the history of Owens College, perhaps at this stage second only to the opening of the College in March 1851.

During the afternoon following the stone laying, the new Court of Governors met for the first time. A number of appointments were made to the Extension College and these included the appointment of the Principal and 10 other professors of Owens College to equivalent chairs in the new institution. The life governors appointed earlier included Sir Thomas Bazley, MP, Thomas Huxley, FRS, and Matthew Arnold, DCL. A new Extension Committee was appointed with Ashton continuing as chair, to lead the fundraising and finalize arrangements with the John Owens Foundation before submitting the application to the Charity Commissioners of England and Wales for their approval.[40]

By the following November a draft agreement was approved by the Trustees of Owens College and submitted to the Charity Commissioners for approval. Final approval was delayed, however, due to lengthy correspondence with the Secretary of the Charity Commissioners about the clause admitting "young persons of either sex." Because Owens's will made specific reference to the education of young males, the Secretary sought two conditions, that the admission of women should not financially jeopardize the admission of "all" men students; and male and female students should be separated "at all times" during teaching.[41]

Further negotiations took place, and it was agreed to delete *all* in the first and *at all times* in the second. The Secretary also insisted that the clause in Owens's will

[38] Ibid., 17.
[39] Ibid., 18.
[40] Ibid., 18–19.
[41] Ibid., 18.

Figure 6.1 New buildings of Owens College in Oxford Road, Manchester.

giving preference to children whose parents were residing in Manchester or South Lancashire should also be adhered to. The Extension Committee, under considerable pressure, felt compelled to comply and the agreement was finally approved in May 1871. This prompted the second parliamentary bill to amalgamate the Owens Extension College with Owens College and form the "Owens College, Manchester," which passed easily through both houses of Parliament and received its royal assent on 24 July 1871, thereby marking another important milestone.[42]

Inauguration of the Oxford Road Buildings

The inauguration of the new buildings on Oxford Road (Figure 6.1) on 7 October 1873 could not come too soon. In the preceding session (1872–1873) the Quay Street buildings had had to accommodate over 1,000 students (334 students in arts, science, and law, 557 evening students, and 112 students in the Pine Street medical buildings), which had stretched the patience of both staff and students.[43] The move to Oxford Road was a major achievement in itself, but it also marked another important milestone in Owens College's determined effort to achieve university status.

[42] Ibid., 19.
[43] Ibid., 20.

The celebrations on 7 October brought together a gathering of notable dignitaries from Manchester and its environs and from London, as well as staff and students. Events were spread throughout the day. The formal opening was held during the day and a soirée was held in the evening when the buildings and departments could be viewed by the public. During the evening a number of lectures were given by the staff. The main ceremony took place in the new chemistry theater with its capacity for 380, and distinguished visitors were seated behind the dais on the platform. The proceedings were presided over by the Duke of Devonshire as chair of governors, and each speaker was cheered as they rose to speak and additional cheers were given when the name "Owens College" was mentioned. In its report of the proceedings, the *Manchester Guardian* notably mentioned the rousing cheers of the medical students who were attending the event.[44]

Among many addresses delivered during the day, Roscoe's lecture, "Original Research as a Means of Education," was particularly notable since it addressed how universities were defined by their academic research, and thereby pointed the direction for Owens College achieving university status.[45] While emphasizing the fundamental need for sound basic knowledge, Roscoe outlined the nature of the research process in relation to the physical sciences, especially chemistry, of which he had firsthand knowledge and experience. However, he also supported the role of research in the liberal arts and humanities. The ability to solve problems was an important transferable skill for professional and commercial work. In a wider context, Roscoe drew attention to the urgent need for science instruction to form part of the curriculum in elementary and secondary schools, as recognized in the recently published reports of the Royal Commission of Scientific Instruction and the Advancement of Science, chaired by the Duke of Devonshire (President of Owens College). This expansion of science education required "the establishment of a systematic method of training science teachers of all grades" and "for this purpose new institutions have to be founded in which the higher branches of the various sciences shall be taught and original research encouraged."[46]

Roscoe also linked original research in chemistry with new opportunities to expand manufacturing and hence national economies:

Among the sixty-three different elements of which the earth, so far as we know, is made up, there are many which have been found only in the minutest quantity. A few only of these rare substances are employed in the arts and manufactures, or are known to play any part in the economy of nature; the rest are substances of interest at present only to the scientific chemist. It would, however, be presumptuous on our part were we to assume that the existence of these bodies is a matter of no moment, for we are constantly learning that substances hitherto supposed to be useless are of the most vital importance.[47]

---

[44] "The Opening of Owens College," *Manchester Guardian*, 8 October 1873, 5.
[45] Henry E. Roscoe, "Original Research as a Means of Education," in *Essays and Addresses by Professors and Lecturers of the Owens College, Manchester* (London: Macmillan, 1874), 21–57.
[46] Ibid., 52.
[47] Ibid., 54.

He used as an example the situation in Germany with the schools of chemistry of Liebig in Munich, Wöhler in Göttingen, Bunsen in Heidelberg, Kekulé in Bonn, Hofmann in Berlin, and Kolbe of Leipzig; indeed, the last named had pointed out that "chemical manufacturers now refuse to take young men into their works unless they have not merely had a scientific education, but also have prosecuted original research."[48] It was this linkage between universities and manufacturing that gave Germany its overwhelming industrial and economic advantage.

At the core of Roscoe's address was the need for government support for science education in general and for universities and original research in particular, not as in France with its centralized and controlling system, but more in the spirit of Germany:

> my aim in the foregoing remarks has been to show that if freedom of enquiry, independence of thought, disinterested and steadfast labour, habits of exact and truthful observation, and of clear perception, are things to be desired as tending to the higher intellectual development of mankind, then original research ought to be encouraged as one of the most valuable means of education. And on this ground alone, and independent of the enormous material benefits which such studies confer on the nation, it is the bounden duty of not only Government, but of every educational establishment, and every citizen of this country who has the progress of humanity at heart, to promote and stimulate the growth of original research among us.[49]

With the presence of so many prominent and influential people, Roscoe's address could be interpreted as a pitch for university status for Owens College and adoption of the German model of education. However, even with the support garnered during the inauguration of the Oxford Road campus, this path would likely prove difficult and still take some years to achieve.

The Path to University Status

The two Acts of Parliament in 1870 and 1871 had laid the foundation for reforming the governance of Owens College by changing it from a charitable trust to a corporate body with a tripartite court of governors, council, and senate. The Oxford Road buildings had greatly expanded accommodations for new departments and the changes brought about by departmental restructuring. There were more well-equipped laboratories for teaching and research. Overall, Owens College could now expect to attract many more students for a wider range of courses, in better accommodations on a larger and more integrated site, though the admittance of women remained a contentious issue. However, even with these achievements, the College's attainment of university status was not a foregone decision. It remained controversial, with many opponents, and would require fortitude and determination on the part of council, the staff, and outside supporters between the 1880s and early 1900s before the University of Manchester as an independent university came into being. Roscoe remained a strong advocate for

[48] Ibid., 37.
[49] Ibid., 57.

university status. The period between 1871 and 1886 (when Roscoe relinquished the Chair of Chemistry after his election as MP for South Manchester) was probably the most intense working period of his life, with his leadership of the chemistry department, his participation in a number of parliamentary inquiries, his extensive consultancy work, and his involvement in the affairs of several professional societies.

The first phase of the Oxford Road building extension was completed in 1872 and included the John Owens Building (the main building) with its lecture rooms, laboratories, and administrative offices, and the chemical laboratories in Coupland Street, to the rear of the main building.[50] Following the site's inauguration the following year, the focus switched to implementing the organizational changes stemming from the two Acts of Parliament.

The Manchester Royal School of Medicine had amalgamated with Owens College in 1872. A new chair in medicine was created and a new purpose-built building was opened in 1874 in Coupland Street, thanks to generous donations. As Roscoe had envisioned when the proposal had originally appeared, a medical school would greatly increase the number of students, would provide an important intellectual linkage between the sciences and the medical school, and would aid Owens College's aspirations of attaining university status. As if to demonstrate its rising importance, further building extensions were opened in 1883 and 1894, with the addition of a dental department in 1884 and a department of public health in 1888.

Transfer of the collections of the Manchester Natural History Society and Manchester Geological Society to Owens College, agreed in outline in the late 1860s, remained to be completed but was dependent on accommodation at the Oxford Road site. The curator of natural history at the museum in Peter Street from 1869 was William Boyd Dawkins, who had studied geology as part of his degree in natural sciences at Jesus College, Oxford. Dawkins had begun cataloging and sorting the collections and had lectured in geology at Owens College. In 1872 Dawkins became the first lecturer in geology at Owens College, following a reorganization of the natural history department, and was appointed professor in October 1874.[51] Unfortunately, space at Oxford Road was at a premium and only part of the geological collections was able to be displayed following transfer of the collections. One condition of the transfer, however, was that the collections must be available for public viewing, and in return the Society agreed to contribute £5,000 toward a new building, to create an endowment fund to pay for the curator's salary and to maintain the collections using proceeds from the disposal of other Society property, including the sale of the Peter Street building.[52] The new building had to wait until 1887 because of the costs and would be designed by Alfred Waterhouse; it was linked to the Beyer Building that housed the geology, zoology, and botany departments.[53]

---

[50] "Owens College Archive," http://archiveshub.jisc.ac.uk/data/gb133-oca.

[51] For Sir William Boyd Dawkins, see *ODNB*.

[52] Samuel J.T.M. Alberti, "Placing Nature: Natural History Collections and Their Owners in Nineteenth-Century Provincial England," *British Journal for the History of Science* 35 (2002): 291–311, on 307.

[53] The museum was extended on several occasions as the collections expanded. Acquisition of the collection from the Petrie excavations in Egypt prompted an extension designed by Paul Waterhouse, Alfred's son, in 1912. "Manchester Museum," https://www.manchester.ac.uk/discover/history-heritage/history/buildings/museum/.

Developments in the curricula during the 1860s and 1870s strengthened the College's aspiration for university status. Two were especially important for strengthening the links between the College and the city: in 1865 Stanley Jevons was appointed Professor of Political Economy (as discussed earlier), and in 1868 Osborne Reynolds was appointed the first Professor of Engineering. For Roscoe, Reynolds's appointment provided an important connection between academia and Manchester's major engineering concerns. The post was "brought about by the eager cooperation of most of Manchester's great engineering firms."[54] Reynolds was only 26 years old when he took up the appointment, having graduated at Cambridge in 1867 and having been elected a Fellow of Queen's College, Cambridge, the same year. He brought energy and drive to the engineering department.[55]

Jevons and Reynolds adapted their teaching and curricula to benefit the industrial development and economy of Manchester. Other science chairs included Geology (1872), Zoology (1879), and a second Chair in Chemistry for Carl Schorlemmer in 1874; arts chairs included Greek for Greenwood in 1869 and Latin for Wilkins in 1869 (thus splitting Greenwood's Classics Chair). In addition, lecturers, assistant lecturers, and demonstrators were appointed to assist the work of the professors as the student numbers steadily increased.

By the mid-1870s the College was already acting as a university. The transformation was underpinned by the development of the Oxford Road site, the merger with the School of Medicine, the extension of the subjects offered, and the growing emphasis on research in both the arts and sciences. What the College still lacked was the authority to set its own examinations and to take full responsibility for its curricula, as it was still tied to the University of London, which provided examinations and degrees to external students from 1858. In 1875 a group comprising Greenwood, Roscoe, Ward, and J. E. Morgan (Professor of Medicine) issued a pamphlet that set out the case for university status.[56] The pamphlet addressed a number of issues: the large number of well-qualified secondary school students seeking higher education, the modern and scientific studies on offer, the need for a university to serve Manchester's commercial and industrial enterprises, and the high cost of studying at Oxbridge.

As part of his advocacy, Roscoe distributed the pamphlet to many of his friends and associates in the academic community to solicit support for the College's aspirations for university status. Responses were in the main supportive, but some reflected concerns that were being expressed in influential circles. Responses from Matthew Arnold, Benjamin Brodie (Professor of Chemistry, Oxford University), Allen Thomson (Professor of Anatomy, Glasgow University), William Thomson (Professor of Natural Philosophy, Glasgow University), and Lyon Playfair were enthusiastic in their support for the College, including setting its own examinations.[57] Allen Thomson's comments went to the heart of the issues at stake:

---

[54] Charlton, *Portrait of a University*, 64.
[55] Ibid., 64.
[56] "Owens College Archive," 8, http://archiveshub.jisc.ac.uk/data/gb133-oca.
[57] "Letters in support of independent university status," OCA/23/11 and OCA/23/3, John Rylands Library, University of Manchester.

In the present state of society in this country, I think that Universities should be placed in our great towns and not in comparatively isolated and or thinly populated districts.... As regards the claims of Owens College to form the nucleus of the new University, I think that these have been so well advocated in your representation as to make them in a great measure irresistible.[58]

On the other hand, several respondents were more cautious, especially about the same body having responsibility for both teaching and examinations. Enforcing standards in medical degrees by restricting the number of medical schools and the number of different examinations had become a sensitive issue nationally. While generally supportive of Owens College becoming a university, both Thomas Huxley (head of the Science Schools, London) and Alexander Williamson (Professor of Chemistry, University College London, and Roscoe's former teacher) focused on the examination question. Williamson was adamant:

In the very able and interesting paper which you sent me it is considered from the point of view of Owens College, and I think that such arguments are entitled to full weight. Many other such schools will doubtless in time arise, and urge similar claims.
    On the other hand, it is argued that there are already too many and too various standards of examination, and that we ought to diminish rather than increase this diversity. In medicine you are, of course, aware that there is an important move in this direction, and I must say that it does seem to me that there are strong motives for desiring a similar tendency towards greater unity in general examinations.[59]

Huxley was even more forceful about examinations:

If the purpose of an examination for a degree, therefore, merely were to determine whether a student is fit to become a teacher in his own University, I should leave the decision of the question entirely to the professors acting as examiners.
    But if a man is to have some social or professional advantages in virtue of his "brand"; if, as in the case of medicine, he is to be allowed to become a registered practitioner, while the man without the "brand" is excluded, then I think society in general has a stake in the question, and has a right to ask for some guarantee that M.B. and M.D., by whatever University conferred, have a common minimum value.[60]

The prospect of Owens College attaining university status attracted a good deal of press attention. Soon after the pamphlet was circulated, a series of articles appeared in the influential scientific journal *Nature* during July 1876. The articles under the byline "The University of Manchester" addressed several concerns. The first article reviewed the current state of higher education nationally before going on to consider

[58] Letter from Professor Allen Thomson to Henry Roscoe, dated 13 May 1876, File GB 133, OCA/23/3, John Rylands Library, University of Manchester.
[59] Letter from Alexander Williamson to Henry Roscoe, dated 26 April 1876. File GB 133 OCA/23/3, John Rylands Library, University of Manchester. Courtesy of University of Manchester.
[60] Letter from Thomas Huxley to Henry Roscoe, undated but probably 1876. File GB 133 OCA/23/3, John Rylands Library, University of Manchester. Courtesy of University of Manchester.

how additional provision required a new direction and adoption of certain princi-ples.[61] The second considered the current deficiency and how it might be remedied.[62] The third addressed how the universities of Britain should be increased.[63] The pro-posal for "the University of Manchester" was reviewed against the theme of each article. Acknowledging the broader educational reforms then in hand and with a par-liamentary commission deliberating on the future of Scottish universities, the article concluded:

> We are convinced that an enlightened government will best complete its efforts in this direction by giving a University Charter to Owens College not, however, as a last and crowning concession, but rather as the first of a series of concessions, all of which, let us hope will, when the time is ripe for them, be frankly and graciously made. Let there be no disguising the fact that Owens College is but the eldest of a large and rapidly increasing family, others of whom may, we hope, in the course of time, make their appearance before the state.[64]

Of the 22 articles appearing in the press, 17 were favorable and 5 less favorable.[65] Nevertheless, the governors of Owens College moved ahead in July 1877 and submit-ted their Memorial for university status to the Privy Council, which was responsible for undertaking a review before its passage on to the queen. Rather than subsiding, opposition continued, mainly due to the aspirations of other large towns to establish universities and the defense of London University as an examining body, but this ten-sion was acerbated by a counter Memorial from Yorkshire College, Leeds.[66]

The Yorkshire College was founded in Leeds in 1874 as the Yorkshire College of Science and aspired to the same status of higher education as Owens College. One of the founding three professors was T. E. Thorpe as the Chair of Chemistry; Thorpe had moved from Owens College to Anderson's College, Glasgow, before this appoint-ment. An Inauguration Ceremony in October 1875 brought together many of the influential and wealthy of the West Riding of Yorkshire, who included the Duke of Devonshire (President of the Yorkshire College, as well as chair of governors at Owens College), Lyon Playfair (MP for Edinburgh and St. Andrews universities, and later MP for South Leeds), and William Forster (the Quaker MP for Bradford and proposer of the 1870 Elementary Education Act). Passionately pleading the case for Yorkshire College, they elicited considerable financial support and a commitment to create a college of higher education embracing science and the arts, with Owens College regu-larly mentioned as the model.[67] In October 1877 the Archbishop of York laid the foun-dation stone of the new College buildings. The Council of the College had appointed

---

[61] "The University of Manchester, I," *Nature* 14 (13 July 1876): 225–226.

[62] "The University of Manchester, II," *Nature* 14 (20 July 1876): 245–246.

[63] "The University of Manchester, III," *Nature* 14 (27 July 1876): 265–266.

[64] "The University of Manchester III," *Nature* 14 (27 July 1876): 266.

[65] "Report on University Status, dated 30 October 1876," OCA/23/5, John Rylands Library, University of Manchester.

[66] "Owens College Archive," http://archiveshub.jisc.ac.uk/data/gb133-oca.

[67] Sir Edward Thorpe, *The Right Honourable Sir Henry Roscoe: A Biographical Sketch* (London: Longmans, Green, 1916), 71.

Alfred Waterhouse as the architect, "whose experience and success in the erection of Owens College seemed to them the highest possible qualification."[68]

It was quickly perceived that Owens College's application for university status would very likely have consequences for Yorkshire College, and correspondence between Roscoe and Thorpe acted as a useful conduit between the two colleges during this hectic period. While the aspirations of emerging colleges in other major cities were recognized by Roscoe and Owens College, they were not in a position at this time to seek university status. With its Memorial submitted, Owens was definitely in the lead. However, Yorkshire College understood the need for urgent action and took an early decision to seek university status. Knowing of this intention, the governors of Owens College included a clause in their application to enable other colleges to enter into incorporation with the University of Manchester. But Yorkshire College was concerned about the name of the university, and a deputation met with the Lord President of the Council (who reviewed Memorials) who, after further deliberations, designated the new university as Victoria University, named in honor of the reigning queen. With both Owens College and Yorkshire College agreeing to the proposed university name (thereby also demonstrating future cooperation between the two colleges), the revised Memorial was signed by the Duke of Devonshire (on behalf of Owens College) and the Archbishop of York (on behalf of Yorkshire College).[69] The Charter for Victoria University was granted by the Queen and finally approved on 20 April 1880.[70] Initially, Owens College was the only constituent college, but there were celebrations in both Manchester and Leeds over the creation of the Victoria University.

Roscoe had always taken an interest in Yorkshire College, even before T. E. Thorpe had become Chair of Chemistry. This was shown on a number of occasions, including the opening of Yorkshire College's new buildings in December 1880 at which Roscoe responded to the toast of "the Victoria University" and expressed the hope that Yorkshire College would soon be one of the incorporated colleges.[71] The 1880 Charter had excluded responsibility for medical examinations (given the concerns over standards and proliferation of degrees), but a supplementary charter was granted in 1883 to allow this. University College, Liverpool, became the second member of the Victoria University in 1884 and Yorkshire College the third in 1887.

As he had done for Owens College, Roscoe was to play an important and influential role in the administration of Victoria University. At a Court meeting to consider whether Greek and Latin should be compulsory, Roscoe spoke forcibly against, citing those students of mathematics or engineering who had never attended a school where Latin was taught.[72] As Thorpe pointed out in his biography of Roscoe, "the 'innovators' won the day by a majority of 2 to 1, and thus effected 'the dethronement, never to rise again, of this mischievous idol.' "[73] When it was agreed to form a Convocation comprising past students of Owens College, Roscoe was put forward as Chairman

[68] Thorpe, *Roscoe*, 75.

[69] "Memorial, adopted by Court of Owens College 22 April 1879, sealed 7 May 1879 and presented 15 May 1879," OCA/23/25, John Rylands Library, University of Manchester.

[70] "Memorial, adopted by Court," 86.

[71] "Memorial, adopted by Court," 88.

[72] Roscoe, *Life and Experiences*, 185.

[73] Thorpe, *Roscoe*, 89–90.

(without his knowledge) but subsequently accepted the nomination and was elected by a large majority. These administrative responsibilities were carried out enthusiastically alongside his considerable workload in the chemistry department, and with his outside commitments.

Admittance of Women to Owens College

Women's admittance to higher education remained a contentious issue through much of the nineteenth century, and opportunities to sit examinations and be awarded university degrees only came about gradually. When Owens College was founded in 1851, public opinion was largely opposed to women studying in higher education.[74] John Owens's will of 1844 under which the College was constituted, had a specific condition that excluded women, "improving and instructing the male sex (not less than 14 years)."[75] However, by the 1860s and early 1870s, provision for school education for girls was expanding, and attitudes toward women studying in higher education were starting to shift not only in society at large, but more importantly within academia. At Owens College, there was mounting support for the admission of women among the professors and governors.

Roscoe was always a strong advocate for women having a full role in Owens College—access to courses, to examinations, and to degrees—and he used his position in the "leadership team" to advance their admission. His advocacy was probably because of the close relationship with his mother and sister following the death of his father when he was just three years old. In an article, "Women at College," published in the *Owens College Magazine* in May 1870, Roscoe took the issue head-on, first addressing the two roles of university education:

> In considering the question of women at College we must bear in mind the two distinct sides to University education. 1. The general one in which school teaching is extended and amplified without any idea of application to the requirement of the afterlife; 2. The special or professional side in which the studies take a given direction with a distinct view to future usefulness.[76]

And then, whether women should be admitted to both or just one:

> For my part I can see no reason why women should be so barred by legal or social enactments from taking any position which their powers or talents entitle them. I believe that the principles of free-trade hold good in intellectual as well as in commercial affairs. Women have an equal right with men in all the benefits which every kind of education can bring. Why, for instance, should women who devote themselves

---

[74] Edward Fiddes, "Introductory Chapter: The Admission of Women to Owens College," in Mabel Tylecote, *The Education of Women at Manchester University 1883 to 1933* (Manchester: Manchester University Press, 1941), 1–16.

[75] Owens's will is quoted in Brian W. Clapp, *John Owens, Manchester Merchant* (Manchester: Manchester University Press, 1965), 173.

[76] H. E. Roscoe, "Women at College," *Owens College Magazine* 2 (May 1870): 129–133, on 130.

with such success to nursing their sick and suffering fellow-creatures be prevented from acquiring that scientific knowledge of disease which will enable them so widely to enlarge their sphere of usefulness.... Then there are hundreds of pursuits in the various branches of commerce, and in the higher and less manual portions of manufacturing industry where properly educated women may as well be employed as men, but for which the necessary preliminary education at present fails.[77]

Finally, returning to higher education and specifically Owens College:

I see no reason why young women should not regularly attend the same lectures and practically work through the same examinations and take the same exhibitions and prizes as young men, and I look forward to the time, not too distant, when we shall wonder why there was so much discussion and so much ill-feeling brought to bear upon the question of admitting women to College.[78]

This unequivocal statement of Roscoe's position remained his core belief and influenced his actions for the remainder of his life. However, within the College, the issue of access for women met some resistance over many years, allowing only incremental improvements. While the Principal dithered, several members of both the Council and Senate, including Roscoe ,felt compelled to pursue the matter, as women were becoming increasingly aggrieved at the lack of progress.

Although the restriction on women becoming students of Owens College had been removed by the 1871 Act, it would take another 12 years before women were fully admitted.[79] Two major issues had arisen from the 1871 Act: the admittance of women crowding out men; and the separation of male and female students in the same classroom. The first never materialized, and the second was never effectively adopted into a policy, leaving a crude symbolic divide of the university buildings marked by Oxford Road: to the west were the main buildings where men were taught, and to the east a house in Brunswick Street for women students. In 1875 a letter appeared in the *Manchester Guardian* mischievously suggesting that classrooms should have a physical wall dividing them, which prompted many other letters voicing similar schemes. It was not until April 1877 that the Court had a formal opportunity to consider admitting women, but with concerns about mixed education and the sentiments of society, it was rejected by 21 votes to 5.[80]

From the 1860s Manchester had become a leading city for advancing women's rights. In January 1867 the Manchester Society for Women's Suffrage was formed to campaign for the rights of women to vote, but when the national campaign began as the National Society for Women's Suffrage in November 1867, it became the Manchester National Society for Women's Suffrage. The Manchester Society's first secretary was Richard Pankhurst (a former student at Owens College), and later his wife Emmeline and daughters Sylvia and Christabel were to take leading roles in the

---

[77] Ibid., 130–131.
[78] Ibid., 133.
[79] Fiddes, "The Admission of Women to Owens College," 7–8.
[80] Ibid., 9.

many campaigns for women's suffrage in Manchester, nationally and internationally. Richard in 1894 joined the Independent Labour Party and the following year stood unsuccessfully for the Manchester constituency of Gorton; much earlier, while a student at Owens College during the period at Quay Street, he had the reputation of being a radical firebrand.[81]

The education of women in Manchester took major steps forward with the formation of the Manchester Association for Promoting the Higher Education of Women in 1869 and the opening of the Manchester High School for Girls in 1874. Both reflected the shift in society's attitudes toward the education of women. However, Owens College tended to shift its position only gradually. In the 1875–1876 session, women were admitted to a list of classes that included Comparative Philology, English Literature, Physics, and Political Economy, but were visitors rather than students and "were denied the privilege of paying fees."[82] However, professors in other subjects (including Roscoe) had also agreed to admit women to their classes; the final selection process remained a mystery.[83]

Outside Manchester and the College, new initiatives for educating women were advancing across Britain. The University of London awarded the first degrees to women in 1878; at the University of Oxford, Somerville College and Lady Margaret Hall were founded in 1879 and women were allowed to attend University lectures and sit University examinations but were not awarded degrees until 1920; and at the University of Cambridge, Girton College was founded in 1869, but the awarding of degrees had to wait until 1948.[84]

A further step in Owens College's embrace of women's education came in 1877 with the establishment of the Manchester and Salford College for Women. The college occupied a house close to Owens College with gloomy and cramped accommodations.[85] While the college was not formally linked with Owens College, several members of the latter's Council and Senate were on the committee, much of the teaching was undertaken by staff from Owens College, and its examinations and certificates were overseen by a Board appointed by the Senate. Such an arrangement was a major step forward, but full integration remained some way in the future.

The 1880 charter for Victoria University had a clear statement on the equality of male and female women students in Clause IV:

> The University shall have power to grant and confer all such degrees and other distinctions as now or at any time hereafter can be granted and conferred by any other University in our United Kingdom of Great Britain and Ireland to and on all persons male or female who shall have pursued a regular course of study in a College in the University and shall submit themselves for examination.[86]

---

[81] Thorpe, *Roscoe*, 91.

[82] Fiddes, "The Admission of Women to Owens College," 9.

[83] The surviving records do not account for the basis of the selection. See also Joseph Thompson, *The Owens College: Its Foundation and Growth* (Manchester: J. E. Cornish, 1886), 494.

[84] "Women Making History: The Centenary," ox.ac.uk/about/oxford.people/womenatoxford/centenary.

[85] Fiddes, "The Admission of Women to Owens College," 9–10.

[86] Ibid., 10.

Students had to have membership in a constituent college (initially only Owens College), but in 1880 women students were still expressly excluded from Owens College. Not surprisingly, this impasse triggered a further press campaign and Council subsequently agreed in April 1883 that membership in Owens College was open to all women students, but with certain conditions. The arrangement was only for a period of five years and with three exceptions—evening classes and for engineering and medicine courses. Evening classes raised social concerns because male and female students would be together in the evening hours to 9 p.m., but the restriction was removed in 1896. By 1897 women were only excluded from engineering and medicine courses: engineering, because it was assumed women would not want to study the subject, though a few years later a women student applied to study engineering and was immediately admitted; and for medicine, although women were first admitted to the Medical Register in 1876 through an act of Parliament, it was only in 1899 that Owens College (then Victoria University) allowed women to study medicine after inquiries had been made at other medical schools as to whether discipline had suffered when men and women studied together.[87] At this time the majority of women students were studying to be teachers, reflecting the limited employment opportunities open to young women.[88] The shackles were finally removed, and women were admitted to all academic subjects at Owens College, but Roscoe for one was probably very disappointed it had taken so long to achieve.

Roscoe's Resignation from Victoria University

Roscoe's election as MP for South Manchester in December 1885 brought his 28-year association with Owens College and Victoria University to an end. His advocacy had been instrumental in overcoming the low student numbers during his early years, in achieving the enlarged campus on Oxford Road, and then securing university status as a college of the federal Victoria University. The final achievement acted as a model for other civic universities. However, Roscoe would be disconnected from the teaching of chemistry, a subject of intense interest since his school days that had led to the influential school of chemistry at Owens College, and had transformed his insight into the important role of science and technical education in the future prosperity of the country.

The Council of Owens College considered Roscoe's resignation following his election as MP on 18 December 1885 and while offering its congratulations:

> accepts with deep regrets the consequent resignation by him of the offices of Professor of Chemistry and Director of the Chemical Laboratories in the Owens College, and invites him to confer with a Committee of the Council as to the time at which the resignation should take effect.[89]

---

[87] Ibid., 14 and 15.
[88] Alex B. Robertson and Colin Lees, "Owens College and the Victoria University, 1851–1903," in *A Portrait of the University of Manchester*, ed. Brian Pullan (London: Third Millennium, 2007), 10–16, on 12.
[89] Roscoe, *Life and Experience*, 240.

It had been Roscoe's hope to be able to combine being an MP while retaining his Chair, but Council must have thought otherwise since they never made such overtures. In acknowledgment of Roscoe's loss to the area, a memorial signed by about 300 of the major manufacturing firms in the region expressed a hope that Roscoe's connection with the College would not be broken.[90] Roscoe officially resigned his Chair the following summer.

His connections with Owens College and Victoria University were not completely severed, because soon after his resignation Roscoe was appointed to a five-year term as a governor of the Victoria University by the Lord President of the Council.[91] In Manchester regularly because of his parliamentary work, Roscoe continued to attend major events and celebrations at Victoria University, where he was always held in high esteem and welcomed by staff and students.

In the post-Roscoe period, the federal Victoria University was accepted by the administration and staff of Owens College as an interim stage on a path toward the fully independent University of Manchester. Even though the curricula and examinations were decided by Owens College (and free of the control of the University of London), with the growth of the Victoria University as charters were approved for University College, Liverpool (1884) and Yorkshire College (1887), the administration became cumbersome and time-wasting, with both the Victoria University and the federal colleges having their own governing bodies. In 1897 the University of Birmingham was created as an independent university without being tied to a federal system. This prompted the constituent colleges of Victoria University to seek the same status. On 15 July 1903 the charter was approved for the independent Victoria University of Manchester alongside Owens College. Parliament approved The Victoria University of Manchester Act on 24 June 1905 that incorporated Owens College with the new Victoria University of Manchester. An aspiration of Manchester going back to the 1830s had finally been achieved, and Roscoe, among many others, would have heartily celebrated the achievement.

[90] Ibid., 241.
[91] Thompson, *Owens College*, 624.

7

Flying the Flag for Chemistry in Cottonopolis

Teaching of Chemistry

As the discoverer of the electron and former Owens student J. J. Thomson argued long ago, "Roscoe's work as a chemist was in teaching and organisation rather than in discovery."[1] Roscoe's great contemporary (and fellow textbook writer) Ira Remsen remarked in his (somewhat inaccurate) obituary of Roscoe: "Roscoe's contributions to chemistry are not numerous" but he "undoubtedly rendered his country a great service in improving the methods of teaching chemistry."[2] The key to understanding Roscoe as a teacher is recognizing his devotion to his own Doctorvater, Robert Bunsen. Bunsen taught chemistry by a method widely accepted in the mid-nineteenth century, namely teaching experimental skills, detection of errors, and powers of observation through the mean of continual drilling in inorganic analysis, first in qualitative inorganic analysis (the detection of different metals) and then in quantitative inorganic analysis (whereby the student measures the exact amount of the metal or non-metal present in a sample).[3] In this Germanic model, this groundwork was followed by a piece of original experimental work under the professor's personal supervision. Carrying out this final stage of chemical training was more problematic in Manchester, as we will see. This model naturally led to both ability and interest in a general chemistry, which was mostly inorganic chemistry but also possessed a strong element of what would later be called physical chemistry. The rapidly growing field of organic chemistry, cultivated by chemists such as August Wilhelm Hofmann and Adolf Baeyer—which was taught at Owens College by Roscoe's German colleague Carl Schorlemmer—was difficult to fit into this framework.

Roscoe was emphatic that taking a chemistry degree at Owens was not a narrow vocational degree—although he disliked the term "liberal education," as he felt it was too closely allied to Classics—and was like any other subject studied there (although he tartly noted that the Professor of Chemistry was paid half the salary of the other professors when he joined Owens College).[4] Hence learning chemistry at Owens College was a training in general scientific principles, teaching both practical skills of broad

---

[1] Sir J. J. Thomson, *Recollections and Reflections* (London: Bell, 1936), 27.

[2] Ira Remsen, "Sir Henry Roscoe (1833–1915)," *Proceedings of the American Academy of Arts and Sciences* 51 (1916): 923–925, both quotes on 924.

[3] By drilling, we mean learning by doing repeated exercises, such as in grammar drills in languages or the drilling of soldiers on the barrack square.

[4] Sir Henry E. Roscoe, *Record of Work Done in the Chemical Department of the Owens College, 1857–1887* (London: Macmillan, 1887), 2; James Francis Donnelly, "Chemical Education and the Chemical Industry in England from the Mid-Nineteenth to the Early Twentieth Century," PhD diss., University of Leeds, 1987, 46.

Henry Enfield Roscoe. Peter J.T. Morris and Peter Reed, Oxford University Press. © Oxford University Press 2024.
DOI: 10.1093/oso/9780190844257.003.0007

application and critical thinking, not a matter of learning numerous facts. Roscoe had a particular dislike for rote learning through books and cramming for examinations. Believing that the professorial method of teaching, as practiced in Germany, was by far the best method of education, he had a low opinion of the tutor-based teaching employed at Oxford and Cambridge. This was probably one reason why he turned down the offer of a Chair at Oxford in succession to his good friend Benjamin Collins Brodie (1817–1880) in 1872, believing that he had a wider scope at Owens.[5] Despite his view, there were tutorial classes in chemistry at Owens College by the mid-1870s, and Roscoe's son Edmund went to Magdalen College, Oxford, where sadly he died of peritonitis at the age of 20 in 1885.[6] As we will see, his positioning of chemistry as a non-vocational degree had advantages from his point of view when engaging with parents from an industrial background.

In contrast to his deprecation of book learning, he strongly supported research as a key element in a chemical education.[7] In Germany this culmination of chemical education in pure research was easy to achieve, as the main degree was the research-based doctorate of philosophy. In England, however, the PhD degree was not common until after World War I when the hitherto usual trip to Germany to take a PhD was no longer considered desirable.[8] Some of the chemists trained by Roscoe took the DSc of the University of London, but this was not a common degree. Hence Owens College lacked any kind of postgraduate training, nor did it have a specific stage in the undergraduate course like Oxford University's Part II, which was introduced in 1916 by Carl Schorlemmer's successor, William Henry Perkin Jr.[9]

It is clear, however, that the most experienced undergraduates could carry out research, and a few of them stayed up to five years in the laboratory in order to do so.[10] Even if the research students at Owens lacked a specific degree-based status, they were not few in number. If one assumes that most research students published a paper (just as most Oxford Part IIs have done), students produced 52 of the 122 papers published by demonstrators and students, either on their own or with a member of staff.[11] Roscoe declared that "I am not going too far when I say that in no

---

[5] Henry Enfield Roscoe, *The Life and Experiences of Sir Henry Enfield Roscoe DCL, LLD, FRS, Written by Himself* (London: Macmillan, 1906), 41–42 and 147–148. For Benjamin C. Brodie, see John S. Rowlinson, Chapter 4, "Chemistry Comes of Age: The 19th Century," in *Chemistry at Oxford: A History from 1600 to 2005*, ed. Robert J.P. Williams, Allan Chapman, and John S. Rowlinson (Cambridge: RSC, 2009), 79–130, on 96–103; *Journal of the Chemical Society, Transactions* 39 (1881): 182–185.

[6] Before he went to Oxford, Edmund took the Junior Class in 1881–1882, the Analytical Chemistry lecture course in 1882, and presumably the qualitative analysis laboratory course, as he registered for the laboratory in 1882. Did his father hope he would take chemistry at Owens College, or was it just to acquaint him with his father's teaching?

[7] Henry E. Roscoe, "Original Research as a Means of Education," in *Essays and Addresses by Professors and Lecturers of the Owens College, Manchester* (London: Macmillan, 1874), 21–57.

[8] Peter J.T. Morris, *The Matter Factory. A History of the Chemical Laboratory* (London: Reaktion, 2015), 115–116; Jack Morrell, Chapter 5, "Research as the Thing: Oxford Chemistry 1912–1939," in Williams, Chapman, and Rowlinson, *Chemistry at Oxford*, 131–186, on 140–141; also see Renate Simpson, *How the PhD Came to Britain: A Century of Struggle for Postgraduate Education* (Guildford, Surrey: Society for Research into Higher Education, 1983).

[9] Morrell, "Research as the Thing," 140–141.

[10] Roscoe, *Record of Work Done*, 10.

[11] Analysis based on the list of papers in ibid., Catalogue III, 45–52.

laboratory in the kingdom has anything like this amount of sound original work been turned out."[12]

There were three distinct groups of students in the chemistry department at Owens. There were the day students. A few of these students took the external Bachelor of Science degree of the University of London, only 18 all told between 1862 and 1883.[13] The average age of all the day students in the early 1860s was significantly lower than today's undergraduates. About four-fifths of the day students were aged between 14 and 17, which reflected the fact that many young people at the time left school at 14 and went out to work.[14] The student fee was 16 guineas per year in the first year and £23 2s in the second year in 1864 (or £1,700 and £2,300 in 2023), which was modest compared with modern university fees.[15] The students tended to be the sons of manufacturers, merchants, and the professional classes. A small minority were sons of shopkeepers and tradesmen.[16] The larger cohort of evening students was a more varied group, as they had daytime jobs, mostly in warehouses, factories, and workshops, but some were teachers. They tended to be slighter older.[17] These remarks apply to the student body as a whole, not just the chemistry students. Indeed, the concept of "chemistry students" is problematic, as students were admitted to Owens College to study a range of subjects, not just one subject, and hence similar to a modern liberal arts college in the United States.

Unfortunately, the class registers are only available from 1875 onward. If we study a cohort of students who first took the entry-level chemistry course from 1875–1876 to 1880–1881 through to their final chemistry courses, matters become clear, albeit for a relatively late period (Table 7.1).[18] A huge number of students, an average of 115 or so a year and comprising around 28% of the entire student body, took the junior lecture course. However, only about 45% went on to take another chemistry course. That many of them failed to pass may be indicated by the large number of students who repeated the class, around a fifth of the total, and some of them even repeated it two or three times.[19] That they failed the course is indicated by the fact that they did not take any more chemical courses (except the analytical chemistry course, which may have occasionally been run concurrently) until their second (or third) attempt. That this conclusion may not be always be true is shown by the case of Charles Kohn, who retook the junior course in 1881–1882, despite coming fourth in the first class in the

[12] Ibid., 17.

[13] Ibid., 17.

[14] Colin Lees and Alex Robertson, "Early Students and the 'University of the Busy': The Quay Street Years of Owens College 1851–1870," *Bulletin of the John Rylands Library* 7 (1997): 161–194, on 164–165.

[15] *The Calendar of Owens College, Manchester, Session 1864–5* (Manchester: Thomas Sowler and Sons, 1864), 25–26.

[16] Lees and Robertson, "Early Students and the 'University of the Busy,'" 167.

[17] Ibid., 164 and 168.

[18] This analysis is based on the class registers for the years 1875–1876 to 1883–1884, John Rylands Library, Manchester. We will seek to publish a complete prosopographical analysis of these registers (and the related laboratory book) in due course.

[19] That they were the same students (and not students with a similar name) was verified by checking their names in the address lists at the front of the class registers, which provides three points of difference: their full name, their parent's name, and their parent's address.

Table 7.1 Numbers of Students in Chemistry Classes and the Laboratory, 1875–1876 to 1883–1884

Year	Day Students	Junior*	Senior**	Analysis	Organic†	Chem Tech†	Chem Phil	Laboratory‡
1875–1876	393	127 (32%)	36	39	22	31	3	77 (77) [79%]
1876–1877	415	117 (28%)	46	37	12	25	4	98 (98) [50%]
1877–1878	418	108 (26%)	34	45	11	18	6	77 (77) [72%]
1878–1879	443	111 (25%)	38	35	12	13	3	81 (81) [58%]
1879–1880	392	112 (29%)	43	37	17	10	10	87 (87) [62%]
1880–1881	417	128 (31%)	36	32	12	27	5	89 (91) [49%]
1881–1882	390	108 (28%)	56	29	25	21	10	108 (108) [50%]
1882–1883	373	105 (28%)	62	33	20	33	11	118 (118) [45%]
1883–1884	400	115 (29%)	50	28	32	35	14	110 (111) [55%]
Total	3641	1033	401	315	163	213	66	845
Average	405	115 (28%)	45	35	18	24	7	

* The figure in brackets is the percentage of day students taking the Junior Class.

** Taking account of any who moved to Organic Chemistry from the Senior Class.

† Total number doing any such course, eliminating any duplication between them.

‡ The figure in brackets is the number of students shown in List I of *Report of Work Done* (per Table 7.1), and the figure in square brackets is the percentage of students taking the Analytical Chemistry and Organic Chemistry course divided by the number in the laboratory.

junior class in 1880–1881.[20] Two-fifths of the group not retake the course and thus either gave up or, having passed, were satisfied taking just one course, perhaps just to have the experience of being taught by Roscoe. Thus, we should not regard these students as bona fide chemistry students. This leaves around 260 students of our initial cohort. Of this group, the largest number (about 45%) took just one other course. Rapidly decreasing numbers took further courses, and only 14 took the full range of courses available.

The junior course comprised lectures in inorganic chemistry, covering the "laws of chemical combination, and a description of the properties and mode of preparation of the non-metallic elementary bodies [i.e., elements], and their most important compounds,"[21] which seems to be equivalent to the modern tenth grade (Year 11 in the UK). The senior class attended lectures on inorganic and organic chemistry which covered "properties of the metals and their compounds, and the composition and relations of the best defined groups of organic bodies [i.e., compounds], and the laws regulating their formation" or roughly twelfth grade (Year 13 in the UK).[22] These

[20] Examination results for the Junior Class in 1880–1881, Laboratory Book, 141, DCH/1/2/3/1, John Rylands Library, Manchester.

[21] *The Calendar of Owens College, 1864–5*, 22.

[22] Ibid., 22.

lectures were invariably lecture-demonstrations, and were probably rather theatrical, rather than simple blackboard and chalk discourses. Roscoe remarked:

> But I would by no means wish to depreciate the value of attendance on a thorough course of Experimental Lectures. By this means the principles and the important facts of the science are brought before the student in a consecutive and systematic manner, and illustrated by experiment, preparation, and diagram, in a way impossible for the pupil himself to make.[23]

In giving his lectures, which he always enjoyed very much, Roscoe had the help of his lecture assistant Joseph Heywood (1826–1893).[24] The senior course was attended by around two-thirds of the core chemistry group in our cohort and had the advantage that they did not involve attendance in the laboratory. About a third of the students who took only one course after the junior course took this course. Despite being the "senior" course, students sometimes took this course in the same year as the junior class, and if taken with the lectures on chemical analysis, they could thereby take three courses in one year. One student, W. S. Blythe, took four courses in 1875–1876 (the fourth one being technological chemistry).[25]

Roscoe's colleague Carl Schorlemmer gave lectures on analytical chemistry, which were later taken over by Carleton Williams.[26] They were a popular option with students who took only two courses (about half of the total), not least because they were usually taken in the same year as the junior class. They were taken by around three-quarters of the core chemistry group as a whole. This course was presumably run in conjunction with the practical course in qualitative inorganic chemical analysis and simple inorganic preparations, as practically everyone who took this course also registered for the laboratory. As Roscoe put it:

> My idea of elementary laboratory teaching is, that to be of any use it must inculcate method and accuracy both in theory and practice. The student must be put on a sound track, and made to understand what he is doing, and why he does it. Moreover, he must gradually gain the power of exact observation, and of logical inference.[27]

Schorlemmer also gave the lectures on organic chemistry. They were usually taken in the second or third year and were not as popular as the senior class or the analysis

---

[23] Roscoe, *Report of Work Done*, 7.

[24] Ibid., 7; Sir Edward Thorpe, *The Right Honourable Sir Henry Enfield Roscoe, PC, DCL, FRS: A Biographical Sketch* (London: Longman, Green, 1916), 101. For a brief obituary of Heywood, see *British Medical Journal* 2, no. 1718 (2 December 1893): 1256.

[25] One suspects he was in a hurry to get back to his father's chemical works; see Peter J.T. Morris and Colin A. Russell, *Archives of the British Chemical Industry, 1750–1914* (Stanford in the Vale, Oxon: British Society for the History of Science, 1988), 24–25. He was not alone in achieving this, at least four other students did so.

[26] *The Calendar of Owens College, Manchester, Session 1872–3* (Manchester: Thomas Sowler and Sons, 1872), 45; *The Owens College, Manchester, Calendar for the Session 1876–7* (Manchester: J. E. Cornish and Thomas Sowler, 1876), 53.

[27] Roscoe, *Record of Work Done*, 7–8.

course. About a quarter of the core chemistry students took this course. These lectures appear to have been historically oriented (partly based on his book *Rise and Development of Organic Chemistry*, published in 1879).[28] He also delivered a course of lectures on chemical philosophy, which combined the history of chemistry with the study of modern theories and the relationship of chemistry to physics. It was the least popular of the courses, being taken by a tenth of the core students, often in their final year if they stayed for more than two years. It was perhaps often regarded as a course for the elite students.

We now come to a problem regarding the class registers. According to *Report of Work Done*, there was a laboratory course on quantitative inorganic analysis. However, there is no trace of this course in the registers. This was because the registers were a record of fees paid. A per diem fee was paid for using the laboratory (and duly recorded), but there is no breakdown of the laboratory courses. The qualitative analysis course and the organic chemical preparations course presumably reflect the enrollment for the corresponding lecture courses, however imperfectly, but there is no such proxy for the quantitative chemical analysis course. One might have thought that the senior course would serve such a role, but it cannot be tied to laboratory registration in the same way as the analysis lectures can be. So, we need to turn to the laboratory instruction itself. The chemistry students registered to take laboratory courses and paid a fee of five guineas (one day a week) to £21 (six days a week) in 1864 to do so.[29] These fees were recorded and thus enabled Roscoe to enumerate the laboratory students in *Report of Work Done*.[30] The second column in Table 7.2 is the total number of day students regardless of the subject studied, and the third column is the number of students who had paid a fee to work in the laboratory.[31]

Colin Lees and Alex Robertson, in their paper on the early years of Owens College, considered the laboratory students to be a distinct group: "who from 1857 were welded into a coherent group by Roscoe in the laboratory ... and were a discrete enough company to be differentiated from 'the College' in the yearly boat race."[32] However, attendance in the laboratory seems to have been a pragmatic matter for most students, rather than to enjoy any esprit de corps. They registered for the laboratory in order to take a given course. The enrollment for the analysis and organic chemistry lecture courses equate to roughly half of all laboratory registrations in the period between 1875–1876 and 1883–1884. If we add an admittedly unknown number for

[28] *The Owens College, Manchester, Calendar for the Session 1880–1* (Manchester: J. E. Cornish and T. Sowler, [1880]), 59–60. Carl Schorlemmer, *The Rise and Development of Organic Chemistry* (Manchester: J. E. Cornish, 1879); a revised version, edited by Arthur Smithells, was published after Schorlemmer's death by Macmillan in 1894.

[29] *Calendar of Owens College, 1864–5*, 35. Five guineas was worth £530 in 2023 and £21 was worth £2,100 according to the Bank of England Inflation Calculator.

[30] Roscoe, *Record of Work Done*, 4.

[31] That the second column refers to *all* day students and not just day students taking chemistry is shown by the table in Lees and Robertson, "Early Students and the 'University of the Busy,'" 163, which gives identical figures for all day students between 1857 and 1869, except for 1867 where it has 170 rather than Roscoe's 173.

[32] Ibid., 180.

Table 7.2 Student Numbers in Chemistry after List I in *Report of Work Done*

Session	Day Students	Lab Students	%
1857–1858	34	15	44
1858–1859	40	23	58
1859–1860	57	24	42
1860–1861	69	21	30
1861–1862	88	22	25
1862–1863	108	34	31
1863–1864	110	38	35
1864–1865	128	49	38
1865–1866	113	41	36
1866–1867	113	37	33
1867–1868	173	44	25
1868–1869	210	57	27
1869–1870	209	51	24
1870–1871	264	60	23
1871–1872	327	66	20
1872–1873	337	66	20
1873–1874	356	78	22
1874–1875	375	71	19
1875–1876	395	77	19
1876–1877	415	98	24
1877–1878	418	77	18
1878–1879	443	81	18
1879–1880	392	87	22
1880–1881	417	91	22
1881–1882	390	108	28
1882–1883	373	118	32
1883–1884	400	111	28
1884–1885	393	120	31
1885–1886	380	118	31
1886–1887	333	101	30
Total	7860	1984	25

the quantitative analysis, we must surely approach three-quarters of the total. An unknown (but small) number of students carried out advanced research in the laboratory. For our cohort, I would surmise this was no more than 20 students over six years. Other occupants of the laboratory in our period were research fellows (including Ludwig Claisen), assistants, and even schoolteachers.

Returning to the quantitative analysis course, Roscoe makes it clear what is obtained from this laboratory work:

> In this he learns by degrees what scientific accuracy means, how exact results can be obtained by careful work, and thus gains in confidence and certainty. Here, too, constant personal supervision on the part of the professor and of the demonstrators is absolutely requisite, as everything depends on the care with which the various operations are carried on, as working from receipts [recipes] without superintendence is really valueless.[33]

Note the emphasis on the supervising role of the professor, which reflects the model set by Bunsen. Roscoe was in fact insistent on this point:

> The personal and individual, attention of the Professor is the true secret of success; it is absolutely essential that he should know and take an interest in the work of every man in his laboratory, whether beginning or finishing his course. The Professor who merely condescends to walk through his laboratory once a day, but who does not give his time to showing each man in his turn how to manipulate, how to overcome some difficulty, or where he has made a mistake, but leaves all this to be done by the Demonstrator, is unfit for his office, and will assuredly not build up a school.[34]

Yet one wonders if the professor protests too much: Did Roscoe always fulfill this ideal himself? In his earlier years in the old laboratory, this was almost certainly the case, as he was only aided by one or at the most two other members of staff who had other duties, such as the teaching of organic chemistry. However, the number of demonstrators grew to three when the new laboratories were opened in 1873, and when the number of students in the laboratory mounted, the professor's ability to supervise every student correspondingly declined.[35] Sir Edward Thorpe later commented:

> Of course, as the number of his pupils increased, and the laboratories became larger and more numerous, it became impossible to give so great a share of individual attention, and much had to be delegated to demonstrators.[36]

Roscoe's external duties, for example as an industrial consultant, grew in this period, as we will see in Chapter 8. In 1876, he was appointed a member of the Royal Commission on Noxious Vapours. Yoshiyuki Kikuchi, on the basis of a laboratory record book in the university archives, has shown that there was a marked shift in 1876 and Roscoe began to rely more on written reports by the demonstrators, which he then signed off.[37] Given the stern ideal set by Bunsen, he probably attempted to

---

[33] Roscoe, *Record of Work Done*, 11.
[34] Ibid., 5.
[35] For a list of lecturer-demonstrators with their dates, see ibid., 6.
[36] Thorpe, *Roscoe*, 100.
[37] Yoshiyuki Kikuchi, *Anglo-American Connections in Japanese Chemistry: The Lab as Contact Zone* (New York: Palgrave Macmillan, 2013), 71; and personal communication from Kikuchi, 11 October 2020, based on his study of a laboratory report book in the Manchester University Archives (DCH/1/2/3/1).

maintain some degree of personal supervision of his students, but he clearly struggled to do so in the later years of his tenure at Owens. If the co-publication of papers is any guide, the supervision of research students seems to have been largely left to the demonstrators and lecturers, who were close to the students in age, with the exception of a few exceptionally gifted students, notably Thorpe and Arthur Harden. The demonstrators tended to be Owens graduates (although not entirely so) and were rather clumsily called demonstrators and assistant lecturers.

The use of inorganic analysis to train chemistry students was well established by the 1870s, but the teaching of organic chemistry was more problematic. One could begin with organic analysis. The first experiment one of the authors (PJTM) did in the Dyson Perrin Laboratory in Oxford a century later was the identification of an unknown, which turned out to be a phenyl ester, as the pungent clouds of monochlorophenol floating above the lab benches soon revealed. But organic analysis was relatively undeveloped in the 1870s, although the Lassaigne sodium test for qualitative elemental analysis had been introduced as far back as 1843.[38] By the 1880s Roscoe and Schorlemmer had developed a course which combined organic analysis with organic preparations and some physical chemistry:

> On entering upon their third session, students usually occupy themselves with Organic Analysis, Vapour Density, Fractional Distillation, &c., together with a series of Organic Preparations made in a systematic manner, and with special reference to the yield obtained.[39]

In his biography of Roscoe, Thorpe gave the total number of chemistry students that passed through the department in Roscoe's tenure as 2,000, and this figure has been cited by Robert Bud and Gerrylynn Roberts.[40] He appears to have based this number on the total number of students in the laboratory, given in the *Record of Work Done* as 1,984 (it is actually given incorrectly as 1,884, but Thorpe may have been using the correct figure). This figure is clearly invalid for a number of reasons. It includes the years 1885–1886 and 1886–1887, when Roscoe was mostly not present in the department. It also includes people who were clearly not chemistry students, such as Ludwig Claisen. Most importantly of all, there is considerable double-counting, as some students registered for the laboratory for more than one year. If we only consider our core chemistry group, the total number of years they registered for the laboratory is 300 years, an average of 1.2 years each. Sixteen of this group did not register for the laboratory at all, and the highest number of years (4) found for this group was taken by a mere seven students;[41] 300 may seem a fairly good proxy for 260 students, but

[38] S. Horwood Tucker, "A Lost Centenary: Lassaigne's Test for Nitrogen. The Identification of Nitrogen, Sulfur, and Halogens in Organic Compounds," *Journal of Chemical Education* 22 (1945): 212–215.

[39] Roscoe, *Report of Work Done*, 14.

[40] Thorpe, *Roscoe*, 108, cited by R. F. Bud and G. K. Roberts, *Science versus Practice: Chemistry in Victorian Britain* (Manchester: Manchester University Press, 1984), 86.

[41] The record in our group, 5 years in the laboratory, is actually held by Robert Frost. However, he retook several courses, which explains why he spent so long in the laboratory. Evidence that he was not good at chemistry is provided by Charles Kingzett, who was irritated that Frost claimed he could not make crystals of calcium hypochlorite, contrary to Kingzett's claim, and then a few days later admitted that he had

the total number of laboratory registrations for this period (the data used by Thorpe) was 848, which is over three times as great. If we assume laboratory registrations are a good proxy up to 1872–1873 and take an average of 43 (based on our cohort) for the remaining 14 years of Roscoe's tenure, we obtain a figure of 840, which seems reasonable, but much lower than 2,000. However, such calculations raise the issue of who was a chemistry student in a period and a college where a modern chemistry degree did not exist. If we were to include all the students who took the junior course (which was taught by Roscoe), the number would be by coincidence indeed around 2,000.

Thus, equipped with an understanding of inorganic chemistry and organic chemistry with a training largely based on qualitative and quantitative inorganic analysis, the student was now equipped to go forth to work in industry or in schools, the destination of a large majority of those who passed through the chemistry department at Owens College. Those who aspire to the higher posts in secondary schools or to teach in colleges and universities—a school of chemistry's highest function, according to Roscoe—stayed on to do some basic research in their fourth year and in some cases even their fifth year.[42] Some of these students were financially assisted by a Dalton Scholarship, some 22 in all during Roscoe's tenure.[43] The scholarships had been established in 1853 by Edward Frankland while he was at Owens, and although originally intended as an award for excellence in research, under Roscoe—the first scholarship was not awarded until 1856—it became a scholarship to fund a research student.[44] It is important to note that these advanced students did not automatically gain a degree as a result, although nine of them did eventually obtain the DSc of the University of London. Rather, it was a stepping stone to taking a PhD in Germany (often at Heidelberg or Bonn) and then a demonstrator-ship at Owens or elsewhere. However, such high-fliers were a tiny minority of the student body.

Technological Chemistry

As we noted in Chapter 1, Manchester was one of the most important industrial centers, not just of the British Empire, but indeed the whole world. While this preeminence was largely based on the textile industry, and cotton in particular, there were also many chemical works in and near Manchester and even further out in Lancashire. One might therefore assume that a training in chemical technology was an important aspect of the chemical syllabus at Owens. Roscoe encountered this assumption, but tried to push back. He did so for several reasons. From a purely pragmatic point of view, he was not an expert on chemical technology, and the teaching of chemical technology would have been both complicated and expensive, involving the acquisition of machinery and the erection of pilot plants, in Roscoe's own words, to "become a

---

now succeeded; see C. T. Kingszett, "Calcic Hypochlorite," *Chemical News* 46 (15 September 1882): 120. It appears he then became a barrister.

[42] Roscoe, *Report of Work Done*, 10. It is interesting to note that none of our cohort who spent four years in the laboratory became a significant chemist.

[43] Ibid., 12.

[44] Roscoe, "Original Research as a Means of Education," 47.

[work]shop."[45] Furthermore, no one had yet broken down chemical technology into its component operations (later called unit operations) in a way which would allow the subject to be taught in a systematic and generalized manner. Roscoe also objected to a focus on chemical technology for more philosophical reasons—which may or may not have been self-serving—he argued that "a sound and thorough training in the scientific principles which underlie all practice" largely based on inorganic chemical analysis was a better foundation for a career in industry than a purely vocational training.[46] Ironically, in one unintended sense, he was correct: some of his students became analysts in works laboratories. By contrast, Roscoe argued that this generalized training would equip his students:

> to take a more intelligent part in the operations of the various manufactures than those who had not had such advantages, but that this education had given an insight into these processes such that those thus trained were able to effect improvements or even to make discoveries of importance.[47]

However, his argument for a generalized training was not simply based on the grounds of mental training or as part of a liberal education, but because he wanted to resist the demands of some parents for a specialized training or even the provision of just the chemistry needed to work in the particular branch of industry that they wish their sons to be employed in, often in a family business. The hardworking W. S. Blythe is a good example of this group. According to Roscoe (and this may be one of Roscoe's characteristic exaggerations):

> The fathers frequently used to come with a story of this kind: "I am a calico printer, or a dyer, or a brewer, and I want you to teach my son Chemistry so far, and only so far, as it is at once applicable to my trade," and when informed that Chemistry as a science must be taught before its applications could be understood, and that his son could not for two or three years at least begin to work upon the subjects directly bearing on his trade, he too often replied that if that were the system he could not afford time for his son to learn on this plan, and that if he could not be taught at once to test his drugs, he should prefer to leave him in the works, where he and his father before him had made a great many commercial successes with no scientific knowledge, and where he saw no reason to doubt that his son would do the same.[48]

These demands created two problems for Roscoe. Even if he had been willing to do so, it would have been impossible to him to provide specific training in say, alkali manufacture, calico printing, synthetic dye manufacture, or pharmaceuticals without at least employing far more staff and using much more space, which would not have been cost effective. The other more immediate problem was to stop the students from leaving when they had learned "just enough chemistry," thus reducing the college's

[45] Ibid., 18.
[46] Ibid., 3; Thorpe, *Roscoe*, 106.
[47] Roscoe, *Report of Work Done*, 3.
[48] Ibid., 9–10.

fee income as well as producing poorly trained chemists, who would not be good for the college's reputation. As it was, comparatively few students stayed on for the whole three-year University of London external degree course. Of the students in our core chemistry group, 30% took four or more courses and only 15% spent three or more years in the laboratory. The idea that one should just learn the minimum and then go into industry was very common at the time, and many students left their universities after one or two sessions, both at Manchester and elsewhere. Hence Roscoe had to construct his course in a way that made it impossible for a student to believe (if he accepted Roscoe's view) that he had really learned chemistry until the end of the second year. At that point, if he was not intending to go into synthetic dyes, medicine, or pharmacy, he might well think the third year devoted to organic chemistry was superfluous. If one takes a cynical view, as an inorganic chemist Roscoe might not have been too concerned about that outcome; at least he would have gotten two years of fees, rather than the one session that their father originally had in mind. Blythe showed that a student could do two years of courses in one year, but he was the exception.

However, Roscoe was also pragmatic. Clearly there was a demand for some form of training in chemical technology, and Owens College had to provide it to satisfy its clientele. Roscoe admitted that it was "very desirable that, supposing him to be intended for a chemical manufacturer, he should attend a course of lectures on some specific subject connected with industrial chemistry."[49] Up to 1880, Roscoe gave a course of lectures on technological chemistry (which he seems to have preferred to the more loaded term "applied chemistry") and Schorlemmer gave others.[50] There were usually at least two courses and sometimes up to four. These courses were intended to teach the "Chemical Principles involved in the most important Chemical Manufactures" and covered such topics as the production of heat and light, water and air (i.e., sanitation), the manufacture of alkalis and acids, the manufacture of glass and porcelain, and dyeing and calico printing. The announcement of these lectures in the calendar was accompanied with the stern warning that "[s]tudents attending this class must be acquainted with the principles of chemical science," in other words, it could only be taken by students who had already taken the courses on pure chemistry.[51] Most of the students taking these courses had done several chemistry courses, but a small number had only taken the junior course, mainly those in the mid-1870s, so things may have been tightened up. About a third of the core chemistry group took these courses, so clearly, they were not regarded as specialized. Roscoe continued to resist the idea that Owens should provide training in industrial practice, arguing that was the job of industry itself. However, he admitted there was a gray area between pure chemistry and industrial practice which perhaps needed to be filled, but he did not know how.[52] Of course we would now recognize that this gap could be and was eventually filled by chemical engineering, with its systematization of industrial practice into its generic components.

[49] H. E. Roscoe, "Remarks on the Teaching of Chemical Technology," *Journal of the Society of Chemical Industry* 3 (1884): 592–594, on 592.

[50] Donnelly, "Chemical Education and the Chemical Industry," 128.

[51] *The Owens College, Manchester: The Calendar for the Session 1878-9.* (Manchester: J. E. Cornish and Thos. Sowler, 1878), 59.

[52] Roscoe, *Life and Experiences*, 188.

With the arrival of Watson Smith, who had worked with Georg Lunge (1839–1923) in Zurich and had a solid background in industrial chemistry, matters began to change.[53] There seems to have been an attempt in the mid-1870s to move into metallurgy, as a metallurgical laboratory was set up under the direction of Roscoe and the hope was expressed that a "course of lectures on Metallurgy will probably be given in the course of the session if a sufficient number of students offer themselves."[54] They presumably failed to do so, as the metallurgical laboratory does not feature in later calendars. Roscoe and Watson developed a Certificate in Technological Chemistry in 1881, which was a four-year course and combined chemistry, mechanical drawing, and geology, among other subjects.[55] Students at Owens began to take the external examinations of the City and Guilds of London Institute from 1882 onward, principally in alkali manufacture, but also in tar distilling and coal tar products, among others.[56] In 1884, Smith became the Lecturer in *Technological* Chemistry with Roscoe's support, but significantly within the new Victoria University, not Owens College itself. One reason for this embrace of chemical technology was the threat of competition from the Manchester Mechanics' Institute, which became the Manchester Technical School in 1883.[57]

Laboratories

The original laboratory and lecture room had been erected in the former garden of the college, where the stables had been in 1851.[58] There was a single lecture room which could hold 150 students and a laboratory which could accommodate 42 students. For several years after his arrival at Owens in 1857, Roscoe struggled to increase the number of students taking chemistry and the quality of the chemistry students. As his reputation burgeoned and the importance of chemistry for industry increased, the number of chemistry students both grew in number and in quality. In 1857–1858, there were 15 students in the laboratory, which was well below the building's capacity.[59] However by 1862–1863, the numbers had more than doubled to 34

---

[53] For Lunge, see P.P.B[edson], "Georg Lunge," *Journal of the Chemical Society, Transactions* 123 (1923): 948–950. For Smith, see P.P.B[edson], "Watson Smith," *Journal of the Chemical Society. Transactions* 117 (1920): 1637–1638.

[54] *The Calendar for the Session 1878-9*, 61. The class list for 1875–1876 records just one student, S. H. Carrington, as working in the metallurgical laboratory. No metallurgical class or attendance in the metallurgical laboratory has been found in any other class register up to 1883–1884.

[55] *The Victoria University Calendar for the Session 1882-3* (Manchester: J. E. Cornish, 1882), 91 and James F. Donnelly, "Getting Technical: The Vicissitudes of Academic Industrial Chemistry in Nineteenth-Century Britain," *History of Education* 26 (1997): 125–143, on 132–133. Also see Donnelly, "Chemical Education and the Chemical Industry," 128–130, and Kikuchi, *Anglo-American Connections*, 72–73.

[56] Roscoe, *Report of Work Done*, 20–21.

[57] Donnelly, "Getting Technical," 132.

[58] Robert H. Kargon, *Science in Victorian Manchester: Enterprise and Expertise* (Manchester: Manchester University Press, 1977), 160. Nonetheless, Roscoe must have told Bunsen (in a now-lost letter) that he was happy with the laboratory, as Bunsen was pleased to hear this news; letter from Bunsen to Roscoe, dated 11 October 1857, Deutsches Museum Archives, HS 908–1033, letter 00919.

[59] Roscoe, *Report of Work Done*, 4. In a letter to Jevons, dated 20 March 1858, Roscoe says he has 13 students in the laboratory, 11 "in my long course of lectures" and 25 in a short evening course, R.D. Collison Black, ed., *Papers and Correspondence of William Stanley Jevons*, volume II, *Correspondence, 1850–1862* (Clifton, NJ: Augustus M. Kelley), Letter 116, 322–325, on 322.

students in the laboratory. The percentage of students in the laboratory relative to the total number of day students decreased from 44% in 1857–1858 to 20% in 1871–1872. This was presumably a reflection of the lack of space in the laboratory (given that the original capacity was only 42), the maximum number appearing to be 66, which is the figure for both 1871–1872 and 1872–1873. Clearly something had to be done. As Roscoe told the Devonshire Commission in 1871, his existing laboratory was "altogether inadequate. The present building was built to accommodate 35 students and now I have 78 working in that place."[60] When the college moved from Quay Street to Oxford Road between 1870 and 1872, the opportunity to build a new laboratory arose.

After visiting the new German and Swiss laboratories with the college principal in the long vacation of 1869, he designed a new laboratory in Oxford Road (Figure 7.1) with the well-established local architect Alfred Waterhouse (1830–1905), who according to Roscoe, "at once understood the special requirements of a laboratory."[61] This laboratory when it opened in 1873 had room for 100 students and at a pinch could hold 200. The lecture theater was designed to hold 380 students.[62] As a former colleague of Bunsen, it is not surprising that his laboratory reflects his master's simplicity:

> There are two long rooms, 30 feet high, lighted both at the side and at the top, which form the chief working laboratories. At the side we have a number of rooms for the various operations: assistants' rooms, library, balance rooms, organic analysis rooms, rooms for spectroscopic work and electrolytic work; also rooms in the basement which are available for special purposes, as also furnace rooms and large store rooms for apparatus and materials. A class-room for smaller classes is also attached, and a room for a mineralogical cabinet. On the second floor, above this portion, is the laboratory and private rooms of the professor, and that is so arranged that he can see what is going on down below. The main building is to be of stone, but I have insisted upon having my building in brick, because I prefer rather to spend the money on the internal arrangements.[63]

Like Bunsen's building in Heidelberg, Roscoe's new laboratory building had two long laboratories side by side. The furnishings at Owens were far less elaborate than in Kolbe's new laboratory building in Leipzig; there were no washbasins, for example, and the bottle racks were very similar to those in a photograph of the Heidelberg laboratory around 1900. Although there were specialized rooms, as Roscoe mentioned in his evidence, the emphasis is clearly on teaching rather than advanced research. The new laboratory building was a decently large edifice which mostly met Roscoe's needs as a teacher, rather than the great chemical palaces of August Wilhelm Hofmann,

---

[60] *Royal Commission on Scientific Instruction* (Command 536, 1872), vol. 1, 500, sc. 7370. The figure he gives is puzzling, as the number given in *Record of Work Done* for 1870–1871 is 60, and all the previous years' figures were lower. He implies that only students worked in the laboratory in this period, an important fact for calculating total student numbers (see above).

[61] *Royal Commission on Scientific Instruction* (Command 536, 1872), vol. 1, 500, sc. 7370; For Waterhouse and his practice, see Colin Cunningham and Prudence Waterhouse, *Alfred Waterhouse 1830–1905: Biography of a Practice* (Oxford: Clarendon Press, 1992).

[62] Education Department, *Reports from Universities and University Colleges Participating in the Grant of £15,000 made by Parliament for "University Colleges in Great Britain"* (London: HMSO, 1896), 191.

[63] *Royal Commission on Scientific Instruction* (Command 536, 1872), vol. 1, 500, sc. 7370.

Figure 7.1 The Quantitative Analysis Chemical laboratory in Oxford Roac.

August Kekulé, and Hermann Kolbe in Germany, with their numerous rooms and a residence for the professor.

One might wonder why Roscoe was not more ambitious in his plans for the new laboratory and why he emulated the plans of Bunsen rather than the more modern designs of Hofmann in Bonn and Berlin and Kolbe in Leipzig. Roscoe followed Bunsen's method of teaching, in which the main professor, rather than the more junior staff, teaches the students, and this style of teaching was reflected in Bunsen's laboratories. The teaching at Owens (and indeed in Britain as a whole at the time) was geared toward undergraduate teaching, and there was no need for the well-equipped research laboratories found in the buildings in Berlin or Leipzig. Inevitably cost was a factor, as a small privately funded provincial university, Owens simply did not have the financial resources available to Hofmann and Kolbe, backed by Prussia and Saxony. It was of course Roscoe's great ambition that British universities should enjoy similar state largesse, but this inevitably lay in the future.

From Owens, this Heidelberg-inspired model spread to other universities in Britain, British schools, and even to Germany. Roscoe later recalled that the laboratories he had:

> designed for Owens College were copied far and wide not only in this country but abroad, and especially at Munich, where [Adolf] Baeyer (Liebig's successor) based his new building on the plans which I sent to him.[64]

In Britain, a similar laboratory was designed at Newcastle by his former student Peter Philips Bedson and the local architect Robert J. Johnson in 1894.[65] Meanwhile, in 1878 Waterhouse and the chemist James Campbell Brown (1843–1910) designed the chemical laboratory at Liverpool.[66] At Leeds, Waterhouse designed the laboratories of the Yorkshire College of Science (now Leeds University) with Roscoe's former assistant Edward Thorpe, which were formally opened in 1885.[67]

Personal Research

Perhaps surprisingly for such a high-profile chemist, Roscoe did not make any major advances in chemistry.[68] This is partly because his main field of research, physical chemistry, was in an intellectual lull during his most productive years. He had moved into this field as a result of his training with Bunsen, and like Bunsen, he was also interested in inorganic chemistry, although mainly in the form of chemical analysis. While these areas were not productive for important intellectual breakthroughs, they

---

[64] Roscoe, *Life and Experiences*, 110.

[65] Personal communication from the late W. Alec Campbell, 1998.

[66] For Brown, see H.B.D[ixon] and J.N.C[ollie], "James Campbell Brown," *Journal of the Chemical Society, Transactions* 99 (1911): 1457–1460.

[67] William H. Brock, "The Chemical Origins of Practical Physics," in *The Case of the Poisonous Socks* (Cambridge: RSC, 2011), 97–113, on 98–99.

[68] Roscoe's research is nicely covered by Jaime Wisniak, "Henry Enfield Roscoe," *Educación Química* 27 (2016): 240–248.

were very useful in applied chemistry, especially for industrial chemistry and for the monitoring of the environment. Had he been an organic chemist, Roscoe would have surely had a greater impact, both on the development of chemistry and on the growth of the synthetic dye industry in Manchester, but he was content to leave this field to his colleague Schorlemmer.

As a close colleague of Bunsen, Roscoe helped to establish spectrum analysis as a mature technique, especially its use in the steel industry. However, he did not introduce any new spectroscopic techniques, nor did he use spectroscopy to discover new elements, unlike his contemporary William Crookes. Another interest Roscoe shared with Bunsen was photochemistry. We have discussed their joint research in this area in Chapter 4. He developed an interesting lecture demonstration in which sealed bulbs of a mixture of hydrogen and chlorine (which were stable in the dark) were subjected to a magnesium flash.[69] If a red or yellow filter was put in front of the bulb, there was no reaction, but if a blue filter was used, there was an immediate explosion, showing that only the light rays at the blue end of the spectrum could initiate the reaction.

Roscoe (with the help of William Dittmar) continued a line of research initiated by Bunsen, namely the relationship between pressure and the solubility of gases in water. He discovered that ammonia and hydrogen chloride did not obey the pressure law below 60°C, but on the other hand, as the concentration of hydrogen chloride in boiling hydrochloric acid was dependent on pressure, hydrochloric acid could not be considered to be a true compound.[70] It was also known that boiling many acids at constant pressure produced mixtures that contained fixed amounts of the acid and water. Roscoe showed that the composition of these mixtures was dependent on pressure and hence could not be considered to be hydrates of the acids.[71] This was a good example of the formation of azeotropic mixtures, but the concept was only introduced as late as 1911 by John Wade (1864–1912) and Richard William Merriman at Guy's Hospital, London.[72] Roscoe also measured the vapor densities of lead and thallium chlorides and showed that there was no decomposition in the vapor phase.[73]

However, Roscoe's most significant achievements were in the field of inorganic chemistry. This partly came about by accident, as these things often do. The occasion for his first encounter with the metal vanadium is best portrayed by Roscoe himself:

In the year 1865 I was asked to visit the copper mines at Mottram, near Alderley Edge, in Cheshire. Copper occurs there together with a large number of other metals, notably cobalt, in the Keuper sandstone, each grain of sand being surrounded by a coating of copper carbonate and oxide of cobalt. The process of extraction consists in treating the

[69] Roscoe, *Life and Experiences*, 132.

[70] Henry E. Roscoe and William Dittmar, "On the Absorption of Hydrochloric Acid and Ammonia in Water," *Quarterly Journal of the Chemical Society of London* 12 (1860): 128–151.

[71] Henry Enfield Roscoe, "On the Composition of the Aqueous Acids of Constant Boiling Point," *Quarterly Journal of the Chemical Society of London* 13 (1861): 146–164.

[72] John Wade and Richard William Merriman, "Influence of Water on the Boiling Point of Ethyl Alcohol at Pressures above and below the Atmospheric Pressure," *Journal of the Chemical Society, Transactions* 99 (1911): 997–1011, on 1004. For Wade, see F. Gowland Hopkins, "John Wade," *Journal of the Chemical Society, Transactions* 103 (1913): 767–774.

[73] Henry Enfield Roscoe, "Note on the Specific Gravity of the Vapours of the Chlorides of Thallium and Lead," *Proceedings of the Royal Society of London* 27 (1878): 426–428.

ore with hydrochloric acid, the copper being afterwards precipitated by metallic iron or zinc. On my visit the manager showed me a dark blue solution which he believed to be a concentrated solution of a copper salt in a peculiar condition, inasmuch as it was not precipitable by zinc. I at once saw that this could not be copper, but that the blue colour was probably due to the presence of the very rare metal vanadium. I asked him to send me some of the solution to test, and found my suspicions confirmed. I inquired in what portion of the ore or residues he found this substance, and after a while obtained from him several tons of a lime precipitate containing this hitherto rare metal.[74]

With his assistant Thomas Thorpe, Roscoe then carried out an extensive investigation of this metal. Vanadium has a complex history. It had originally been discovered in a supposed ore of lead (actually vanadinite) in 1801 by the Spanish mineralogist Andrés Manuel del Río y Fernández (1764–1849) in Mexico City. He was struck by the numerous colors produced by the compounds of this element and called it "panchromium." Later, in 1805, he was persuaded that panchromium was just an impure sample of chromium and withdrew his claim. The Swedish chemist Nils Gabriel Sefström (1787–1845) rediscovered the element in iron ores in 1831 and called the element "vanadium" after the Norse goddess of beauty, Vanadis, because of the beautiful colors of its compounds.[75] It was then studied by the great Swedish chemist Jöns Jakob Berzelius (1779–1848), who claimed to have isolated the metal and determined that the highest oxide of vanadium was a trioxide (VO_3). He believed the atomic weight of vanadium to be 68.5 (in modern terms).[76]

Roscoe was able to show that the highest oxide was in fact a pentoxide and that vanadium was chemically related to phosphorus and arsenic, not chromium as Berzelius had assumed.[77] Furthermore, the vanadium chloride of Berzelius and later chemists was in fact an oxychloride similar to phosphorus oxychloride. He was able to isolate the metal for the first time, but not without difficulty.[78] Roscoe's Bakerian lecture on vanadium was delivered in December 1867, before Dimitri Mendeleev (1834–1907) introduced his periodic table. Mendeleev placed vanadium in group V with phosphorus and arsenic, but also with niobium and tantalum. Vanadium and the other transition metals were later separated out to become group Vb (or 5 in modern usage). Roscoe determined the atomic weight of vanadium to be 51.4, which

[74] Roscoe, *Life and Experiences*, 141–142. For the history of the Mottram mines, see G. Warrington, "The Copper Mines of Alderley Edge and Mottram St. Andrew, Cheshire," *Journal of the Chester Archaeology Society* 64 (1981): 47–73; and "Mottram Mines," https://mottramstandrewpc.org.uk/the-village/mottram-mines/, which has images of mottramite vanadium ore.

[75] Mary Elvira Weeks, *The Discovery of the Elements*, 6th ed., enlarged and revised (Easton, PA: Journal of Chemical Education, 1960), 353–364. Also see Lyman R. Caswell, "Andrés del Río, Alexander von Humboldt, and the Twice-Discovered Element," *Bulletin for the History of Chemistry* 28 (2003): 35–41.

[76] James F.W. Johnston, "Report on the Recent Progress and Present State of Chemical Science," in *Report of the First and Second Meetings of the British Association for the Advancement of Science at York in 1831 and Oxford in 1832* (London: John Murray, 1833), 414–529, on 470. Also see J. Berzelius, "Ueber das Vanadin und seine Eigenschaften," *Annalen der Physik* 98, no. 5 (1831): 1–67.

[77] Henry E. Roscoe, "Researches on Vanadium," Part I (Bakerian Lecture), *Proceedings of the Royal Society of London* 16 (1868): 220–228. See Roscoe, *Life and Experiences*, 142–144; Thorpe, *Roscoe*, 123–130; and Weeks, *Discovery of the Elements*, 361–363.

[78] Roscoe, "Researches on Vanadium," Part II, *Philosophical Transactions of the Royal Society* 159 (1869): 679–692.

is fairly close to the modern figure of 50.9. He argued that Berzelius had obtained a figure of 68.5 because his reduction of the pentoxide to the metal was incomplete, and was actually the molecular weight of vanadium suboxide (vanadium(II) oxide), further complicated by the presence of phosphorus, which is very difficult to remove.[79] Soon after his Bakerian lecture, Roscoe gave a Friday evening discourse at the Royal Institution. The assistant, while preparing for the lecture, came across a sample of ammonium vanadate sent by Berzelius to Faraday in 1831. Edward Frankland, who was covering for the Fullerian Professorship at the Royal Institution at the time, as Faraday had died three months earlier, kindly sent him part of this sample.[80] Roscoe was able to show that it contained a considerable amount of phosphoric acid, which he claimed would have utterly confounded the Swedish chemist's investigation of the element. Phosphorus and vanadium are often found together in ores as they have the same formulae for their compounds (i.e., they are isomorphic).

Roscoe later studied the chemistry of niobium in the same sub-group as vanadium in the mid-1870s, having obtained the mineral columbite from Greenland. Again, he was able to isolate the metal and also showed that niobium trichloride unusually reacted with carbon dioxide to form niobium oxytrichloride and carbon monoxide.[81] In between these group Vb researches, he prepared both tungsten pentachloride and uranium pentachloride, and also tungsten pentabromide.[82] It is interesting to note how his research after 1870 may have been guided by Mendeelev's new periodic table. Vanadium and niobium were in Mendeleev's group V (Vb) and tungsten and uranium were in group VI (VIb). Roscoe and Schorlemmer discussed the periodic law—giving some credit to John Newlands (1837–1898)—and reproduced a periodic table in part II of the second volume of their *Treatise on Chemistry* in 1879, then proceeded to discuss the existence of undiscovered elements.[83] Mendeleev himself certainly regarded Roscoe as a leading supporter of the periodic table in the eighth edition (1905) of his *Principles of Chemistry*: "I consider Roscoe, de Boisbaudran, Nilson, Winkler, Brauner, Carnelley, Thorpe, and others who justified the application of the periodic law to chemical actuality as the true confirmers of the periodic law."[84] It would be fair to say that Roscoe's colleagues Thomas Carnelley and Borislav Brauner played a more important role (see below). Roscoe warmly welcomed Mendeleev to the 1887 meeting of the British Association for the Advancement of Science in Manchester and

[79] Roscoe "Researches on Vanadium," Part I, 223–224.

[80] For Frankland, see *ODNB* and Colin A. Russell, *Edward Frankland: Chemistry, Controversy, and Conspiracy in Victorian England* (Cambridge: Cambridge University Press, 1996). As Russell points out on 221, Frankland was not the Fullerian Professor at this time, as is often erroneously stated. Odling replaced Faraday in 1868.

[81] Henry E. Roscoe, "Note on Metallic Niobium and a New Niobium Chloride," *Chemical News* 37 (18 January 1878): 25–26. This variety of Columbite is now called Columbite-(Fe).

[82] H. E. Roscoe, "On the Study of Some Tungsten Compounds," *Chemical News* 37 (23 February 1872): 90. H. E. Roscoe, "On a New Chloride of Uranium," *Journal of the Chemical Society* 27 (1874): 933–935.

[83] H. E. Roscoe and C. Schorlemmer, *A Treatise of Chemistry*, vol. 2: *Metals*, part II (London: Macmillan, 1879), "Periodic Law of the Elements," 506–512, and "On the Existence of Undiscovered Elements," 514–515. For Newlands, see *ODNB* and Eric R. Scerri, *The Periodic Table: Its Story and Its Significance*, 2nd ed. (New York: Oxford University Press, 2020), 81–90.

[84] As translated in Yu I. Solov'ev, "DI Mendeleev and the English Chemists," *Journal of Chemical Education* 61 (1984): 1069–1071, on 1070, citing D. I. Mendeleev, *The Periodic Law* (Moscow: Academy of Sciences Press, 1958), 315 (in Russian).

Figure 7.2 Dimitri Mendeleev and Harry Roscoe in a group photograph taken during the British Association for the Advancement of Science Meeting in Manchester in September 1887. Front row, left to right: N. A. Menshutkin; Mendeleev; Roscoe; back row: C. Martius; T. Carnelley; W. J. Russell; C. Schorlemmer W. Odling.

Image courtesy of the Royal Society of Chemistry Library.

had a series of photographs taken of Mendeleev with himself, Schorlemmer, and other leading international chemists (Figure 7.2).[85] In 1878, the Swiss-American chemist Marc Delafontaine (1837?–1911) claimed the discovery of a new element, philippium, in samarskite ore from North Carolina, but Roscoe was able to show that this supposed element was a mixture of yttrium and terbium four years later.[86]

Finally, we may note Roscoe's research on perchloric acid and its derivatives with Schorlemmer in the early 1860s.[87] He was able to prepare the fiercely reactive pure perchloric acid and its monohydrate. He was keen to analyze ethyl perchlorate, which had first been made by the American chemist Clark Hare (1816–1905), the son of Robert Hare, and the Danish-American chemist Martin Boyè (1812–1909) in 1840.[88] Perhaps wisely, Hare then became an eminent lawyer. When Roscoe was filtering a small amount of ethyl perchlorate, there was a terrific explosion which punched a hole in the base of the filter stand and pierced his left hand with hundreds of glass fragments from the shattered test-tube. As he later remarked: "if I had had my hand under the test-tube it would have bored a hole in it just as it did in the filter-stand."[89]

Astronomy and Solar Physics

Most academic scientists in the early twenty-first century restrict themselves to one scientific discipline and in fact are usually only concerned with a small part of that discipline. While a process of specialization was already underway in the mid-nineteenth century, there was still considerable overlap between chemistry and physics (for example in the case of atomic spectroscopy). One of the most active areas of atomic spectroscopy in the 1860s was solar physics and astronomy. Indeed, the rise of atomic spectroscopy was one major driver for the conversion of astronomy, up to then largely concerned with positional astronomy, into the modern field of astrophysics.[90]

---

[85] For a discussion of the identity of the chemists in these photographs, see Peter J.T. Morris, "Who Are the Chemists in the Picture of Henry Roscoe and Dimitri Mendeleev?," *RSC Historical Group Newsletter* 79 (Winter 2021): 29–37 (electronic version), https://www.rsc.org/globalassets/03-membership-community/connect-with-others/through-interests/interest-groups/historical/newsletters/historical-group-newsletter—-winter-2021.pdf.

[86] Henry E. Roscoe, "A Study of Some of the Earth-Metals Contained in Samarskite," *Journal of the Chemical Society, Transactions* 41 (1882): 277–282. Marco Fontani, Mariagrazia Costa, and Mary Virginia Orna, *The Lost Elements: The Periodic Table's Shadow Side* (New York: Oxford University Press, 2015), 119–121.

[87] Henry Enfield Roscoe, "On Perchloric Acid and Its Hydrates," *Proceedings of the Royal Society of London* 11 (1862): 493–503.

[88] Clark Hare and Martin H. Boyè "On the Perchlorate of the Oxide of Ethule or Perchloric Ether," *Transactions of the American Philosophical Society* 8 (1843): 73–76. For Clark Hare, see William Draper Lewis, "John Innes Clark Hare," *American Law Register (1898–1907)*, new series, 45 (December, 1906): 711–717, which does not contain any mention of his chemical activities or his father. For Martin Boyè, see his entry in Wyndham D. Miles, *American Chemists and Chemical Engineers* (Washington, DC: American Chemical Society, 1976), 44–45; Edgar Fahs Smith, "Martin Hans Boyè, Chemist, 1812–1909," *Journal of Chemical Education* 21 (1944): 7–11; "Martin Hans Boye," *Science* 39 (19 March 1909): 448–449.

[89] Henry Enfield Roscoe, "Note on Perchloric Ether," *Journal of the Chemical Society* 15 (1862): 213–215. Quotation from Roscoe, *Life and Experiences*, 108.

[90] A. J. Meadows, *Early Solar Physics* (Oxford: Pergamon Press, 1971); John B. Hearnshaw, *The Analysis of Starlight: Two Centuries of Astronomical Spectroscopy* (Cambridge: Cambridge University Press, 2014); Barbara J. Becker, *Unravelling Starlight: William and Margaret Huggins and the Rise of the New Astronomy*

Solar eclipses have attracted attention for millennia, but for most of that time as a fearful portent and a visual wonder, rather than for strictly scientific purposes. As such, any observers might be gratified (or alarmed) if there was a solar eclipse in their locality, but they would not travel to observe them in foreign parts.[91] The study of solar eclipses was transformed by two technical developments: solar photography and solar spectroscopy. The chemist (and head of the eponymous printing firm) Warren de la Rue (1815–1889) developed the photoheliograph, a telescope which could take photographs of the Sun, in 1856.[92] A pressing issue at the time was the relationship between the prominences seen in solar eclipses—did they originate from the Sun or the Moon?[93] In order to take photographs of a solar eclipse, de la Rue and a party of astronomers set off to northern Spain on *HMS Himalaya*, a ship borrowed from the Royal Navy. The group viewed the eclipse of 18 July 1860 from the village of Rivabellosa and took over 40 photographs, including two of totality.[94] Attention now turned to the possible use of atomic spectroscopy. Jules Janssen (1824–1907) observed the total eclipse of 18 August 1868 from Guntur, India, and discovered the yellow line of helium in the chromosphere of the Sun, which was also observed by Norman Lockyer (1836–1920) two months later from the more prosaic location of West Hampstead.[95]

A scientific party was then assembled to observe the solar eclipse of 22 December 1870, which included Lockyer, George Darwin (1845–1912), the Manchester photographer Alfred Brothers (1826–1912), Roscoe, and Roscoe's former assistant, Thorpe.[96] They took the train via the Brenner Pass and Rome (which had been taken from the Papal States by the Italian army a few months earlier) to Naples. There they embarked the HM Dispatch Boat *Psyche* to travel to Catania in Sicily. En route, the ship became embedded on a submerged rock, but fortunately near the shore, and everyone and their equipment were taken onto dry land at Acrireale, at the foot of Mount Etna,

(Cambridge: Cambridge University Press, 2011); Ileana Chinnici, *Decoding the Stars: A Biography of Angelo Secchi, Jesuit and Scientist* (Leiden: Brill, 2019).

[91] For a general account of Victorian solar eclipse expeditions, see Alex Soojung-Kim Pang, *Empire and the Sun: Victorian Solar Eclipse Expeditions* (Stanford, CA: Stanford University Press, 2002).

[92] Peter D. Hingley, "The First Photographic Eclipse?," *Astronomy & Geophysics* 42, no. 1 (2001): 1–18. For Warren de la Rue, see *ODNB*; David Le Conte, "Two Guernseymen and Two Eclipses," *Antiquarian Astronomer* 4 (January 2008): 55–68; Le Conte, "Warren De La Rue: Pioneer Astronomical Photographer," *Antiquarian Astronomer* 5 (February 2011): 14–35. For the early use of the photoheliograph at Kew Observatory, see Lee T. Macdonald, "'Solar Spot Mania': The Origins and Early Years of Solar Research at Kew Observatory, 1852–1860," *Journal for the History of Astronomy* 46 (2015): 469–490, and *Kew Observatory and the Evolution of Victorian Science, 1840–1910* (Pittsburgh: University of Pittsburgh Press, 2018), 90–100.

[93] For the history of the scientific study of solar eclipses up to 1870, see A. J. Meadows, *Science and Controversy: A Biography of Sir Norman Lockyer, Founder of Nature*, 2nd ed. (London: Macmillan, 2008), 51–66.

[94] Hingley, "The First Photographic Eclipse?"

[95] Neither Janssen nor Lockyer immediately realized it could be a new element; this came later. Biman B. Nath, *The Story of Helium and the Birth of Astrophysics* (New York: Springer, 2013). For Janssen, see E.B.K[nobel], "Pierre Jules Cesar Janssen," *Monthly Notices of the Royal Astronomical Society* 68, no. 4 (1908): 245–249; and for Lockyer, see *ODNB* and Meadows, *Science and Controversy*.

[96] For an overview of the December 1870 expeditions, see Meadows, *Science and Controversy*, 66–68. For Roscoe's own account, see Roscoe, *Life and Experiences*, 158–161. For Angelo Secchi's expedition to Augusta, 60 kilometers south of Mount Etna (he did not see totality because of clouds), see Chinnici, *Decoding the Stars*, 140–147.

about nine miles from Catania. As Roscoe later remarked, "It was certainly a well-ordered shipwreck, for no one even wet their shoes."[97] Roscoe led Darwin, Edward Bowen (1836–1901) of Harrow School, and George Seabroke (1848–1918) of Rugby School on an expedition halfway up Mount Etna.[98] However, clouds rolled in as the eclipse started and it began to snow. Roscoe's party was left enshrouded in gloom, while Thorpe at Catania and Brothers at Syracuse saw the eclipse perfectly. Perhaps chastened by this experience, Roscoe did not travel to see any further eclipses, but Lockyer observed the solar eclipse of 12 December 1871 from Bekul (now Bekal) in southwest India and used a prismatic camera to observe the solar chromosphere.[99] He subsequently led a number of eclipse expeditions and might be considered to be one of the first "eclipse-chasers."

In the 1870s Lockyer developed a theory of atomic dissociation which arose from his study of spectra.[100] It was clear that cool stars contained molecules, but hotter stars such as our Sun only displayed atomic lines in their spectra. The spectra of even hotter stars seemed to only contain the lines of atomic hydrogen. Lockyer suggested that just as molecules were decomposed at high temperatures into atoms, in the very high temperatures found in stars (and the temperatures of stars were thought to be higher at the time than they really are) atoms might be dissociated into sub-atomic units, or heavier elements might be dissociated into hydrogen atoms. Roscoe became concerned at the damage that such speculations could do to Lockyer's reputation, and he was probably also worried about the possible impact on the reputation of the scientific reform movement given Lockyer's important role in that movement. His criticisms of Lockyer's ideas and experimental results fell into two categories. He censured Lockyer for making facile statements about the temperatures of electric arcs and sparks (both of which were used to produce spectra in the laboratory), and of stars, when it was still far from certain what their actual temperatures were. Roscoe then showed that Lockyer's claims to make hydrogen from other elements were vitiated by the presence of chemical impurities. For example, in the case of sodium metal, the hydrogen could be occluded in the metal or produced by traces of sodium hydroxide. While Lockyer continued to believe in his hypothesis and carried out further research to prove it, Roscoe's intervention did much to prevent his ideas creating a major scientific controversy. In December 1879, Roscoe reported to Bunsen that Lockyer had given up his attempts to decompose all the elements.[101] We now know that Lockyer was in fact partly correct, even if his experiments were ill-founded. Atoms are indeed dissociated in stars, but only to produce nuclei and electrons in the form of a plasma, not the complete destruction of the atomic nucleus, which requires the extremely high temperatures of the Big Bang.

Roscoe was also a close friend of Joseph Baxendell (1815–1887), an estate agent by trade who became a keen amateur astronomer, then astronomer to the

[97] Roscoe, *Life and Experiences*, 160.
[98] For Bowen, see *ODNB*; and for Seabroke, see "George Mitchell Seabroke," *Monthly Notices of the Royal Astronomical Society* 79, no. 4 (1919): 231–233.
[99] Meadows, *Science and Controversy*, 68–71.
[100] See Meadows, *Science and Controversy*, Chapter VI "What Is an Atom," 134–173, for a detailed account of this controversy, and 147–151 therein for Roscoe's involvement.
[101] Letter from Roscoe to Bunsen, dated 17 December 1879, letter 18, Heid. Hs. 2741 III A-41, University of Heidelberg library.

Manchester corporation in 1859, and finally meteorologist to the Southport corporation.[102] Baxendell and Roscoe were both members of the Manchester Literary and Philosophical Society and joint secretaries in the 1860s and early 1870s. Baxendell was a keen observer of variable stars, but shared an interest in solar physics with Roscoe. They showed that the so-called chemical intensity of sunlight did not correlate with the visible intensity with a change in altitude of the Sun; the chemical intensity being much greater, relatively speaking, when the Sun was lower in the sky.[103]

Textbooks

A good educationalist can extend the impact of their teaching beyond their immediate pupils to a much wider circle of students across the globe—and indeed across time—by writing textbooks.[104] Through their textbooks, they can influence the teaching of chemistry and how it is understood by its participants. Major chemists such as Antoine Lavoisier (1743–1794), Mendeleev, and Louis P. Hammett (1894–1987) actually transformed the nature of chemistry through their textbooks.[105] However, most leading chemists are not textbook authors, preferring to focus on their research and lecturing. There no textbooks associated with the names of Emil Fischer (1852–1919), and Robert Burns Woodward (1917–1979). On the other hand, there are many chemists who are only known to a wider circle through their textbooks, such as the organic chemistry textbooks of Wilfred Hickinbottom (1896–1979), Ivor Finar (1912–1984), and Jerry March (1929–1997).[106]

Roscoe built up a close relationship with the publisher Macmillan & Company, which was founded in 1843 by two Scottish brothers, Daniel Macmillan (1813–1857) and Alexander Macmillan (1818–1896).[107] Daniel was the grandfather of the later prime minister Harold Macmillan, who worked for the firm whenever he was not in political office. Daniel died before Roscoe was involved with the firm, but in his autobiography Roscoe said that Alexander was "a good and valued friend ... [who] had a racy Scottish humour which made him an excellent companion and a charming host."[108] Roscoe wrote *Lessons in Elementary Chemistry* for Macmillan's series of school textbooks in 1866 at a time when the teaching of chemistry in schools was increasing.[109] It is written

[102] For Baxendell, see *ODNB* and Kargon, *Science in Victorian Manchester*, 74–77.

[103] Thorpe, *Roscoe*, 114–115; Henry Enfield Roscoe and Joseph Baxendell, "Note on the Relative Chemical Intensities of Direct Sunlight and Diffuse Daylight at Different Altitudes of the Sun," *Proceedings of the Royal Society of London* 15 (1867): 20–24.

[104] Anders Lundgren and Bernadette Bensaude-Vincent, eds., *Communicating Chemistry: Textbooks and Their Audiences, 1789–1939* (Canton, MA: Science History Publications/USA, 2000).

[105] For Hammett, see F. H. Westheimer, "Louis Plack Hammett," *Biographical Memoirs of the National Academy of Sciences* 72 (1997): 136–149.

[106] For Hickinbottom, see *Chemistry in Britain* 16 (1980): 218; for Finar, see *Chemistry in Britain* 22 (1986): 142; and for March, see his death notice in *New York Times*, December 28, 1997, Section 1, 29.

[107] Rosemary T. Van Arsdel, entry on Macmillan family in the *ODNB*; Charles Morgan, *The House of Macmillan (1843–1943)* (London: Macmillan, 1944). Also see A. J. Sanders, "A Long Life's Relationship: Archibald Geikie, Alexander Macmillan and His Publishing House," *Geological Society, London, Special Publications* 480 (2018): 139–148.

[108] Roscoe, *Life and Experiences*, 155.

[109] Henry Enfield Roscoe, *Lessons in Elementary Chemistry* (London: Macmillan, 1866).

in a very clear style without any of the floweriness we associate with mid-Victorian writing and looks surprisingly modern. Indeed, it differs only slightly in its style and only moderately in its content from the textbooks used in schools a century later, such as the inorganic and organic chemistry textbooks of P. John Durrant (1901–1989), which were used by both of the authors at school.[110] This modern look was enhanced by the use of the metric system throughout the book. The book was enduring: the seventh edition was published by Macmillan as late as 1900.[111] The most important aspect of this textbook, however, was the extraordinary number of translations published in a short space of time. At the suggestion of Lothar Meyer, Schorlemmer translated *Elementary Chemistry* into German and it was widely used in that country despite having a foreign author.[112] Translations into Russian, French, Italian, Hungarian, Polish, Swedish, Greek, Japanese, and Urdu followed.[113] Roscoe and his contemporaries regarded the translation into an Indian language as something quite remarkable.[114]

No doubt wishing to capitalize on the growing interest in science, and inspired by the work of James Wilson (1836–1931) at Rugby School to introduce science into the school curriculum, Alexander Macmillan asked Roscoe in 1870 to join forces with the biologist Thomas Huxley and the physicist Balfour Stewart to edit a series of elementary primers on science.[115] Roscoe then proceeded to write the primer on chemistry, the second in the series after an introductory volume by Huxley, in 1872.[116] This is very much a chemistry of everyday life, drawing on materials found in the home or in nature, accompanied by simple experiments. Like his hero Faraday, Roscoe opens the book with the burning of a candle. The experiments, illustrated by the characteristic line drawings of the period, would be considered to be more advanced (and even alarming) today than they clearly were in the 1870s. The preparation of oxygen, for example, involves the heating of mercury(II) oxide in a test-tube with a spirit lamp. The chemistry is more elementary than in *Lessons in Elementary Chemistry* and less overtly didactic, although they would be considered to be excessively pedagogic for

---

[110] For example, P. J. Durrant, *General and Inorganic Chemistry*, 3rd ed. (London: Longmans, Green, 1964) and P. J. Durrant, *Organic Chemistry*, 9th impression (London: Longmans, Green, 1963). Although he became Vice-Master of Selwyn College, Cambridge, there does not seem to be a scholarly obituary of Durrant, but see the obituary in the *Daily Telegraph*, 16 March 1989, 27.

[111] Henry E. Roscoe, *Lessons in Elementary Chemistry: Inorganic and Organic* (London: Macmillan, 1900).

[112] Roscoe, *Life and Experiences*, 149.

[113] As listed in Roscoe, *Report of Work Done*, 29. It has not been possible to trace the Russian, French, Greek, or Swedish versions. In the case of the Greek and Swedish versions, there may be confusion between *Elementary Chemistry* and *Chemistry* (i.e., the primer) as there are Greek and Swedish translations of the latter.

[114] Roscoe, *Life and Experiences*, 150. Sadly, it has not been possible to trace a copy of this Urdu version.

[115] Ibid., 151; Morgan, *The House of Macmillan*, 70–71. A primer is a book, usually small in size, to enable children to learn the basic rudiments of a subject; so-called because it is the first book they use in the subject. The term is now only used for elementary books of Classical grammar. For Wilson, see *ODNB*. It seems that Roscoe saw this series as being in competition with the science books produced by the Society for Promoting Christian Knowledge (SPCK), but whether this competition was ideological or commercial is not entirely clear. Roscoe to Thomas Huxley dated 24 January 1875, 25.277–278, Huxley papers, Imperial College Archives. Alexander Macmillan himself believed "that religion and science were allies, not foes"; Morgan, *House of Macmillan*, 70. The SPCK science books were mostly secular in content by the 1870s. For a different view, which assumes there was an ideological aspect, see Ruth Barton, *The X Club: Power and Authority in Victorian England* (Chicago: University of Chicago Press, 2018): 309 and 311.

[116] H. E. Roscoe, *Chemistry* (London: Macmillan, 1872). Huxley's *Introductory* primer, despite being the first in the series, did not appear until 1880, a delay which was a source of considerable anxiety to Roscoe, as shown by his letter to Huxley in 1875.

a young audience nowadays. To Roscoe's discomfiture, secondary schools were soon using the primer to teach chemistry to their older students, rather than the more obviously suitable *Lessons in Elementary Chemistry*.[117] The last edition of the primer appeared in 1904. *Chemistry* was soon translated into an even wider range of languages than its predecessor, including German, Italian, Icelandic, Polish, Lithuanian, Japanese, Bengali, Turkish, Malayalam, and Tamil.[118] Roscoe also published *Inorganic Chemistry for Beginners* (assisted by Joseph Lunt) in 1893.[119] He then wrote *Inorganic Chemistry for Advanced Students* with Arthur Harden in 1899.[120]

A far more ambitious publication was the multi-volume *Treatise of Chemistry*, which Roscoe wrote with his colleague Schorlemmer from 1877 onward.[121] Roscoe wrote the parts on inorganic chemistry and Schorlemmer the historical sections and the parts on organic chemistry. The first volume was well received, but it was a hideously complicated undertaking. It was a typically convoluted Victorian publication, divided not only into volumes, but also parts of volumes. As Roscoe later noted, the organic chemistry required more frequent revision than the inorganic chemistry, which greatly complicated any revision of the treatise.[122] Even the inorganic parts were being revised in the late 1880s while the later volumes were still appearing. The project came to a juddering halt in 1892 when Schorlemmer died, while the organic chemistry volumes were still being written. Macmillan must have been keen on the *Treatise of Chemistry*, probably with libraries in mind, as Roscoe produced a completely new edition between 1911 and 1913 with the help of several chemists. Another revised version produced by the dye chemist and Owens graduate John C. Cain (1871–1921) appeared in the early 1920s after Roscoe's death, but seems to have quickly run out of steam.[123] By this time it would have struck many chemists

[117] Roscoe, *Life and Experiences*, 153.

[118] Roscoe, *Report of Work Done*, 29. Greek and Swedish translations have also been found. The Tamil translation is mentioned in Roscoe, *Life and Experiences*, 151. These translations are all still in existence except, it seems, the Turkish and Malayalam versions. The strongest single holding, but not a complete set, is in the University of Manchester Library, presumably Roscoe's personal collection. The others are in their respective national libraries, except for the Lithuanian version which is in the Polish national library. For the translation of *Chemistry* into Japanese and Hebrew, see Yona Siderer, "Translations of Roscoe's Chemistry Books into Japanese and Hebrew: Historical, Cultural and Linguistic Aspects," *Substantia* 5, no. 2 (2021): 41–54; and for the translation into Greek, see Kostas Gavroglu, "The Transmission and Assimilation of Scientific Ideas to the Greek Speaking World ca. 1700–1900: The Case of Chemistry," in *The Making of the Chemist: The Social History of Chemistry in Europe 1789–1914*, ed. David Knight and Helge Kragh (Cambridge: Cambridge University Press, 1998), 289–304, on 300–301.

[119] Sir Henry Roscoe assisted by Joseph Lunt, *Inorganic Chemistry for Beginners* (London: Macmillan, 1893). There was a second edition in 1912. Joseph Lunt (1866–1940) had an interesting career. A member of the British Astronomical Association, he left the Lister Institute to become an assistant at the Royal Observatory, Cape of Good Hope, where he remained until his retirement in 1926. He discovered no less than seven Index Catalogue (IC) objects, namely IC 2621, IC 4670, IC 5170, IC 5171, IC 5181, IC 5201, and IC 5224. See German *Wikipedia* and S2A3 Biographical Database of Southern African Science, https://www.s2a3.org.za/bio/Biograph_final.php?serial=1734.

[120] Sir Henry Roscoe and Arthur Harden, *Inorganic Chemistry for Advanced Students* (London: Macmillan, 1899). There was a second edition in 1910.

[121] See the list of publications at the end of this volume for the complete bibliographic details.

[122] Roscoe, *Life and Experiences*, 154.

[123] Sir H. E. Roscoe and C. Schorlemmer, *A Treatise on Chemistry*, 5th ed., completely revised by Dr. J. C. Cain (London: Macmillan, 1920–1923). Cain's death in 1921 would not have helped. For Cain, see J.F.T[horpe], "John Cannell Cain," *Journal of the Chemical Society, Transactions* 119 (1921): 533–537.

as anachronistic to write a "Treatise on Chemistry" which left out physical chemistry. Probably to his frustration, Roscoe's textbooks were influential in inverse proportion to the level of chemistry in them, the primer being by far the most widely used.[124]

Roscoe was also associated with the series of books on practical chemistry as practiced at Owens College. The first edition of *The Owens College Junior Course of Practical Chemistry* was written by Francis Jones, chemistry master at Manchester Grammar School, under the supervision of Roscoe, in 1872 and was published by Macmillan.[125] It is comprised of mostly simple inorganic chemistry experiments and inorganic group analysis, a genre still familiar to most O level chemistry students in England as late as the 1960s.[126] It was joined by *The Owens College Junior Course of Practical Organic Chemistry* written by Julius B. Cohen in 1887, which became *Practical Organic Chemistry for Advanced Students* in 1900 and simply *Practical Organic Chemistry* in the second edition of 1908.[127] Jones's textbook remained influential until Arthur I. Vogel (1905–1966) published his *Textbook of Qualitative Inorganic Analysis* in 1937, which remained in print, duly revised and with changes in the title, until 1999.[128] Cohen's textbook was also widely used, but was eventually displaced by *Practical Organic Chemistry* by Frederick Mann (1897–1982) and Bernard Saunders (1903–1983), first published in 1936 and still being printed (as its fourth edition of 1970) in the 1970s.[129] Vogel first published the equally influential *Textbook of Practical Organic Chemistry* in 1948.[130]

[124] Roscoe, *Life and Experiences*, 153.

[125] Francis Jones, *The Owens College Junior Course of Practical Chemistry* (London: Macmillan, 1872). Rather oddly, there appear to be two different frontispieces for this book, one with Jones as the sole author with a preface by Roscoe, held by Harvard University Library (obtained from the Barnard Club at some point) and the British Library, and another frontispiece with Roscoe given as the first author followed by Jones, held by the Radcliffe Science Library (RSL), Oxford. A possible explanation lies in the fact that the RSL copy was accessioned in 1875; was Roscoe added as an author when it was reprinted? Yet both frontispieces have the same date, 1872, and there is no indication of a second impression in the RSL copy. The book was still being published under Jones's name as late as 1903.

[126] One of the authors (PJTM) took a chemistry A level along similar lines, the very last time it was available, in January 1974. He had to go to a laboratory in central London and carry out a set of inorganic experiments as part of the examinations. In England and Wales up to 1986, O[ordinary] levels were taken about age 16 and A[dvanced] levels were taken around the age of 18.

[127] Julius B. Cohen, *The Owens College Junior Course of Practical Organic Chemistry* (London: Macmillan, 1887); Cohen, *Practical Organic Chemistry for Advanced Students* (London: Macmillan, 1900). There was a third edition in 1924 and it was last published in 1937.

[128] Arthur I. Vogel, *A Textbook of Qualitative Inorganic Analysis* (London: Longmans, Green, 1937). The third edition was published by Longmans, Green under the new title of *A Text-Book of Qualitative Chemical Analysis Including Semimicro Qualitative Analysis* in 1945, and the fourth edition by Longmans, Green as *A Text-Book of Macro and Semimicro Qualitative Inorganic Analysis* in 1954. For Vogel, see *Chemistry in Britain* 2 (1966): 548.

[129] F. G. Mann and B. C. Saunders, *Practical Organic Chemistry* (London: Longmans, Green, 1936). For Mann, see I. T. Millar, "Frederick George Mann," *Biographical Memoirs of Fellows of the Royal Society* 30 (1984): 408–441; and for Saunders, see *Chemistry in Britain* 20 (1984): 917.

[130] Arthur I. Vogel, *A Textbook of Practical Organic Chemistry Including Organic Analysis* (London: Longmans, Green, 1948). There was a fifth edition of 1989 and it was published in Delhi as late as 2004.

Other Publications

In addition to his textbooks, Roscoe wrote two other major books which had a strong historical flavor. In May and June of 1868, Roscoe gave a series of lectures on spectrum analysis to the Society of Apothecaries to an audience comprising Society of Apothecaries members, their apprentices, office bearers of the Royal College of Physicians and Surgeons, and members of the Medical Councils in London.[131] They were also attended by professional and scientific celebrities and they were reported, for example, in the British Medical Journal.[132] The society was under pressure to justify its role in the medical profession, and after the death of its Professor of Chemistry William Brande (1786–1866) in 1866, the council of the society decided to hold three series of six lectures, one on chemical philosophy, one on materia medica and therapeutics, and one on botany and morphology.[133] William Odling (1829–1921) was approached to give the first series of lectures, but declined.[134] Roscoe was then invited and accepted, but the other two series of lectures were later abandoned as all the money put aside for the lectures had been spent on the chemistry lectures. Roscoe then published the lectures more or less as given, but with appendices containing new or additional material in Spectrum Analysis: Six Lectures in 1869 (Figure 7.3).[135] They contained a good deal of historical material, from the time of Newton to the latest work of William Huggins (1824–1910) and William Allen Miller (1829–1921) on astronomical spectroscopy.[136] As Ian Rae has pointed out, Roscoe presents the history of atomic spectra as a linear progression which ignores the contributions of many scientists who did not fit this narrative, which was biased in favor of Bunsen, Kirchhoff, Huggins, and Miller and, for example, overlooked the contributions of Sir George Stokes (1819–1903).[137] Of course this is always the problem when trying to present a coherent account of scientific developments, and it is hardly surprising that Roscoe gave the most credit to the scientists he knew best. He was not, after all, trying to write a complete history of spectrum analysis, but rather to present the technique as a major scientific advance.

While one might have thought that Roscoe would have concentrated on the chemical applications of atomic spectroscopy given his audience, in fact he gives much space

---

[131] Anna Simmons, "The Chemical and Pharmaceutical Trading Activities of the Society of Apothecaries, 1822–1922," PhD diss., Open University, Milton Keynes, 2004, 92–93.

[132] "Professor Roscoe's Lectures on Spectrum Analysis: Lecture One," British Medical Journal 1, no. 384 (9 May 1868): 460; "Lecture Two," British Medical Journal 1, no. 385 (16 May 1868): 488; "Lecture Three," British Medical Journal 1, no. 387 (30 May 1868): 541; "Lecture Four," British Medical Journal 1, no. 389 (13 June 1868): 594. The last two lectures were not reported; perhaps the journal realized by then that the lectures would have no medical content.

[133] For Brande, see ODNB and Journal of the Chemical Society 19 (1866): 509–511.

[134] For Odling, see ODNB; Rowlinson, "Chemistry Comes of Age," especially 103–113; and J. E. Marsh, "William Odling," Journal of the Chemical Society, Transactions 119 (1921): 553–564. For his involvement with the development of the periodic table, see Scerri, Periodic Table, 90–95.

[135] Henry E. Roscoe, Spectrum Analysis: Six Lectures Delivered in 1868 before the Society of Apothecaries of London (London: Macmillan, 1869).

[136] For Huggins, see ODNB and Becker, Unravelling Starlight; for Miller, see ODNB and Journal of the Chemical Society 24 (1871): 617–620.

[137] Ian D. Rae, "Spectrum Analysis: The Priority Claims of Stokes and Kirchhoff," Ambix 44 (1997): 131–144, on 136–137. For Stokes, see ODNB and Mark McCartney, Andrew Whitaker, and Alastair Wood, eds., George Gabriel Stokes: Life, Science and Faith (Oxford: Oxford University Press, 2019).

Figure 7.3 Cover of H. E. Roscoe, *Spectrum Analysis: Six Lectures* (1869).

over to the solar and astronomical uses of spectrum analysis, including the study of the solar chromosphere by Lockyer and Janssen, stellar spectra by Huggins and Miller, and the spectral classification of stars by Father Angelo Secchi.[138] Indeed, Roscoe laments in his foreword that our knowledge of the Sun was increasing so quickly that he could not cover it all in the book. Although it was not written in the style of a textbook, it was widely used as an introduction to atomic spectroscopy. A second edition and third edition quickly followed, and Roscoe wrote a fourth edition with his physics colleague Arthur Schuster in 1885.[139] However, rather remarkably, the format of the book was not altered although the material was updated.

Roscoe had a great interest in the two great Manchester scientists of the nineteenth century, John Dalton (1766–1844) and James Joule (1818–1889); he felt that both science and industry owed an incalculable debt to both men.[140] After he came to Manchester he joined the Manchester Literary and Philosophical Society, which was accommodated in Dalton's former house. He contributed a paper on Dalton to its *Memoirs*.[141] Roscoe later discovered some of Dalton's notebooks in the society's possession which had been donated by W. C. Henry, the son of Dalton's friend William Henry and an inept biographer of Dalton, although they were always stored in the society's premises. It is not clear when Roscoe became aware of the significance of the documents, but probably in the early 1890s.[142] Working on these manuscripts with his student Arthur Harden, he radically changed his view of how Dalton arrived at his atomic theory.

Roscoe and Harden published the results of their research in *A New View of the Origin of Dalton's Atomic Theory* in 1896.[143] This is a relatively short book of 191 pages, most of which is a reproduction of Dalton's diary, letters, and lectures. Only the long opening chapter, entitled "The Genesis of Dalton's Atomic Theory," is the original work of Roscoe and Harden. Thanks largely to the view put forward by Dalton's early supporter Thomas Thomson and accepted by Henry in his biography, it had been assumed by most scholars up to the 1890s that Dalton had arrived at the concept of atoms of definite and characteristic weight from the law of multiple proportions in an inductive

---

[138] For Secchi, see Chinnici, *Decoding the Stars*.

[139] The second edition was published by Macmillan in 1870 and a third edition in 1873. The fourth edition, "revised and considerably enlarged by the author and by Arthur Schuster," was published by Macmillan in 1885. A German translation was made by Carl Schorlemmer and was published by F. Vieweg of Brunswick in 1870.

[140] For Dalton, see ODNB; Frank Greenaway, *John Dalton and the Atom* (London: Heinemann, 1966); and Arnold Thackray, *John Dalton: Critical Assessments of His Life and Science* (Cambridge, MA: Harvard University Press, 1972). For Joule (whose name should be pronounced "Jowl," not "Jool"), see *ODNB* and Donald S.L. Cardwell, *James Joule: A Biography*. (Manchester: Manchester University Press, 1989).

[141] He gave a lecture on Dalton's instruments as part of the Special Loan Exhibition at the South Kensington Museum on 3 June 1876, *Chemical News* 32 no. 863 (9 June 1876): 242. H. E. Roscoe, "Some Remarks on Dalton's First Table of Atomic Weights," *Memoirs of the Manchester Literary and Philosophical Society* 5, Series 3, (1876): 269–275; "Lecture on John Dalton Delivered to the Boys of Eton College," probably in the 1890s, is an appendix to Roscoe, *Life and Experiences*, 395–404.

[142] These manuscripts were badly damaged in an air-raid in 1940 and were assumed lost. However, they were rediscovered in 1990, and although the damage to the manuscripts is considerable, it has been possible to conserve them and they are now stored in the John Rylands Library. Diana Leitch and Alfred G. Williamson, *The Dalton Tradition* (Manchester: John Rylands University Library, 1991).

[143] Henry E. Roscoe and Arthur Harden, *A New View of the Origin of Dalton's Atomic Theory: A Contribution to Chemical History* (London: Macmillan, 1896).

manner. However, it appeared clear to Roscoe and Harden from his manuscripts that Dalton had obtained the concept of atoms of a specific mass from his work on the gases of the atmosphere, and it was this idea which led to the law of multiple proportions. In this way, his concept of atoms had arisen from purely physical considerations and in an hypothetico-deductive approach, rather than by induction. Roscoe and Harden also disagreed with the generally accepted idea that Dalton had believed that equal volumes of gases held equal numbers of atoms (or molecules). They concluded that:

> Dalton was led in 1803 to the conception of atoms of characteristic weight and size by his investigations on the diffusion of gases. From these same investigations he also received the incentive to determine the relative weights and sizes of the atoms of different substances.[144]

Modern scholarship, beginning with the work of Leonard Nash (1918–2013), has laid more emphasis on Dalton's work as a natural philosopher on the solubility of different gases in water, an idea first put forward by George Wilson (1818–1859) in his obituary of Dalton in 1845.[145] The perhaps naïve belief that the atomic theory arose out of a logical inductive or hypthothetico-deductive method employed by a chemist has been replaced by a less dogmatic view that Dalton, as a natural philosopher and meteorologist, had a particular Newtonian view of the properties of gases which had not occurred to chemists. With the help of these manuscripts, Roscoe also wrote *John Dalton and the Rise of Modern Chemistry* for the Century Science Series edited by Roscoe and published by Cassells in 1895.[146] It is a relatively short work of 212 pages which was particularly suitable for older schoolchildren and chemistry students.

Roscoe's Colleagues

For many years, Roscoe ran the chemistry department with only a limited number of staff. When he arrived in 1857, his friend Frederick Guthrie (1833–1886) was the assistant.[147] Roscoe brought his private assistant William Dittmar (1833–1892) with him to Manchester. Roscoe's close association with his two assistants was not sufficient to prevent both of them leaving the department within a few years. Guthrie went to Edinburgh University in 1859 before serving as Professor of Chemistry and Physics at the Royal College of Mauritius between 1861 and 1867. He then switched to physics

---

[144] This is the conclusion they reached in *A New View*, but made more explicitly in H. E. Roscoe and Arthur Harden, "The Genesis of Dalton's Atomic Theory," *The London, Edinburgh, and Dublin Philosophical Magazine and Journal of Science* 43 (1897): 153–161, on 161.

[145] This discussion of more recent work is based on Arnold W. Thackray, "The Origin of Dalton's Chemical Atomic Theory: Daltonian Doubts Resolved," *Isis* 57 (1966): 35–55, and Alan J. Rocke, "In Search of El Dorado: John Dalton and the Origins of the Atomic Theory," *Social Research: An International Quarterly* 72 (2005): 125–158. Also see Leonard K. Nash, "The Origin of Dalton's Chemical Atomic Theory," *Isis* 47 (1956): 101–116. For Wilson, see *Quarterly Journal of the Chemical Society* 13 (1861): 169–170.

[146] Sir Henry E. Roscoe, *John Dalton and the Rise of Modern Chemistry* (London: Cassells, 1895). It was published in the same year by Macmillan in New York, but also in London, which seems to be an unusual clash.

[147] Roscoe, *Life and Experiences*, 104 and 106.

Figure 7.4 Carl Schorlemmer.

completely. He became a lecturer at the Royal School of Mines in 1869, a Fellow of the Royal Society two years later, and subsequently a professor.[148] Dittmar also moved to Edinburgh University in 1861, returned to Owens in 1873, but almost immediately moved to the Andersonian University in Glasgow.[149]

Dittmar suggested another German chemist, Carl Schorlemmer (Figure 7.4), as his successor as Roscoe's assistant.[150] Schorlemmer had also trained as a pharmacist, then went to Heidelberg and Giessen, where Heinrich Kopp awakened his interest in the history of chemistry. He mostly worked on the chemistry of hydrocarbons and dyes. He came from a humble background (his father was a joiner), and he became a friend of Karl Marx and Friedrich Engels; he was one of only 13 known individuals at Marx's funeral in 1883. He was a member of the First International (International Workingmen's Association) and of the German Social Democratic Party (SPD). Despite their ostensible political differences, Roscoe and Schorlemmer were very close. According to Roscoe: "we were very much attached to each other; indeed I do

[148] *ODNB*; G.C.F[oster], "Frederick Guthrie," *Nature* 35 (4 November 1886): 8–10. The date given in the latter for his fellowship of the Royal Society is incorrect.
[149] A.C.B[rown], "William Dittmar," *Nature* 45 (24 March 1892): 493–494.
[150] *ODNB*; Karl Heinig, *Carl Schorlemmer, Chemiker und Kommunist ersten Ranges* (Leipzig: BSB B. G. Teubner, 1974); A.H[arden], "Carl Schorlemmer," *Journal of the Chemical Society, Transactions* 63 (1893): 756–763.

not think during the whole of that time we ever had a disagreement."[151] They taught chemistry in concert, they also carried out joint research on perchloric acid and wrote the *Treatise of Chemistry* together. Schorlemmer's death of a lung disease at only 57 in 1892 was a great blow to Roscoe. Schorlemmer became the official assistant after Dittmar's departure, then the senior assistant, and finally the first Professor of Organic Chemistry in 1874.[152] He was succeeded by William Henry Perkin, junior (1860–1929), who later established a world-class school of organic chemistry at Oxford.[153]

Schorlemmer's successor as Roscoe's private assistant was Thomas Edward Thorpe (1845–1925), who had been a local clerk before joining the chemistry department.[154] He helped Roscoe with his important research on vanadium in the mid-1860s. After taking a chemistry degree at Owens between 1863 and 1867 (initially as an evening student), Thorpe then took his PhD at Heidelberg, before returning briefly to Owens. After teaching at Anderson's University between 1870 and 1874, he moved to the Yorkshire College of Science in Leeds (which became part of the Victoria University in 1887) and then the Normal School of Science in London (which later became Imperial College) in 1885. In 1894 he became the first Government Chemist, with a remit to create a new laboratory in the Strand, London, to replace the existing inadequate facilities.[155] When he retired as the Government Chemist, he was knighted and became Sir Edward Thorpe, despite having been known as Tom Thorpe for his whole life. Thorpe's successor as the assistant was Francis Jones (1845–1925), who had first met Roscoe in Heidelberg and became the Science Master at Manchester Grammar School in 1872.[156]

Within a few years, after a number of short-term appointments, the staff had increased to comprise three demonstrators and assistant lecturers (see Table 7.3).[157] W. Carleton Williams (1850–1927) was the first of this new cohort in 1873.[158] He was educated at Owens College and at the Universities of Heidelberg and Bonn. He resigned in 1883 and went to Firth College, Sheffield, where he became Professor of Chemistry in succession to Thomas Carnelley. He resigned in 1904 at the relatively early age of 54 to give the newly created University of Sheffield a free hand to appoint someone new. Matthew Moncrieff Pattison Muir (1848–1931), always known as M. M. Pattison Muir, was educated at the University of Glasgow and briefly Tübingen University.[159] He joined the chemistry department from Anderson's University in

[151] Roscoe, *Life and Experiences*, 107.

[152] *Calendar of Owens College, 1864–1865*, 34 (laboratory assistant); *The Calendar of Owens College, 1872-3*, 42 (senior assistant); Roscoe, *Record of Work Done*, 6 (professor in 1874).

[153] Jack Morrell, "W. H. Perkin, Jr., at Manchester and Oxford: From Irwell to Isis," *Osiris*, 2nd Series, VIII, *Research Schools: Historical Reappraisals* (1993): 104–126; *ODNB*; A. John Greenaway, Jocelyn F. Thorpe, and Robert Robinson, *The Life and Work of Professor William Henry Perkin* (London: Chemical Society, [1932]).

[154] *ODNB*; P.P.B[edson], "Sir Edward Thorpe," *Journal of the Chemical Society (Resumed)* 129 (1926): 1031–1050.

[155] Peter W. Hammond and Harold Egan, *Weighed in the Balance: A History of the Laboratory of the Government Chemist* (London: HMSO, 1992), 123–163.

[156] H.B.B[aker], "Francis Jones," *Journal of the Chemical Society (Resumed)* 129 (1926): 1020–1021.

[157] Roscoe, *Record of Work Done*, 6.

[158] P. Phillips Bedson, "William Carleton Williams," *Journal of the Chemical Society (Resumed)* (1927): 3202–3205.

[159] R. S. Morrell, "M. M. Pattison Muir," *Journal of the Chemical Society (Resumed)* (1932): 1330–1334.

Table 7.3 Roscoe's Colleagues in the Chemistry Department at Owens, 1857–1885

Name	Birth	Death	Birthplace	UG Education	DSc?	PG Education	On Staff	Later Career
Guthrie, Frederick	1833	1886	London	UCL	No	Heidelberg, Marburg	1855–1859	Edinburgh, RSM
Roscoe, Henry Enfield	1833	1915	London	UCL	No	Heidelberg	1857–1886	, Parliament
Dittmar, William	1833	1892	Darmstadt	Heidelberg	No	none	1859–1861	Edinburgh, Andersonian
Schorlemmer, Carl	1834	1892	Darmstadt	Heidelberg, Giessen	No	none	1861?–1892	none
Thorpe, Thomas Edward	1845	1925	Manchester	Owens	Yes	Heidelberg, Bonn	1869–1870	Andersonian, YCS, NSS, LGC
Jones, Francis	1845	1925	Edinburgh	Edinburgh	No	Heidelberg	1870–1872	MGS
Smith, Henry Arthur		1872			No		1872–1872	Died in post
Stone, Daniel	1820	1873	Manchester	Edinburgh	No		1872–1873	MRSM, analytical chemist
Williams, W. Carleton	1850	1927	Salford	Owens	Yes	Heidelberg, Bonn	1873–1883	Firth
Grimshaw, Harry	1851	1904	Manchester	Owens	Yes	none	1873–1875	Chemical manufacturer
Muir, M. M. Pattison	1848	1931	Glasgow	Glasgow	No	Tübingen	1875–1877	Cambridge
Wilkinson, Oswald	1854	1931	Stockport		No		1875–1876	Analyst and consultant, manager J&P Coats dyehouse
Carnelley, Thomas	1854	1890	Manchester	Owens	Yes	Bonn	1875–1880	Firth, Dundee, Aberdeen
Hannay, James B.	1855	1931	Glasgow		No		1878–1879	Inventor
Bedson, Peter Phillips	1853	1943	Manchester	Owens	Yes	Bonn	1879–1882	Armstrong College

Name	Born	Died	Birthplace	University	PhD abroad	PhD location	Years	Career
Smith, Watson	1845	1920	Stroud	Owens	No	Heidelberg, ETH	1880–1889	UCL, Editor JSCI
Baker, Harry	1859	1935	London	Owens	Yes	Heidelberg	1882–1888	Aluminium Co. Castner–Kellner Co.
Smithells, Arthur	1860	1939	Bury	Glasgow, Owens	Yes	Munich, Heidelberg	1883–1885	YCS/Leeds
Bott, William Norman	1861	1935	Wiesbaden	Heidelberg	No		1883–1886	Government analyst, Singapore, mining engineer
Cohen, Julius Berend	1859	1935	Eccles	Owens	No	Munich	1885–1891	YCS/Leeds
Bailey, George Herbert	1852	1924	Barnard Castle	London	Yes	Owens, Heidelberg	1885–1909	Owens, British Aluminium Co.
Harden, Arthur	1865	1940	Manchester	Owens	Yes	Erlangen	1888–1897	Lister Institute

1873. Just as Roscoe had focused on vanadium, Muir concentrated his research on bismuth. He left Owens to become the praelector in chemistry at Gonville and Caius College, Cambridge. Unable to do advanced research at Cambridge because of a lack of postgraduate students, Muir became a leading historian of chemistry. Harry Grimshaw (1851–1904) was educated at Owens College and was then appointed a demonstrator and assistant lecturer in chemistry in 1873.[160] He resigned after two years to set up a chemical company called Grimshaw Brothers, with his brother in Clayton, Manchester, mainly to supply chemicals to the cotton industry. He then became involved with the Recovered Rubber Works in collaboration with Thomas Rowley. This was a pioneering business at a time when natural rubber was still scarce—yet alone reclaimed rubber—the firm described itself as the "oldest and original Founders of this important industry."[161]

Grimshaw was replaced by Thomas Carnelley (1854–1890), who had acted as a private assistant to Roscoe in 1872–1874, before studying at the University of Bonn; he took a Doctor of Science degree at the University of London in 1876 rather than a German PhD.[162] He had been appointed a demonstrator and assistant lecturer at Owens College in 1875 and also became Principal of the North Staffordshire School of Science, at Hanley, during the same period. When Firth College was set up in Sheffield in 1879, he became the first Professor of Chemistry there. He moved to Dundee in 1882 and then to Aberdeen in 1888, before he sadly died as a result of an internal abscess at the young age of 36 in 1890. While he was at Owens, Carnelley became very interested in the Periodic Law established by Mendeleev and developed a number of mathematical formulations for an elements atomic weight and physical properties.[163] However, these formulae were purely empirical and he gained no insight into the atomic structure of the elements.

Peter Phillips Bedson (1853–1943) was educated at Owens College and studied under August Kekulé at Bonn before becoming a demonstrator and assistant lecturer in 1879.[164] Three years later he was appointed Professor of Chemistry at Durham College of Physical Science (later Armstrong College and now the University of Newcastle) in Newcastle-upon-Tyne. He built the new laboratories at Newcastle in the same style as Roscoe's laboratories at Owens College. He was interested in occupational hazards of white lead manufacture and coal mining, and made a particular study of coal. His research suffered from a lack of postgraduate students, and the college became a military hospital in World War I. Bedson retired soon after the war ended, but lived to be 90, dying in 1943 in the middle of World War II. He was probably the last survivor of Owens College's chemistry department who served under Roscoe.

[160] Obituary in *India-Rubber Journal* (1 February 1904): 129, and portrait in *India-Rubber Journal* (15 February 1904): 160. Entry on Grimshaw Brothers in Grace's Guide https://www.gracesguide.co.uk/Main_Page; some of the records of this firm are held by the Science Museum. Also see *India Rubber World* 24(2) (1 May 1901): 229.

[161] Advertisement for the Recovered Rubber Works, Clayton, Manchester, late Grimshaw Brothers, *India-Rubber and Gutta-Percha Trades Journal* (24 September 1897): v.

[162] "Thomas Carnelley," *Journal of the Chemical Society, Transactions* 59 (1891): 455–461.

[163] Ibid. and Jaime Wisniak, "Thomas Carnelley," *Educación Química* 23 (2012): 465–473, on 466–469.

[164] J. A. Smythe, "Peter Phillips Bedson," *Journal of the Chemical Society (Resumed)* (1944): 40–41.

Arthur Smithells (1860–1939), having been educated at Owens College, Munich, and Heidelberg, was appointed demonstrator and assistant lecturer in 1883.[165] He replaced Thorpe as Professor of Chemistry at Yorkshire College of Science in 1885, eventually becoming Pro-Vice-Chancellor of its successor, the University of Leeds. His particular interest was the chemistry of the flame, and thanks to his efforts, the Livesey Professorship of Coal Gas and Fuel Industries was established at Leeds in 1910, with William A. Bone (1871–1938) as its first occupant.[166] Julius B. Cohen (1859–1935) was a close friend of Smithells at Owens, and after an unhappy year at the Clayton Aniline Company, he joined Smithells in Munich, where he took his PhD.[167] He then became a demonstrator and assistant lecturer at Owens College in 1885. He rejoined Smithells at Leeds in 1890 and remained there for the rest of his career. Cohen had a wide range of interests in organic chemistry and laid the early foundations for the development of organic reaction mechanisms, for example, studying the effect of substituents in the benzene ring and steric hindrance on the path of reactions.

One of Roscoe's last assistant lecturers was perhaps the most bizarre. William Bott (1861–1935) joined the department in 1883 shortly after taking his PhD at Heidelberg.[168] After leaving Owens in 1886, he went to Göttingen, before returning briefly to Owens before he became a government analyst in the Straits Settlements (now Singapore) in 1890, but he left around 1903 to set up a mining engineering partnership in London under the name of William Norman Bott (or Norman-Bott).[169] The firm seems to have been most active in West African mining, but Norman-Bott was one of the first to point out the importance of the new oilfields of Sakhalin in Russia in 1903.[170] At the beginning of World War I, he was charged with failing to register under the Alien Registration Order.[171] At first he claimed to have been born in Manchester in 1865 or 1867, but it then transpired that when he had applied for a job in the Straits Settlements in 1889, he had stated, correctly, that he had been born in Wiesbaden in 1861. It was then discovered that he had been naturalized in 1885 (while at Owens) and his case was dismissed. Why he could have not said all this in the first place is hard to fathom. Indeed, had he been born in 1867,

---

[165] ODNB; H. S. Raper, "Arthur Smithells," Obituary Notices of Fellows of the Royal Society 3 (1940): 96–107.

[166] ODNB; A. C. Egerton, "William Arthur Bone," Obituary Notices of Fellows of the Royal Society 2 (1939): 586–611.

[167] H. S. Raper, "Julius Berend Cohen," Obituary Notices of Fellows of the Royal Society 1 (1935): 502–513. He was the first cousin of the father of the chemical historian Otto Theodore Benfey, to whom he bears a strong resemblance.

[168] The life of William Norman Bott has been pieced together from various sources, notably the records of the Deutsche Chemische Gesellschaft collated by Ernst Homburg (personal communications, 10 and 16 September 2021), and the application of William Harold Richard Allen to be a fellow of the Chemical Society in 1904 (Proceedings of the Chemical Society 20, no. 285, 226). He studied chemistry at Heidelberg in 1881–1882, as Wilhelm Bott of Wiesbaden; see Rudolf Werner Soukup und Roland Zenz, Eine Bibliothek als beredte Zeugin eines umfassenden Wandels des wissenschaftlichen Weltbilds. Teil II: Ansätze einer Rekonstruktion des wissenschaftlichen Netzwerks Bunsens unter besonderer Berücksichtigung von Bunsens Privatbibliothek, 144, http://www.rudolf-werner-soukup.at/Publikationen/Dokumente/Bunsenbibliothek_Teil_2_Rekonstruktion.pdf. For his academic records and PhD examination, see H IV 102–100, Heidelberg University Archives.

[169] The London Gazette, 3 July 1903, 4223 and 14 April 1911, 2997.

[170] Vagit Alekperov, Oil of Russia: Past, Present and Future (Minneapolis: East View Press, 2011), 118.

[171] "W. N. Bott, Late of the Straits, in Trouble," Malaya Tribune, 16 October 1914, 5. The Chemist and Druggist, 5 September 1914, 42 and 12 September 1914, 36.

he would have only been 16 when he was appointed an assistant lecturer! After the war, he moved to Congleton in Derbyshire, where he set up a chemical works in an old textile mill at Bath Vale.[172] He died in Congleton in 1935 and the year of birth on the death certificate was given as 1869.[173]

In 1879, the former assistant of the Professor of Technical Chemistry at the Federal Polytechnic (ETH), Zurich, Georg Lunge (himself a student of Bunsen), Watson Smith (1845–1920), was appointed Lecturer of Chemical Technology at Owens.[174] He thus brought applied chemistry to the forefront of the chemistry department's teaching. However, he strove to make chemical technology a new department within the university, which put him at odds with Roscoe, who enjoyed the support of the chemical manufacturers. He moved to University College London (UCL) in 1889 to take up a similar position, but he was only a lecturer and again he was subordinate to the Professor of Chemistry, William Ramsay (1852–1916).[175] This unhappy situation ended with the abolition of the post at UCL in 1894. Another assistant lecturer with a strong connection with the chemical industry was Harry Baker (1859–1935), who studied chemistry under Roscoe at Owens before taking his PhD with Bunsen in Heidelberg.[176] He joined the department as a demonstrator and assistant lecturer in 1882, but he resigned in 1888 to join the Aluminium Company (see Chapter 8) in Oldbury and worked with Hamilton Young Castner (1858–1899).[177] When Castner's aluminum process was undermined by the new electrolytic process, he developed an electrolytic process for sodium hydroxide, only to discover that the Austrian Carl Kellner (1851–1905) had developed a similar process. The two parties joined forces to form the Castner–Kellner Company in 1895 and Baker moved to Runcorn to work for the new company. He retired soon after the company became part of Imperial Chemical Industries (ICI) in 1926.

Arthur Harden (1865–1940) became an assistant lecturer under Roscoe's successor, Harold Baily Dixon, but as he was educated at Owens under Roscoe and Julius Cohen, and collaborated with Roscoe on *Inorganic Chemistry for Advanced Students* in 1899, he deserves to be mentioned here. He was born in Manchester and did his PhD in Erlangen under Otto Fischer (1852–1932).[178] At Manchester, he taught the history of chemistry and worked with Roscoe on Dalton. At this stage, he was not greatly involved in research, but had an interest in photochemical induction, and in particular the photochemical reaction between carbon monoxide and chloride, which he had taken over from Roscoe. In 1897, he replaced Joseph Lunt, another Roscoe student,

---

[172] Royal Geographical Society, London, *A List of the Honorary Members, Honorary Corresponding Members and Fellows* (London: Royal Geographical Society, [1921]), 65 (listed as a fellow). For the Bath Vale works, see the Archaeological Dataservice website, https://archaeologydataservice.ac.uk/library/browse/issue.xhtml?recordId=1151780&recordType=GreyLitSeries.

[173] Familysearch genealogical website, https://www.familysearch.org/en/.

[174] P.P.B[edson], "Watson Smith," *Journal of the Chemical Society, Transactions* 117 (1920): 1637–1638.

[175] For Ramsay, see *ODNB* and Morris Travers, *The Life of Sir William Ramsay* (London: Edward Arnold, 1956).

[176] E. Otho Glover, "Harry Baker," *Journal of the Chemical Society (Resumed)* (1936): 539.

[177] For Castner, see *ODNB*.

[178] *ODNB*; F. G. Hopkins and C. J. Martin, "Arthur Harden," *Obituary Notices of Fellows of the Royal Society* 4 (1942): 2–14.

at the Jenner Institute of Preventive Medicine in London, at a time when Roscoe was closely involved with the institute (see Chapter 11). This appointment led to Harden taking up biochemistry and, in particular, alcoholic fermentation (he wrote a major textbook on the topic in 1911).[179] He discovered the importance of a coenzyme (now usually called a cofactor) in fermentation, which he called "co-zymase" but is now known as nicotinamide adenine dinucleotide (NAD$^+$) in 1906.[180] He then discovered the importance of phosphates in fermentation. His findings were to find wider application throughout biochemistry. He won the Nobel Prize in Chemistry in 1929 with the German-Swedish biochemist Hans von Euler-Chelpin (1873–1964) for their independent but related work on alcoholic fermentation.[181] Harden retired from the Lister Institute in the following year.

Most of the staff at Owens in Roscoe's time were educated at Owens before taking higher training at a German university. A small minority had been undergraduates at Glasgow or Edinburgh. Strikingly, none of the staff had been at Oxford or Cambridge, and apart from Roscoe and Guthrie, none of them had been educated in London. It might be argued that it was a weakness that Owens employed mostly its own graduates (and nearly all of them holders of the Dalton chemical scholarship) as its junior staff. Perhaps the department might have flourished more if it had employed a wider range of chemistry graduates. There seems to have been two reasons for this. Roscoe had a particular way of teaching chemistry, and he trusted his own graduates to follow his pattern of teaching. Being a junior member of staff seems to have been regarded as part of the training of the most able students before they became members of staff at other universities. It is striking that none of the junior staff in Roscoe's time (Bailey was at the very end of his tenure) was retained at Owens. Just as none of them came from Oxford or Cambridge, none of them went to Oxford, and only one, Pattison Muir, went to Cambridge. Only two went to London colleges: Thorpe went to the Normal School of Science (later Imperial College), and Watson Smith had an unhappy period at UCL. Manchester was not yet the elite waiting room of British chemistry. By contrast, seven went to a northern or Scottish university. Early on, two went to the Andersonian in Glasgow (Dittmar, Thorpe), but the two Yorkshire colleges predominate. Three went to the Yorkshire College (later Leeds)—Thorpe, Smithells, and Cohen—and two—Williams and Carnelley—went to Firth College (later Sheffield). Carnelley also went to two Scottish universities (Dundee and Aberdeen), but he was the exception. Bedson went to Armstrong College (later Newcastle). Hence Roscoe's protégés played an important role in developing chemical education in Leeds, Sheffield, and Newcastle. Given Roscoe's involvement with the chemical industry, it is not surprising that five of them went into the chemical industry, mining, or dyeing. In terms of their stature in the British chemical community, the two preeminent chemists were Thorpe and Harden; the latter being a Nobel laureate. In the second rank, we

[179] Arthur Harden, *Alcoholic Fermentation* (London: Longmans, Green, 1911). There was a second edition in 1914, a third edition in 1923, and a fourth edition in 1932, all published by Longmans, Green.

[180] Robert E. Kohler, "The Background to Arthur Harden's Discovery of Cozymase," *Bulletin of the History of Medicine* 48 (1974): 22–40.

[181] For Euler-Chelpin, see the biography on the Nobel Foundation website, https://www.nobelprize.org/prizes/chemistry/1929/euler-chelpin/biographical/.

would enroll Carnelley, Smithells, Bedson, and Cohen. While this shows the importance of Roscoe's school, it is not that impressive given that he taught in Manchester for nearly three decades. Perhaps the problem was the quality of the college's intake, as shown by the failure of students to pass even the junior course as late as the early 1880s.

Roscoe's Students and Research Fellows

We are fortunate that Roscoe himself provides tables of his leading students. One is a list of teachers of importance, and the other is a list all the students who became "teachers of their science."[182] Interestingly, he makes no distinction between university teachers and schoolteachers. Most of the men recorded in the first list were also on the staff of the college, namely Thorpe, Carnelley, Smithells, Williams, Bedson, and Jones. There are two other men in this list. Charles Romley Alder Wright (1844–1894) was lame because of a hip problem. He became Lecturer of Chemistry at St. Mary's Hospital Medical School in 1871 and played a leading role in the formation of the Institute of Chemistry (later the Royal Institute of Chemistry) in 1877.[183] He died of diabetes mellitus, then untreatable, in 1894. William Marshall Watts (1844–1919) studied at Heidelberg as well as Owens College and then spent two years at the University of Glasgow, before becoming the science master at Manchester Grammar School in 1868.[184] He then went to Giggleswick School in Yorkshire in 1872 and was replaced by Jones. He published *Index of Spectra* in the same year.[185] He retired from Giggleswick School in 1904. Following in Roscoe's footsteps, he published *An Introduction to the Study of Spectrum Analysis* in the same year.[186]

Of the men on the longer list, only a few British chemists are noteworthy. George Newth (1851–1936) taught at the Royal College of Chemistry (latterly Imperial College) until his retirement in 1909.[187] He is best remembered for his textbook of inorganic chemistry rather than any chemical research.[188] Sydney Young (1857–1937) was born in Farnworth near Widnes and was educated at Owens College before studying under Rudolph Fittig (1835–1910) at Strasbourg.[189] He became an assistant to Ramsay at University College, Bristol, in 1882 and succeeded him as the Professor of Chemistry. He became Professor of Chemistry at Trinity College Dublin in 1904,

[182] Roscoe, *Report of Work Done*, 24–25.

[183] "Charles Romley Alder Wright," *Journal of the Chemical Society, Transactions* 67 (1895): 1113–1115.

[184] "W. Marshall Watts and the Tamblyn-Watts family," https://bearalley.blogspot.com/2007/07/w-marshall-watts-and-tamblyn-watts.html.

[185] The publishing history of the *Index of Spectra* is complex. It seems that the first edition was published in London by H. Gilman, but there is no copy in the Internet Archives. The revised edition of 1889 was published by Abel Heywood in Manchester and William Wesley in London.

[186] W. Marshall Watts, *An Introduction to the Study of Spectrum Analysis* (London: Longmans, Green, 1904).

[187] "George Samuel Newth," *Journal and Proceedings of the Institute of Chemistry of Great Britain and Ireland* 60 (1936): 258.

[188] G. S. Newth, *A Text-Book of Inorganic Chemistry* (London: Longmans, Green, 1894), which then went almost immediately into a new edition in 1895, reaching a fifth edition by 1897. The ninth edition was published in 1902 and as late as 1923 there was a new and enlarged edition which was reprinted in 1926.

[189] *ODNB*; W.R.G. Atkins, "Sydney Young," *Obituary Notices of Fellows of the Royal Society* 2 (1938): 370–379.

Figure 7.5 Harry Roscoe with his staff and students at Owens College, 1882–1883, from a copy once owned by Alexandre Claparède. Back row, left to right: Harry Baker, Watson Smith, Arthur Harden, William Blair Syme, Conrad Gerland, Gibson Dyson, William Ray, Saville Shaw, Alexandre Claparède, Charles Henry Keutgen, Duncan Scott Macnair, Hermann Alfred Rademacher, Frederick Jackson, Horace Edward Brothers, Ludwig Claisen, Willy Bott. Middle row: Carl Schorlemmer, Henry E. Roscoe, Charles A. Burghardt (mineralogist). Front row: George Herbert Bailey, Norman Erskine Maccallum, Allen Duval, Augustus Schlösser (later Mallinson). Bibliothèque de Genève Icon M 1971–79.

and after living through a turbulent period in Irish history, retired in 1928. He was President of the Royal Irish Academy between 1921 and 1926. Young was a physical chemist who shared Roscoe's interest in fractional distillation and azeotropic mixtures. He died in Bristol.

By contrast, the foreign chemists who came as Berkeley fellows in 1881 became leading figures in their home countries. Ludwig Claisen (1851–1930) was born in Cologne and studied chemistry at the University of Bonn before coming to Owens College.[190] He then went to Munich in 1885, before becoming a Professor of Organic Chemistry at Aachen, Kiel, and finally Berlin, where he worked closely with Emil Fischer. He died in Bad Godesberg. His particular interest was the condensation of carbonyl compounds which created carbon-carbon bonds. Claisen is best known today for his four name reactions: the Claisen-Schmidt condensation (1881), the Claisen condensation (1887), the Claisen reaction (1890), and the Claisen rearrangement (1912). The reader will note that he did not publish any of these reactions while he was in Manchester, and his research while he was there does not seem to have resulted in any significant publications. He published only one paper in English, on the action of hydrochloric acid on hydrogen cyanide in the presence of ethyl alcohol, with Francis E. Matthews (1862–1929), who discovered the sodium polymerization of butadiene in 1910.[191] Owens College appears to have had a greater impact on the other Berkeley fellow, the Czech chemist Bohuslav Brauner (1855–1935). Brauner was born in Prague and studied under Bunsen at Heidelberg, who dismissed his queries about the periodic table and the position of the rare earths in it.[192] By contrast, Roscoe encouraged Brauner to isolate the rare earth elements and determine their atomic weights. During his short stay in Manchester, he used spectroscopy to show that the supposed element didymium was actually a mixture of two elements, but he did not isolate them. Carl Auer von Welsbach (1858–1929) separated didymium into praseodymium and neodymium in 1885.[193] Brauner also predicted the existence of a rare earth element between neodymium and samarium. This gap was confirmed by Henry Moseley's X-ray spectra in 1914 and was eventually filled by the synthetic element promethium in 1945.[194]

[190] DB; German Wikipedia; R. Anschütz, "Ludwig Claisen: Ein Gedenkblatt," Berichte der Deutschen Chemischen Gesellschaft 69, no. 7 (1936): A97–A170.

[191] L. Claisen and F. E. Matthews, "On the Action of Haloïd Acids upon Hydrocyanic Acid," Journal of the Chemical Society, Transactions 41 (1882): 264–268. For Matthews, see W.R.H[odgkinson], "Francis Edward Matthews," Journal of the Chemical Society (Resumed) (1929): 2970–2972.

[192] S. I. Levy, "Brauner Memorial Lecture," Journal of the Chemical Society (Resumed) (1935): 1876–1890; J.H[eyrovsky], "Prof. Bohuslav Brauner," Nature 135 (1935): 497–498; also see Soňa Štrbáňová, "Nationalism and the Process of Reception and Appropriation of the Periodic System in Europe and the Czech Lands," in Early Responses to the Periodic System, ed. Masanori Kaji, Helge Kragh, and Gabor Palló (New York: Oxford University Press, 2015) 121–149.

[193] For Welsbach, see Roland Adunka and Mary Virginia Orna, Carl Auer von Welsbach: Chemist, Inventor, Entrepreneur (Cham: Springer, 2018).

[194] P. M. Heimann, "Moseley and Celtium: The Search for a Missing Element," Annals of Science 23 (1967): 249–260; Eric Scerri, A Tale of Seven Elements (Oxford: Oxford University Press, 2013), chapter 9, "Element 61—Promethium." For Henry Moseley (1887–1915), see J. L. Heilbron, H. G. J. Moseley: The Life and Letters of an English Physicist, 1887–1915 (Berkeley: University of California Press, 1974) and Roy MacLeod, Russell G. Egdell, and Elizabeth Bruton, eds., For Science, King and Country: The Life and Legacy of Henry Moseley (London: Uniform Press, 2018).

It is striking that Roscoe in his autobiography does not refer to the presence of either Claisen or Brauner at Owens, but mentions two of his Japanese students, in whom he clearly took great pride, perhaps because he saw them as his contribution to the modernization of Japan. These students were Shigetake Sugiura (1855–1924) (called Suguira in *Life and Experiences*, he seems to have used this variation sometimes while he was in England) and Yoshimi Hiraga (1857–1943) (called Kiraga in *Life and Experiences*).[195] Sugiura initially came to England to learn agricultural chemistry at the Royal Agricultural College at Cirencester in 1876, but finding English agriculture to be completely different from Japanese practice, he went to Owens College to study chemistry.[196] After two years there, he transferred to the Royal College of Chemistry in London to work under Edward Frankland. In his autobiography, Roscoe claims that Sugiura had left because he had been ashamed to win the second-place prize in his second year after gaining the first place a year earlier. Sugiura denied this was the case.[197] He argued that he went to the Royal College of Chemistry to broaden his experience before going back to Japan, having considered going to Germany instead. Given the unreliability of *Life and Experiences*, Sugiura's explanation is more likely to be correct and Roscoe's story probably stemmed from Western stereotypes about the Japanese being afraid to lose face.[198] However, soon afterward he had some kind of nervous breakdown (called neurasthenia at the time), and after a seaside stay at Hastings failed to cure the problem, he returned to Japan. From 1882 and 1885 he was Principal at the Preparatory School attached to Tokyo University and gradually turned his attention away from chemistry to ethics and nationalism. Having eventually recovered his health, he became a tutor to the future Emperor Hirohito and his future wife, Princess Nagako (later the Empress Kojun, who died as recently as 2000). Like Sugiura, Hiraga hailed from the chemistry department at Tokyo University and went to Owens College in 1878, just after Sigiura had left.[199] During his three-year stay in Manchester, he was apprenticed to a local dyeing company and this experience shaped his future career. He became an engineer and was appointed Director of the Osaka Prefecture Industrial Research Institute in 1903. He then became involved with the Osaka Orimono Kaisha (Osaka Fabric Company) and was made president of the company in 1911. Hiraga also published two school textbooks on chemistry and a monograph on dyeing in the 1880s.[200]

Ironically, Roscoe does not mention the most important of his Japanese students, namely Toyokichi Takamatsu (1852–1937).[201] Once again having studied chemistry

---

[195] Roscoe, *Life and Experiences*, 114–115.
[196] Joji Sakurai, "Shigetake Sugiura," *Journal of the Chemical Society (Resumed)* 129 (1926): 3246–3248. Also see Jun Uchida, "From Island Nation to Oceanic Empire: A Vision of Japanese Expansion from the Periphery," *The Journal of Japanese Studies* 42 (2016): 57–90.
[197] In Sakurai, "Shigetake Sugiura."
[198] Personal communication from Yoshiyuki Kikuchi, 26 August 2020.
[199] Personal communication from Yoshiyuki Kikuchi, 25 September 2020.
[200] *Kagaku shoho* [Elementary Chemistry] (Tokyo: Hiraga Yoshimi, 1883), written with Jugo Sugiura; *Senshokujutsu tekiyō* [Outlines of Dyeing Techniques] (Tokyo: Inouesokichi, 1886); *Senshokuhō: Jitchi kanben yūeki* [Dyeing Methods: Simple and Beneficial in Practice] (Fukuoka: Kaishōdō, 1887); *Shōgakkōyō rika. Kagaku hen* [Elementary School Science: Chemistry] (Tokyo: Fukyūsha, 1887).
[201] Kikuchi, *Anglo-American Connections*, 70–74, and personal communication from Yoshiyuki Kikuchi, 9 October 2020.

at Tokyo University, Takamatsu went to Owens College in 1879, a year after Hiraga, and they overlapped for two years. Takamatsu became close to Watson Smith, who had just joined the staff at Owens, and they published four papers in the *Journal of the Chemical Society*.[202] On the advice of Roscoe, and in common with most of the other talented students at Owens, Takamatsu went to Germany in 1881 to work in Hofmann's laboratory in Berlin for a year. Thanks to his experience in Berlin, he became an expert on dye chemistry and was appointed head of the department of applied chemistry at Tokyo University when the department was split in 1883 (thus achieving what his mentor Watson Smith failed to do). He became a director of the Tokyo Gas Company between 1909 and 1914 before being appointed Director of the Government Industrial Research Institute in Tokyo (*Tokyo Kōgyō Shikenjo*) in 1915. He stayed in this office for nine years, until his retirement, and simultaneously gave policy recommendations to the government for the promotion of chemical industries during World War I as a core member of the Chemical Industry Study Commission instituted by the Ministry of Agriculture and Commerce in October 1914. Takamatsu published a college textbook of chemistry which was produced by the Department of Education in three parts between 1890 and 1894, which was largely based on Roscoe's two elementary chemistry textbooks, but also drew from other Western textbooks, such as Ira Remsen's *Organic Chemistry* of 1885.[203] By a quirk, this textbook became influential in the then Japanese protectorate of Korea, when Japanese-educated Inpyo Hong published a textbook based on Takamatsu's volume in 1907.[204]

Was Roscoe a "Chemist Breeder"?

Jack Morrell published a famous paper in 1972, in which he argued that certain criteria showed by Justus von Liebig (1803–1873) at Giessen and Thomas Thomson (1773–1852) in Glasgow created successful research schools ("chemist breeders" in Morrell's colorful phrase), whereas Thomas Charles Hope (1766–1844) in Edinburgh did not.[205] Morrell mentioned "the important pioneering British research schools run by Thomson, Hofmann, and H. E. Roscoe" with the implication that they too were "chemist breeders," which can be judged to be successful according to the same criteria. We will assess Roscoe's school in terms of Morrell's four criteria, but we would

---

[202] Watson Smith and T. Takamatsu, "On Pentathionic Acid," *Journal of the Chemical Society, Transactions* 37 (1880): 592–608; "On Phenylnaphthalene," *Journal of the Chemical Society, Transactions* 39 (1881): 546–551; "Sulphonic Acids Derived from Isodinaphthyl (ββ-dinaphthyl)," *Journal of the Chemical Society, Transactions* 39 (1881): 551–554; "On Pentathionic Acid (Part II)," *Journal of the Chemical Society, Transactions* 41 (1882): 162–167.

[203] *Kagaku kyōkasho* [Chemistry Textbook] (Tokyo: Monbushō Sōmukyoku, 1890f)

[204] Jongseok Park and Byung-Hoon Chung, "British Chemist Henry E. Roscoe's Unintended Contribution to Korean Chemistry in 1907," *Journal of Chemical Education* 92 (2015): 593–594.

[205] Jack B. Morrell, "The Chemist Breeders: The Research Schools of Liebig and Thomas Thomson," *Ambix* 19 (1972): 1–46. For Liebig, see William H. Brock, *Justus von Liebig: The Chemical Gatekeeper* (Cambridge: Cambridge University Press, 1997); and for Thomson, see *ODNB*. For Hope, see *ODNB* and Robert G.W. Anderson, "Thomas Charles Hope and the Limiting Legacy of Joseph Black," in *Cradle of Chemistry: The Early Years of Chemistry at the University of Edinburgh* ed. Anderson (Edinburgh: John Donald, 2015), 147–162.

argue that Morrell's statement is in fact incorrect: Roscoe did not set up, nor operate a research school. As we have seen, it was not possible to do a postgraduate degree at Owens and there was therefore no postgraduate research, strictly construed. Nonetheless, the most able students were encouraged to stay on and do some original research. The junior staff of the college also did their own research. Crucially, however, this research was either a collaboration between a student and a demonstrator or the demonstrator's own line of research. It was not part of an overall research program guided by the director. Roscoe did publish some coauthored papers, notably with Bunsen, but he did not publish papers with his students or his junior colleagues, with the notable exception of Thorpe.

Let us overlook this inconvenient fact and subject Roscoe to the tests Morrell presents in his seminal paper. To summarize, Morrell has four sets of criteria:

1. There should be a director, who has a large program of chemical research to carry out and has a growing or established national (or even better, an international) reputation. This director should be powerful within their institution and have access to at least adequate funding. Crucially, this director should possess charisma, thus maintaining control of their group and being able to maintain morale while overcoming any passing difficulties.
2. The school will have a regular supply of able students, who will flourish under the director's regime. These best and most ambitious of these students, in Morrell's words, will "first [be] disciples of their master and later as apostles who would diffuse his work and methods by establishing their own schools modelled on his."[206]
3. The director and their students will develop "a set of relatively simple, fast and reliable experimental techniques could be steadily applied by both brilliant and ordinary students to the solution of significant problems in a new or growing field of enquiry."[207]
4. The school will make itself known to the chemical world, through its publications, as "[p]ublication was vital to the success of any ambitious research school."[208] Hence the director should either have easy access to a leading journal (perhaps as its editor) or even have their own journal (as Liebig did).

It is evident that these criteria are very much based on the case of Liebig himself, and hence by definition Liebig will inevitably fulfill them. A more systematic approach would have been to have studied a large number of research schools across the whole of the nineteenth and twentieth centuries and in several different countries in order to establish the criteria which are common to the most successful ones using statistical analysis. Nevertheless, Morrell's paper has been widely accepted, and we will apply the above criteria here.

As a director, Roscoe was very powerful within Owens college and had a national reputation and to some degree an international reputation. He was a strong leader and

[206] Morrell, "The Chemist Breeders," 4.
[207] Ibid., 5.
[208] Ibid., 5.

organizer, important traits for a research leader. He was a bluff northerner (despite being born in London) and it is more debatable whether he had charisma. It would be more accurate to say that he was respected and liked by those who knew him well. But he certainly had the ability to overcome major difficulties and maintain morale, as his development of the chemistry department shows. As we have seen, the number and quality of the students were both low initially, but a steady stream of students became available, partly at least thanks to Roscoe's growing reputation. It must be emphasized that these were purely undergraduate students. We have examined the later career of his students, and several of them took up academic positions in the new colleges set up across northern England and also in Scotland. They had much less impact on the "golden triangle" of Oxford, Cambridge, and London, with the marked exception of Thorpe. The crucial question here is whether Roscoe's approach to chemistry was carried forward to a later generation. His successor at Manchester, Harold Baily Dixon (1852–1930), was a Londoner with a Mancunian father who was educated at Christ Church, Oxford, and became a fellow of Trinity College, Oxford, before taking Roscoe's Chair at Manchester.[209] Although Dixon and Roscoe had similar interests, the element of any kind of "apostolic succession" is lacking. Roscoe influenced the design of the laboratories and the style of teaching in Leeds, Sheffield, and Newcastle-upon-Tyne. Nonetheless, it is hard to ascertain any lasting tradition of "Roscoian chemistry" even in the north of England. Roscoe's approach was largely overtaken by the growing importance of organic chemistry, changes in physical chemistry stemming from the work of Jacobus van't Hoff, Svante Arrhenius, and Wilhelm Ostwald, and a shift from practically oriented chemistry to a more theoretically based approach to teaching chemistry.

In terms of ease of publication, Roscoe had good access to the memoirs of the Manchester Literary and Philosophical Society, but this ceased to be a serious scientific journal in his lifetime. He also published papers in the *Journal of the Chemical Society* (*JCS*) and would not have encountered any difficulties in being published there. For his more physical papers (for example, his papers on photochemistry with Bunsen) he tended to favor the *Philosophical Transactions*. He also produced papers which appeared in the proceedings of the Royal Institution. His colleague Schorlemmer generally published his papers in the *JCS* or the proceedings of the Royal Society. His junior colleagues and students generally published their research in the *JCS* or the Manchester memoirs. The number of papers produced was high, but not especially high: in the 29 and a half years between the summer of 1857 and the end of 1886, Roscoe published 61 papers (of which 20 papers were coauthored); Schorlemmer (in a shorter period) published 54 papers (of which 16 papers were coauthored). Their junior colleagues and students published 122 papers (not including papers with Roscoe or Schorlemmer, which were few in number).

However, the key point is that there was no overarching research program. Roscoe worked on a number of topics, most notably on exotic metals, but he did not publish papers with his students and colleagues on these topics. Nor did he develop any specific chemical technique (apart from his work with Bunsen on photochemistry,

[209] H.B.B[aker] and W.A.B[one], "Harold Baily Dixon," *Journal of the Chemical Society (Resumed)* (1931): 3349–3368.

which owed more to Bunsen than Roscoe) or any specialized chemical apparatus. Indeed, the nearest Owens College had to a research program was Schorlemmer's work on the aliphatic hydrocarbons on one hand, and his work on synthetic dyes with Richard S. Dale, the son of the dye manufacturer John Dale, on the other.[210] Nor did Roscoe have access to significant funding for his research school, not least because he had no postgraduate students. Hence while Roscoe (and to an almost equal degree Schorlemmer) produced a significant number of papers by themselves and trained a large number of undergraduates, some of whom went on to lead chemistry departments in other parts of northern England—after additional mentoring as junior members of staff at Owens—it would be inaccurate to call Roscoe a "chemist breeder" in the sense defined by Morrell.

However, we could turn this round, and argue—given the evidence presented in this chapter—that Roscoe was a "chemist breeder" despite not meeting many of the criteria in Morrell's paper. We can go further and declare that Roscoe had a program that was not a research program, but rather a model of how to train chemists. Morrell, given his focus on Liebig and Giessen, assumes that the training of chemists was shaped by the professor's research program—they had to be able to do the tasks it required. If the professor does not have such a research program or one which does not require any novel skills, their focus is going to be on the future employment of their students. This focus is not just for the benefit of the students themselves, but a professor whose students consistently obtain good positions after graduation is clearly going to be a professor whose courses will be in high demand and thus be respected. Morrell assumed that respect stemmed from the successful completion of a major research program, but surely it is more likely that professors as a whole are more likely to be respected for the quality of their teaching and hence the employability of their students? Hence Roscoe is closer to the situation discussed by Ernst Homburg in his paper on Friedrich Stromeyer than the one analyzed by Morrell.[211]

In the 1820s degree-level chemistry was a minority activity carried on by an elite group of chemists who mostly became academics. Fifty years later, however, chemistry was much more widely taught in universities and colleges, and only relatively few graduates had both the desire and the ability to become academics. Most graduates moved out of chemistry; for example, they took up relatively low-level positions in industry or government, or went into school teaching at a time when the teaching of science in schools was being promoted—not least as a result of the educational reforms championed by Roscoe. Roscoe had stressed the importance of his students occupying higher-level positions in industry, directing research or running companies in an enlightened way, not least to show that his graduates could earn high salaries, but the reality was that they became analysts in rudimentary works laboratories. Medical students also became an important group for the chemistry department after the Royal Manchester School of Medicine became part of Owens College in 1872, followed by the opening of the new medical school building in Oxford Road (again designed by

---

[210] Carsten Reinhardt and Anthony S. Travis, *Heinrich Caro and the Creation of Modern Chemical Industry* (Dordrecht: Kluwer, 2000), 61.

[211] Ernst Homburg, "The Rise of Analytical Chemistry and Its Consequences for the Development of the German Chemical Profession (1780–1860)," *Ambix* 46 (1999): 1–32.

Waterhouse) in 1874.[212] There were 166 medical students at Owens College in 1875–1876 and no less than 323 just 10 years later.[213] This compares with 395 day-students in 1875–1876 and 380 in 1885–1886.

Conclusion

From unpromising beginnings when he arrived in 1857, Roscoe created one of the largest and most prestigious chemistry departments in Britain. This was a major achievement which is universally acknowledged. However, we have to be clear about the nature of his achievement. He did not create a major research program of his own, nor did the department become a powerhouse of chemical research, partly for the lack of postgraduate students, although its output was—in British terms at least—impressive. In effect, Roscoe farmed out the postgraduate training of his most promising students to German universities, while welcoming most of them back to Owens College for further vocational training as academic teachers before they finally left for other colleges and universities. The undergraduate training at Owens was very much a continual drilling in inorganic chemical analysis, supplemented by lectures which explained the theoretical background to this analysis. This method was not unique to Owens or even novel; it was a method which was in general use in the mid-nineteenth century. In Roscoe's case, it stemmed from Bunsen's teaching at Heidelberg, which can be traced back to the teaching of Stromeyer in the early nineteenth century. The teaching of organic chemistry fit awkwardly into this scheme, as Roscoe admitted, but it was eventually shoehorned into a similar system of teaching through experimental exercises.

The only novel aspect of Roscoe's teaching was an emphasis on original research, which should not be confused with Henry Armstrong's later concept of "heuristic" teaching which was incorporated at all levels of teaching. By contrast, original research at Owens, however highly it was prized, was only undertaken by an elite group of students who stayed on after the basic three-year course. Roscoe did not fashion an elite school of chemistry as Manchester became in the first half of the twentieth century (when it was called the waiting room of Oxford and Cambridge), but he created something perhaps even more significant, namely a mechanism for producing thousands of well-trained if relatively undistinguished chemists to meet the growing demand for chemists in schools, government, and industry. These Roscoe-trained chemists also replaced people who had been self-taught in chemistry or who had learned aspects of chemistry through empirical practice in industry. Much of the demand for lower-paid chemists was met through learning chemistry in evening classes (such as those held at Owens College) well into the twentieth century.

---

[212] Stella V.F. Butler, "A Transformation in Training: The Formation of University Medical Faculties in Manchester, Leeds, and Liverpool, 1870–84," *Medical History* 30 (1986): 115–132, especially 120. Also see James Hopkins, "The Disciplinary Development of University Buildings: Medicine and Manchester," *Baltic Journal of Art History* 16 (2018): 101–114; Hopkins, "The (Dis)assembling of Form: Revealing the Ideas Built into Manchester's Medical School," *Journal of the History of Medicine and Allied Sciences* 75 (2020): 24–53.

[213] Roscoe, *Record of Work Done*, 4.

8

Roscoe's Engagement with Industry

When Roscoe began his studies at University College London (UCL) in 1848, London was already in a state of heightened expectation of the Great Exhibition in Hyde Park, where the spectacular displays in its "Crystal Palace" in 1851 would attract over six million visitors from across the world and confirm Britain's position as the world's leading industrial nation. Chemical manufacturers took full advantage to show their wide range of products alongside displays of production processes.[1] Chemistry was generally accepted for much of the nineteenth century as the science most relevant to industry.[2]

Roscoe's likely first introduction to the industrial application of chemistry was by Alexander Williamson, his Chemistry Lecturer at UCL. Williamson did not have any direct working experience in the industry, but after completing his PhD with Liebig at Giessen in 1845 he undertook research in Paris and became friends with some of the outstanding French chemists, including Jean-Baptiste Dumas, Adolphe Wurtz, August Laurent, and Charles Gerhard, who were advancing chemistry for the economic benefit of France. Williamson arrived back in London in 1849 to take up the appointment of Professor of Analytical and Practical Chemistry in the Birkbeck Laboratory at UCL. In the chemistry course that Williamson adopted, students developed an understanding of basic chemistry through a series of practical exercises, alongside a program of lectures, before studying the principles of chemical technology in the manufacture of a range of commodities.[3] Later, from 1868, Williamson arranged regular visits to chemical manufactories so students could relate how principles learned at UCL underpinned large-scale industrial processes.

Working with Robert Bunsen over an extensive period in Heidelberg, Roscoe came to appreciate how collaboration between university chemists and industrial firms led to improved working of industrial processes. Bunsen had used his innovative gas analysis techniques to investigate the efficiency of German iron furnaces and had found that about 50% of the fuel (effectively the gases reducing the iron oxide to iron) was allowed to escape, making the process inefficient.[4] Later, Bunsen and the British chemist Lyon Playfair used the same analysis techniques for their study of British iron furnaces funded by the British Association for the Advancement of Science, in which

---

[1] Peter Reed, *Entrepreneurial Ventures in Chemistry: The Muspratts of Liverpool, 1793–1934* (Farnham, Surrey: Ashgate, 2015), 85.

[2] J. F. Donnelly, "Getting Technical: The Vicissitudes of Academic Industrial Chemistry in Nineteenth-Century Britain," *History of Education* 26 (1997): 125–143, on 125.

[3] Takaaki Inuzuka, *Alexander Williamson: A Victorian Chemist and the Making of Modern Japan*, trans. Haruko Laurie (London: UCL Press, 2021), 16–17.

[4] Henry Enfield Roscoe, "Bunsen Memorial Lecture," *Journal of the Chemical Society* 77 (1900): 513–554, on 519–520.

Henry Enfield Roscoe. Peter J.T. Morris and Peter Reed, Oxford University Press. © Oxford University Press 2024.
DOI: 10.1093/oso/9780190844257.003.0008

the loss of the reducing gases was found to be 81.5%.[5] While the German iron founders acted promptly to revise their operations quite quickly, the British prevaricated over a six-year period, avoiding the improvement of their production.

When Roscoe took up his appointment in late 1857, Owens College was facing financial difficulties linked to low student numbers. Even in Manchester, a major center for industrial and commercial enterprise, few owners and managers of major businesses were convinced of the value of a university education, having had no experience of such education and its benefits. Roscoe's riposte was direct: working as a consultant, he demonstrated how scientific expertise could improve industry and commerce, keeping their operations up to date and working more efficiently and economically. Surprisingly, such an enterprising initiative brought criticism from outside Owens College. It was considered in contravention of the spirit of John Owens's bequest and also inappropriate for a university professor who should concentrate on teaching his students, for which he was paid by the College.

Alongside the explosives and soap industries, the Leblanc soda industry was a major strength of the British chemical industry during the nineteenth century, at least until the early 1880s.[6] The Leblanc soda industry adopted the process for the production of soda (sodium carbonate) from salt, invented by the French chemist Nicholas Leblanc in 1788–1789, rather than relying on vegetable sources that were prone to supply fluctuations, especially during periods of war.[7] With Britain's abundant supply of the two main raw materials, namely salt and coal, the Leblanc industry (as it became known) established itself on a large scale in several areas of the country, including Tyneside, Glasgow, and southwest Lancashire (with access to Cheshire salt and local coal).[8] As home and international demand for soda increased rapidly because of its use in soap making, production levels grew rapidly during the 1860s and 1870s. But the process produced two major pollutants: hydrogen chloride gas (or acid gas, as it was often called) and sulfur waste (see Figure 8.1 for details of the Leblanc process).[9] The acid gas was released from tall chimneys, much to the annoyance of communities living near the works, while the sulfur waste (given the name "galligu" because of its black viscous nature and evil smell) was dumped as waste on land adjoining the works or at sea. The sulfur waste played an important economic role in the Leblanc process; most of the expensive sulfur (initially imported from Sicily but from the 1840s obtained in the form of pyrites) which was used to produce sulfuric acid (for the first stage of the process) ended in the sulfur waste. By the 1850s hundreds of thousands of tons of sulfur waste had been dumped, and in an effort to regenerate the sulfur several

[5] Robert Bunsen and Lyon Playfair, "Report on the Gases Evolved from Iron Furnaces, with Reference to the Theory of the Smelting of Iron," in *Report of the Fifteenth Meeting of the British Association for the Advancement of Science* (London: John Murray, 1846), 142–186.

[6] H. W. Richardson, chapter 9, "Chemicals," in *The Development of British Industry and Foreign Competition 1875–1914*, ed. Derek H. Aldcroft (London: George Allen and Unwin, 1968), 274–306, on 281.

[7] Reed, *Entrepreneurial Ventures in Chemistry*, 27–29.

[8] M. H. Matthews, "The Development of the Synthetic Alkali Industry in Great Britain by 1823," *Annals of Science* 33 (1976): 371–382, on 373; also see Kenneth Warren, *Chemical Foundation: The Alkali Industry in Britain to 1926* (Oxford: Clarendon Press, 1980), 44.

[9] Lyon Playfair referred to acid gas as "[t]he monster nuisance of all"; see A. E. Dingle, "The Monster Nuisance of All: Landowners, Alkali Manufacturers, and Air Pollution, 1828–64," *Economic History Review* 35 (1982): 529–548, on 529.

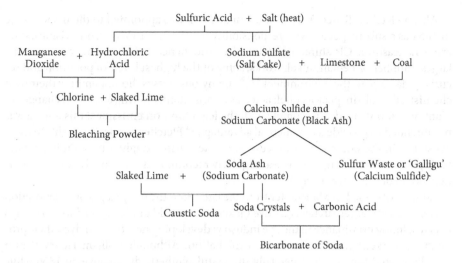

Figure 8.1 A schematic diagram of the Leblanc process.

From: Peter Reed, *Acid Rain and the Rise of the Environmental Chemist in Nineteenth-Century Britain*, 92, published by Routledge (© 2014), by permission of Taylor & Francis Group.

chemists, including the German-born and trained Ludwig Mond, invented processes that were partially successful.

Although the production of Leblanc alkali in Britain began on Tyneside and Clydeside on a small scale during the first two decades of the nineteenth century, it was the Muspratt works in Liverpool that from 1823 rapidly grew its production to satisfy the demands of local soap makers. Almost synchronous with the building of the works were the nuisance accusations as a constant stream of acid gas descended on local neighborhoods, and the threat of legal action.[10] By the 1850s Muspratt and other Leblanc manufacturers on Merseyside were forced to relocate to the more isolated site on the banks of the River Mersey that would grow into the town of Widnes (about equidistant from Liverpool and Manchester). By 1860 Widnes had become a major center for the Leblanc soda industry.[11] The relocation, however, did not remove or reduce the pollution; it grew instead as demand for soda (and associated chemicals) rose steadily, forcing Parliament to approve the Alkali Act in 1863. The terms of the legislation were strictly defined: each works was permitted to release no more than 5% of the acid gas (so at least 95% had to be absorbed); the Alkali Inspectorate was established to enforce the terms; all Leblanc soda works had to register with the Inspectorate; the Chief Inspector was to submit an annual report; and the Chief Inspector could seek financial penalties in the county courts.[12]

---

[10] Reed, *Entrepreneurial Ventures in Chemistry*, 37–38.

[11] D.W.F. Hardie, *A History of the Chemical Industry in Widnes* ([Liverpool]: Imperial Chemical Industries, 1950), facing 128.

[12] Peter Reed, *Acid Rain and the Rise of the Environmental Chemist in Nineteenth-Century Britain: The Life and Work of Robert Angus Smith* (Farnham, Surrey: Ashgate, 2014), 107.

Alfred Fletcher, Roscoe's fellow student at UCL, was appointed to the Inspectorate in 1864 as a sub-inspector with responsibility for the No. 1 (Western Division), covering Lancashire, Cheshire, and Flintshire, and hence Widnes. This area had the largest number of Leblanc alkali works, some of the highest levels of pollution and recurring damage to the environment.[13] As many businesses did not employ their own chemists, the sub-inspectors often acted as a consultant, offering advice on changes to plant operations needed to comply with the legal limit on emissions; this advice was not intended to provide a commercial advantage.[14] Fletcher regularly sought Roscoe's assistance in the enforcement process, and when firms sought advice on improving their plant or operation, Roscoe was regularly recommended as an advisor (being independent of the Inspectorate).

Roscoe witnessed in Manchester from the late 1850s the early stages of collaboration between chemical researchers and dye makers that would gradually lead to Germany's overwhelming dominance in the dye industry: developing new types of dyes, their production, marketing, and commercial exploitation. Although William Henry Perkin had discovered the first commercially successful synthetic dye, mauve, in 1856 while a student at the Royal College of Chemistry, Britain eventually lost the opportunity to create a major dye-making industry. By the late 1860s Germany was shifting its focus from improving production to searching for new colors.[15] This transition is illustrated by the German chemist Heinrich Caro, who had trained as a calico printer before relocating to work between 1859 and 1866 for Roberts, Dale and Company in Manchester, major suppliers of chemicals to the textile industry, who were in fierce competition with Simpson, Maule and Nicholson.[16] Caro investigated new dyes and processes while seeking out printers and dyers in Manchester (Roscoe's father-in-law Edmund Potter become a customer), and also met with Roscoe and his colleagues, including Carl Schorlemmer, in the chemistry department at Owens College in Quay Street.[17] Caro's external consultants included Peter Griess (who was working at Samuel Allsopp's Brewery in Burton-on-Trent) and Edward Schunck. Informal collaborations and discussions took place among German dyers who included Carl Alexander Martius (Wilhelm Hofmann's assistant at the Royal College of Chemistry in London) and other chemists at the local German Clubs, the Thatched House Inn, and the chemistry laboratory at Owens.[18] Roscoe had witnessed at first hand the benefit of such collaboration and kept in regular contact with Caro after he returned to Germany at the end of 1866.[19]

Back in Germany, Caro was hired by BASF to head its research division, and from 1873 began an informal collaboration with Adolf Baeyer's university laboratories, first in Strasbourg and then in Munich (from 1875).[20] In 1876 BASF and Baeyer signed a

---

[13] Peter Reed, "Alfred Fletcher's Campaign for Control of Black Smoke in Britain, 1864–1896: Anticipating the 1956 Clean Air Act," *International Journal for the History of Engineering and Technology* 91 (2021): 27–48.

[14] Reed, *Acid Rain*, 123–125.

[15] John Joseph Beer, *The Emergence of the German Dye Industry* (Urbana: University of Illinois Press, 1956), 64.

[16] Carsten Reinhardt and Anthony S. Travis, *Heinrich Caro and the Creation of Modern Chemical Industry* (Dordrecht: Kluwer, 2000), 52.

[17] Ibid., 59.

[18] Ibid., 81–82.

[19] Ibid., 127.

[20] Ibid., 59, 160, and 177.

formal contract that included work undertaken by not only Baeyer but also his assistants and co-workers.[21] For BASF, the collaboration provided valuable research information that led to the commercial exploitation of new groups of dyes, while Baeyer (and his team) gained valuable insight into novel molecules and reactions that were ripe for further research; finding a manufacturing pathway for synthetic indigo was probably the apogee of their collaboration.[22] As competition between German dye companies intensified, strong and exclusive collaborations developed between individual companies and universities, as in the case of Caro and Baeyer.[23] With state and government grants allowing the appointment of several chairs in chemistry (and other senior research workers), there was capacity not only for teaching but also for research collaboration.[24] However, in parallel, companies were employing large numbers of university-trained chemists to help develop better production processes, leading in the 1880s and 1890s to the creation of major industrial research laboratories, including the central research laboratory which oversaw the departmental research laboratories, as the control of patents became a crucial issue.[25]

Roscoe was fully aware of these developments (as were other British scientists, including Raphael Meldola).[26] Lack of government funding for universities and reliance on private donations (in the case of Owens College until 1890) prevented Roscoe and Owens College from taking the same strategy, much to the annoyance of several leading industrialists, including the German-trained but Manchester-based dye maker Ivan Levinstein, who on many occasions challenged Roscoe about the lack of research support from Owens College in finding new dyes.[27] Industrial research laboratories only materialized in Britain in the 1890s. The United Alkali Company (UAC), formed in 1890 by amalgamating all 47 chemical works associated with the Leblanc soda industry, in the same year established a Central Laboratory, modeled on the German dye manufacturers' research laboratories but on a much smaller scale.[28] Instead of focusing its limited technical resources on new technologies, far too much effort was wasted trying to improve the obsolete Leblanc process. UAC's final undoing was their rejection of the emerging electrolytic manufacture of alkali and chlorine that in 1895 would turn into a major new industry, with Roscoe as a technical advisor and board director.[29]

Many aspiring American chemists trained at German universities and then returned to establish chemistry departments based on laboratory-based learning, an emphasis on research and collaboration with industry.[30] Johns Hopkins University in

[21] Ibid., 59 and 186.

[22] Ibid., 59 and 186–187.

[23] Ibid., 65.

[24] Ibid., 65.

[25] Ibid., 219–256.

[26] Raphael Meldola, "The Scientific Development of the Coal-Tar Colour Industry," *Journal of the Society of Arts* 34, no. 1749 (28 May 1886): 759–771.

[27] H. B. Charlton, *Portrait of a University 1851–1951* (Manchester: Manchester University Press, 1951), 150.

[28] Peter Reed, "Making War Work for Industry: The United Alkali Company's Central Laboratory during World War One," *Ambix* 62 (2015): 72–93, on 73 and 75.

[29] Reed, "Making War Work," 77–78.

[30] Edward H. Beardsley, *The Rise of the American Chemistry Profession, 1850–1900* (Gainesville: University of Florida Press, 1964), 1.

Baltimore, Maryland, became a unique case in the United States when founded in 1876 as a private research university; established through the benefaction of the American entrepreneur Johns Hopkins, the university's mission was to integrate teaching and research and thereby train PhDs in line with the German model.[31] The first Professor of Chemistry was Ira Remsen, who drew on his experiences at the University of Munich and the University of Göttingen for the laboratories and research programs.[32] Although he understood the important link between German universities and industry, Remsen believed in pure scientific research and "disdained the commercial and industrial world"; with the chemistry department at John Hopkins achieving a notable output of PhD students, most were encouraged to teach rather than work in industry.[33] Many other German-trained chemists, including Roscoe, adjusted their approach to the opportunities and constraints afforded by the cultural, economic, and political context.

As well as being well-organized and articulate, Roscoe was a very genial and social person who built a remarkable network of friends and associates in Manchester and in London. His Manchester network grew out of his association with the Manchester Lit and Phil, the Manchester and Salford Sanitary Association, and the regional office of the Alkali Inspectorate, and through his consultancy work. In London his network drew on membership in several Royal Commissions, giving evidence to numerous parliamentary inquiries, work on behalf of the Royal Society, membership in the Chemical Society, and as a Member of Parliament. Roscoe's regular attendance at BAAS annual meetings in different towns broadened his networks and increased his influence. These networks were to play an important role in the formation of the Society of Chemical Industry in 1881. Mond had sought out Roscoe's support, knowing he could bring together leading chemists from both academia and industry; they had first met at the Norwich meeting of the BAAS in 1868 when Mond delivered a lecture on recycling the sulfur in the sulfur waste from the Leblanc process, and he came to respect Roscoe's practical knowledge of chemistry and admired his accomplishments at Owens College.[34] Mond had been reluctant to recruit non German-trained chemists, but made it known that he welcomed chemists from Owens College, which had become "a competitor of the German universities as a place for chemical education."[35] Roscoe's student Albert William Tangye served as a director of Brunner Mond & Company between 1919 and 1926.[36]

---

[31] Owen Hannaway, "The German Model of Chemical Education in America: Ira Remsen at Johns Hopkins (1876–1913)," *Ambix* 23 (1976): 145–164, on 145.

[32] William Albert Noyes and James Flack Norris, "Biographical Memoir of Ira Remsen (1846–1927)," *National Academy of Sciences of the United States Biographical Memoirs* 14 (1931): 207–257; also see Hannaway, "German Model," 145–164.

[33] Hannaway, "German Model," 157.

[34] J. M. Cohen, *The Life of Ludwig Mond* (London: Methuen, 1956), 158. See also Ludwig Mond, "On the Manufacture of Sulphur from Alkali Waste in Great Britain," in *Report of the Thirty-Eighth Meeting of the British Association for the Advancement of Science Held in Norwich in 1868* (London: John Murray, 1869), 40.

[35] W. J. Reader, *Imperial Chemical Industries: A History*, vol. 1: *The Forerunners 1870–1926* (London: Oxford University Press, 1970), 93.

[36] Cohen, *Ludwig Mond*, 159–160.

The foregoing provides the context for Roscoe's varied engagements with chemical industry over the period from 1857 when he arrived in Manchester to his death in December 1915. Roscoe was regularly sought out to join (and often lead) collaborative ventures because of his organizational ability, his sound business acumen, and his aptitude for working with others across academia and industry.

Consultancies

From the middle of the nineteenth century and against the backdrop of industrial and urban expansion, chemists (often from academia) acted as consultants to safeguard society's demands for safe air, water, and food.[37] This was also the period when few qualified chemists were employed in the chemical industry and only a small number of the entrepreneur-owners of chemical companies were qualified chemists.[38] Installation of new plant or modification of existing plant was overseen by chemists acting in a consultancy role. Roscoe's consultancies while at Owens College began in the 1860s and were undertaken, at least initially, to advocate the link between the College and the industrial and commercial community in Manchester and the surrounding area, including southwest Lancashire (St. Helens, Widnes, and Runcorn), a major center for the alkali industry, as part of a campaign to increase student numbers. Improvements in laboratory facilities at Owens College and better instrumentation allowed Roscoe to broaden his consultancies to embrace work for the Alkali Inspectorate, local authorities, local boards of health, and many individual businesses. Once student numbers had stabilized and plans for the Oxford Road site were underway, Roscoe viewed his relationship with industry and commerce as a two-way process: Owens College through Roscoe enabled businesses to be more productive and economic (and less polluting); and Roscoe sought out his business connections for financial support to enhance the College's facilities and staffing, in particular but not exclusively, for the science and engineering departments. Besides these altruistic benefits, Roscoe probably gained financially, although records have not survived.

Surviving records show that Roscoe's first consultancy was in 1865 as a result of visiting the copper mines at Mottram near Alderley Edge (Cheshire), where the Keuper sandstone had been found to contain copper and other metals, including cobalt.[39] Upon examining a solution of the metals, Roscoe suspected the presence of vanadium, a rare metal at the time. His later laboratory investigations confirmed the metal as vanadium and so began a long series of investigations to elucidate the metal's physical and chemical properties (see Chapter 7).

---

[37] Viviane Quirke and Peter Reed, "Chemistry, Consultants, and Companies, c. 1850–2000: Introduction," *Ambix* 67 (2020): 207–213, on 208.

[38] James Donnelly, "Consultants, Managers, Testing Slaves: Changing Roles for Chemists in the British Alkali Industry, 1850–1920," *Technology and Culture* 35 (1994): 100–128, on 102; and Donnelly, "Defining the Industrial Chemist in the United Kingdom, 1850–1921," *Journal of Social History* 29 (1996): 779–796, on 780.

[39] Henry Enfield Roscoe, *The Life and Experiences of Sir Henry Enfield Roscoe DCL, LLD, FRS, Written by Himself* (London: Macmillan, 1906), 141.

A substantial proportion of Roscoe's consultancies involved the analyses of water and air samples, as his surviving letter books from April 1865 show. Rapid population growth in urban areas was accompanied by a steady rise in demand for fresh clean water, forcing local authorities to urgently seek out new sources, often natural wells. Before making a decision on the water's use, assessing any impurities via thorough chemical and microscopic analysis was vital. Sources then required regular analysis in case the nature and level of any impurities had changed. Roscoe rarely inspected the sources, but was sent samples for analysis, and his reports always made reference to the water's suitable use. For example, in a letter to Henry Roberts (Engineers Office, Widnes) dated 26 May 1879, samples from Stocks Well and from Netherley were said to be of good quality, and letters dated 27 November 1879 and 28 May 1880 confirmed that the water was still good quality. On the other hand, a letter to W. W. Wright (Lostock Hall Water), dated 22 May 1881, reported that the water was not usable for locomotives or drinking, probably because the water was very hard or contained other mineral salts that could cause locomotive boilers to explode or would harm human health.[40]

The most important analysis Roscoe undertook was on water from Thirlmere, a natural lake in the Lake District, some 60 miles north of Manchester. It was under consideration as a source to meet Manchester's rising demand. In 1878 Roscoe was commissioned by Manchester Corporation to analyze and compare water samples from Manchester Corporation Water Company and Thirlmere Lake as part of their planning phase. Roscoe concluded: "I am therefore of the opinion that for chemical, sanitary and industrial reasons the Thirlmere supply is of the greatest importance for Manchester and Lancashire District."[41]

He produced his positive report against the backdrop of bitter opposition concerned over the likely disfiguring of a spectacular natural area that was led by the Thirlmere Defence Association, whose members included Octavia Hill, John Ruskin, and William Morris (among many other high-profile campaigners). Nevertheless, a parliamentary bill was finally approved in 1879 for the damming of Thirlmere Lake, thereby converting the lake into a reservoir.[42] Because of substantial construction work, the first supply of water did not arrive until 1894, supplying Manchester with a daily average of about 8 million gallons of water. By 1899 the average daily demand of the city had risen to 32.5 million gallons, reflecting the steady rise in population and the increasing demands of industry and commerce.[43]

As mentioned earlier in the chapter, the release of chemical nuisances from industrial works became an increasing public concern during the second half of the nineteenth century, and their regulation formed an important part of Roscoe's consultancy

---

[40] Letter to W. W. Wright (Lostock Hall Water), dated 22 May 1881, "Henry Roscoe Copybook of Letters (1878–1883)," MSS CH R108, John Rylands Library, University of Manchester. See also Colin A. Russell and John A. Hudson, *Early Railway Chemistry and Its Legacy* (Cambridge: RSC, 2012).

[41] Report to Chairman of the Water Committee of Manchester Corporation, dated 14 March 1878, "Henry Roscoe Letter Book (1866–1878)," MSS CH R107, John Rylands Library, University of Manchester. Courtesy of the University of Manchester.

[42] Advertisement—"Thirlmere Defence Association," *Carlisle Patriot*, 19 October 1877, 1.

[43] "Manchester's Water Supply," *Manchester Courier and Lancashire General Advertiser*, 6 January 1900, 17.

portfolio. Poor operation of chemical plant and the proximity of polluting works to residential housing generated rising concerns over damage to property and health. In one case, Roscoe was approached by solicitors in Hyde (Lancashire) about the tar distillation plant at Hyde Chemical Works. After inspecting the area around the works, Roscoe concluded that the works was not operated satisfactorily.[44] Whether the case was taken to court is not known. In another case, Roscoe was asked to inspect the workings of the Macclesfield Bone Works, where carcasses were boiled to separate the bones from the flesh (with the release of very unpleasant odors) before their use for the manufacture of fertilizers.[45]

Advances in instrumentation that included spectroscopes and photometers broadened Roscoe's consultancies. With the gas main network expanding through major towns and cities, questions were raised about the presence of sulfur in town gas and its effect on the gas's illuminating power. In 1868 Roscoe was asked to carry out photometric measurements of the gas supplied by the Manchester Gas Company.[46] In another investigation in 1878, William Crookes, editor of *Chemical News* and discoverer of thallium in 1861, sent a sample of thallium for spectroscopic and chemical analysis, and Roscoe was able to confirm the abundant presence of lead, as Crookes had suspected.[47]

Roscoe was a regular participant at the annual meetings of the British Association for the Advancement of Science (BAAS), held in a different location each year. When the BAAS held its 1861 meeting in Manchester, Roscoe was appointed one of the three local secretaries responsible for organizing visits to places of scientific interest and making arrangements for meeting venues. While the meetings "afforded him [Roscoe] an excellent opportunity of showing his organizing powers and business aptitudes," Roscoe took full advantage by not only networking with scientists and business leaders and promoting the benefits of Owens College, but presenting a report on the chemical industry in South Lancashire and giving a talk to the Chemistry Section (Section B).[48]

The report, *On the Recent Progress and Present Condition of Manufacturing Chemistry in the South Lancashire District*, provided an up-to-date overview of one of Britain's major chemical manufacturing centers.[49] The principal chemicals produced were sulfuric acid, soda, bleaching powder, dyestuffs, and disinfectants, which formed a vital part of Britain's international trade as well as supplying the

[44] Letter to Messrs. J. and J. Hibbert, Solicitors, dated December 1873, "Henry Roscoe Letter Book (1866–1878)," MSS CH R107, John Rylands Library, University of Manchester.

[45] Letter dated 22 May 1874, "Henry Roscoe Letter Book (1866–1878)," MSS CH R107, John Rylands Library, University of Manchester.

[46] Letter from Manchester Gas Company, dated March 1868, "Henry Roscoe Letter Book (1866–1878), MSS CH R107," John Rylands Library, University of Manchester.

[47] Letter to William Crookes, dated 2 December 1878, "Henry Roscoe Letter Book (1866–1878)," MSS CH R107, John Rylands Library, University of Manchester.

[48] Sir Edward Thorpe, *The Right Honourable Sir Henry Enfield Roscoe, PC, DCL, FRS: A Biographical Sketch* (London: Longman, Green, 1916), 38.

[49] E. Schunck, R. Angus Smith, and H. E. Roscoe, "On the Recent Progress and Present Condition of Manufacturing Chemistry in the South Lancashire District," in *Report of the Thirty-Third Meeting of the British Association for the Advancement of Science Held in Manchester in 1861* (London: John Murray, 1862), 108–128.

home market. The authors were Roscoe, Robert Angus Smith (the Manchester chemical consultant and air quality expert), and Edward Schunck (Manchester-based dye researcher) who, using their wide knowledge of these chemicals and their manufacture, reported on the progress over the previous decade, the new processes, and the modifications made to old processes; a concise statistical overview confirmed the importance of the region's chemical industry. Since BAAS members were more interested in the science than the economics, the report focused less on the latter. Such reports fulfilled one of the BAAS's declared aims; namely, to highlight and promote major local industries for participants unfamiliar with the city or its environs where the meeting was held. For historians, these reports have subsequently provided a "state of the industry" snapshot for the location and the particular year. Roscoe's paper to the Chemistry Section, "On Perchloric Acid and Its Hydrates," met another objective of the BAAS meetings: to highlight recent research by local scientists.

The Manchester and Salford Sanitary Association (MSSA) had been formed in October 1852 to aid the work undertaken by Manchester and Salford local authorities to improve poor sanitary conditions. Roscoe joined several other scientific members of the Manchester Literary and Philosophical Society, including Robert Angus Smith and Frederick Crace Calvert, to conduct surveys and give talks in support of the MSSA's work. With Parliament passing the Adulteration of Food and Drink Act in 1860 to reduce occurrence of food adulteration, the MSSA published a report on the quality of food available in the conurbation in August 1863, based on chemical analyses by Roscoe, Crace Calvert, and Schunck. The report drew attention to the frequent occurrence of food adulteration, though fortunately only rarely dangerous to health.[50] The 1860 act proved largely ineffectual and was replaced in 1873 by the Sale of Food and Drugs Act.[51]

When working as an advisor and assessor for Fletcher and the Alkali Inspectorate, Roscoe was drawn into a challenging situation, with its tension between environmental control and commercial enterprise. Fletcher's private diaries reveal that Roscoe's expertise was used on a regular basis over the period 1866–1870, but only as his heavy responsibilities at Owens College allowed.[52] Roscoe's role was essentially twofold: to recommend what permanent changes were needed to be made to the plant or the operating procedures in order for the works to comply with the regulations; and to act as an independent assessor to estimate the damages that might be sought by the Inspectorate in a court case against an offending works. It was in the most serious cases with recurring noncompliance and where the financial damages sought might be quite high, or in an extreme situation in which the Inspectorate might issue a closure notice, that Roscoe's independent input was sought. In a few instances, the Inspectorate recommended Roscoe to the business owner (or their lawyer) as an independent expert who could advise on the necessary modifications to plant or

---

[50] "Manchester and Salford Sanitary Association," *Manchester Guardian*, 9 April 1864, 6.
[51] *An Act for Preventing the Adulteration of Articles of Food or Drink*, P.P. 1860 (103); Sale of Food and Drugs Act, P.P. 1875 (63).
[52] Diaries of Alfred Fletcher (held in the private papers of the Fletcher family).

operations to comply with the legal emission limit and thus avoid closure. It was not unknown for Roscoe to act on occasions as an independent arbiter in disputes be-tween the Inspectorate and a manufacturer over the level of damages to be sought. When the Inspectorate took legal action in the courts, Roscoe would have to be in attendance, ready to present his evidence or his assessment of damages (together with their justification) should the judge require them; with the court's unpredictable time-table, Roscoe became frustrated by the waste of time he could ill-afford. This probably accounts for why Roscoe's consultancy work for the Alkali Inspectorate was limited to this four-year period.[53]

Roscoe's consultancy work on behalf of Owens College was not immune from crit-icism. Some thought that such work took Owens College away from its original pur-pose (and the terms of John Owens's will). They accused Roscoe of using his academic knowledge for material gain rather than for the benefit of education, since teaching staff were not well paid compared with other professions. Much of the criticism appeared in the columns of *The Times* over several weeks, although strangely the ar-ticle that started the debate, "The Endowment of Education in Its Economic Aspects," first appeared in the *Theological Review*.[54] The issue was brought to national attention in an editorial in *The Times* of 28 December 1874.[55] This editorial acknowledged that "it was economically sound to devote public money to the support of learning and science," since without it higher education was not sustainable and "learned and sci-entific men are ... compelled to maintain themselves by engaging in educational and commercial pursuits, except in the few fortunate cases in which they happen to be possessed of private means." The article went on to state that allowing subjects such as chemistry and physics to align with commercial activities would advance them disproportionately when compared to other disciplines; research undertaken in the physical sciences would take precedence over that in the arts and humanities because of the commercial benefit it would bestow. The editorial further suggested that the best students would be drawn to commerce rather than academia because of its finan-cial rewards.

In his response, the Owens College Principal pointed out how the College was able to support teaching in all disciplines from their endowments, and research formed an important activity for every discipline and not just for the sciences with their opportunities for commercial activities (as with consultancies) that can reward staff financially.[56] His consultancies had enabled Roscoe to demonstrate directly to leaders in industry and commerce (possible employers of Owens College's students) how training in higher education would benefit firms' innovation and economic sustainability.

[53] Fletcher's diaries have 28 entries for Roscoe working with him between August 1866 and August 1870.
[54] Michael Sanderson, *The Universities in the Nineteenth Century* (London: Routledge and Kegan Paul, 1975), 109. See also "The Political Economy of Endowments," *The Times*, 28 December 1874, 3.
[55] "The Political Economy of Endowments," *The Times*, 28 December 1874, 3.
[56] "Owens College, Manchester," *The Times*, 6 January 1875, 7.

Appointment to the 1876 Royal Commission
on Noxious Vapours

Roscoe was appointed to the Royal Commission on Noxious Vapours in July 1876. Set up by the Disraeli government, the Commission, under the chairmanship of Henry Bruce, 1st Baron Aberdare, the former Home Secretary and Lord President of the Council, was to review the working of the Alkali Works Act 1863 (and its later amendments in 1868 and 1874) for the control of pollution from industrial works. While the Alkali Inspectorate's ability to enforce the legal limits set for these chemicals were generally well received by Parliament, some members were skeptical whether the Inspectorate's enforcement was thorough enough, while others (aided by public campaign groups) sought vociferously to have other noxious chemicals blighting residential areas added to the list of regulated chemicals.

Roscoe was one of three scientific Commissioners appointed to the Commission. The others were Alexander Williamson, Professor of Chemistry and Practical Chemistry at UCL (and Roscoe's mentor), and Frederick Abel, President of the Chemical Society, Lecturer in Chemistry at the Royal Military Academy at Woolwich and chemist to the War Office. The other Commissioners were Algernon Percy, Earl Percy (later 6th Duke of Northumberland), William Brodrick (8th Viscount Midleton), Wilbraham Egerton (later 1st Earl of Egerton), James Stevenson (owner of the Jarrow Chemical Company, one of Britain's largest chemical works), and William Hornby (a retired vice-admiral).[57] About 200 witnesses gave evidence to the Commission, including several working for the Alkali Inspectorate (Robert Angus Smith, Alfred Fletcher, and other sub-inspectors), chemical manufacturers, landowners, land agents, representatives of Local Boards, medical officers of health (including John Simon, former Medical Officer to the Local Government Board), town clerks, residents, and gardeners. The first interview (with the Chief Inspector, Robert Angus Smith) took place in London on 3 August 1876, and others followed in Liverpool, Tynemouth, Newcastle-upon-Tyne, and Swansea, with the final meeting in London on 3 November 1877. Although meetings were planned well in advance, not all the Commissioners attended every meeting. Roscoe attended most meetings, even though his available time was severely restricted by his ongoing work on Owens College's extension and university status (see Chapter 6).

Roscoe's nomination to this Royal Commission was his first such appointment, although he had given evidence to several parliamentary inquiries previously. Even so, his experience as a consultant and assessor for Alfred Fletcher on behalf of the Alkali Inspectorate and his independent advice to manufacturers brought valuable insight into the working of the existing legislation and where revisions must be beneficial. Angus Smith gave evidence at the first and last meetings of the Commission, and Roscoe's questioning was broadly directed at three issues: the desirability of giving the Local Government Board (the government body with oversight of the Alkali Inspectorate) the legal authority to reduce the emission level year on year for each vapor under regulation; estimating how many additional staff would be required if the

---

[57] *Report of the Royal Commission on Noxious Vapours.* P.P. C. 2159 (1878), 1.

additional vapors were placed under regulation; and assessing the nature and number of "persistent" emissions above the legal limit that were undermining the work of the Inspectorate. The final report was published in August 1877 and the recommendations included: setting limits for the acids of sulfur and nitrogen; the deposit of sulfur waste so as to cause a nuisance was deemed an offense; and that manure works, ammonia works, tar distilleries, coke ovens, cement works, cobalt works, lead works, and dye works manufacturing coal-tar derivatives should be subject to inspection. The most important section related to the adoption of the "best practicable means" available for preventing escapes, rather than waiting until the final technical means was developed, so placing more responsibility back on the manufacturer.[58]

The Commission's recommendations were subsequently debated by Parliament and finally incorporated into revised legislation, which became the Alkali, etc., Works Regulation Act 1881.[59] As Roy MacLeod has pointed out, this legislation strengthened the Alkali Inspectorate's control:

> first, by extending the fixed standards to sulphuric acid and nitric acid; second, by placing more than a dozen new kinds of works under the "best practicable means" test pending the formulation of fixed standards; and third, by giving the Local Government Board power to expand inspectorial control over cement and salt works by provisional order, as soon as suitable means of regulation could be devised.[60]

Even though further parliamentary committees and legislation proved necessary in controlling new toxic chemicals, the 1876–1878 Commission proved an important milestone in Parliament's determination to protect the British population from harmful noxious chemicals. Roscoe, alongside the other Commissioners, had helped reframe important legislation, while his role as a Commissioner raised his personal standing and enhanced Owens College's reputation at a time when it was seeking university status.

Formation of the Society of Chemical Industry

After the formation of the Chemical Society (1841), the Pharmaceutical Society (1841), the Institute of Chemistry (1871), and the Society of Public Analysts (1874), further professional splintering occurred in 1881 when the Society for Chemical Industry (SCI) was formed.[61] The SCI sought to foster interest in all aspects of industrial chemistry for those working in the industry and others sharing an interest in the scientific and technological advances influencing the industry. Earlier, local societies had been formed in the major manufacturing regions, primarily for those working

[58] *Report of the Royal Commission on Noxious Vapours*. P.P. C. 2159 (1878), 36–37.

[59] *Alkali, &c., Works Regulation Act 1881*. P.P. 119 (1881).

[60] Roy M. MacLeod, "The Alkali Acts Administration, 1863–1884: The Emergence of the Civil Scientist," *Victorian Studies* 9 (1965): 85–112, on 107.

[61] Peter Reed, "Learning and Institutions: Emergence of Laboratory-Based Learning, Research Schools and Professionalization," in *A Cultural History of Chemistry in the Nineteenth Century*, ed. Peter J. Ramberg (London: Bloomsbury Academic, 2022), 191–215, on 211.

in the chemical industry, but in 1881 the aspiration for a national society emerged, with Roscoe taking a leading role in bringing together chemists in both industry and academia.[62]

Regional societies had been formed during the 1860s and 1870s in two major manufacturing centers in Britain: Tyneside and southwest Lancashire (St. Helens, Runcorn, and Widnes). Tyneside had the Newcastle Chemical Society (founded in 1868) as well as the Tyne Chemical Society; southwest Lancashire had the Faraday Club (formed in 1875). Their members were drawn from across different sectors of the industry, though most were proprietors of the major companies who were most likely to gain from the meetings and social events. A further initiative took place in November 1879 when the consulting chemist John Hargreaves, after meeting a representative of the Tyne Chemical Society, organized a meeting in Widnes of those working in the local chemical industry with the purpose of forming a Chemical Society for South Lancashire. Hargreaves was apparently unaware of the Faraday Club, but their members were invited to the second meeting. The formation of a regional society for south Lancashire gained momentum when Ludwig Mond of the ammonia-soda firm Brunner, Mond agreed to chair the meeting on 29 January 1880. Although unable to attend, Roscoe gave his enthusiastic support: "it will give me much pleasure to assist those interested in its formation in every way in my power."[63] From this point, many leading industrialists (and others supporting the formation of such a society) saw Roscoe as the one person with the vision and organizing ability to bring together the disparate groups across industry and academia.

Roscoe's leadership became evident at the next meeting, which was held at Owens College on 19 April 1880. With Roscoe in the chair, several manufacturers spoke out that the society should not be confined to a region, but should be for the whole of the United Kingdom, and should be named the Society of Chemical Engineers. While there was general agreement about a national society, some of those attending objected to the proposed name because many working in the industry did not see themselves as chemical engineers and would feel excluded. This disagreement was deferred to a later meeting at Owens College, but not before a large committee had been formed that included Ludwig Mond, James Liebig Muspratt (chemical manufacturer and grandson of James Muspratt), Ferdinand Hurter (industrial chemist), Peter Spence (chemical manufacturer), Robert Angus Smith (Alkali Inspectorate), Edward Schunck (dye researcher), Thomas Edward Thorpe (Yorkshire College), and John Hargeaves (industrial chemist), with Roscoe as Chairman and George Davis as honorary General Secretary.

The Committee's full report on the purpose and proposed activities of the society was presented to the inaugural meeting of about 60 people held at the premises of the Chemical Society in Burlington House, London, on 4 April 1881. Among the objectives set out in the report were: "interchanging ideas respecting improvements in the various processes"; "to publish information ... by means of a journal or otherwise"; to hold annual general meetings; and to organize local sections. There remained

[62] "The Society of Chemical Industry," *Journal of the Society of Chemical Industry*, Jubilee volume (July 1931): 9–13.
[63] Ibid., 10.

PRESIDENT

Prof. H. E. Roscoe, F.R.S.

VICE-PRESIDENTS

Prof. F. A. Abel, F.R.S.	Jas. Young, F.R.S.
Prof. A. W. Williamson, F.R.S.	F. H. Gossage.
I. Lowthian Bell, F.R.S.	E. K. Muspratt.
W. H. Perkin, F.R.S.	H. Lee Pattinson.
Dr. C. W. Siemens, F.R.S.	Walter Weldon.
Dr. Angus Smith, F.R.S.	Dr. G. D. Longstaff.

COMMITTEE

Capt. W. de W. Abney, F.R.S.	Jas. Mactear.
Prof. Chas. Graham.	H. Sprengel.
W. Crookes, F.R.S.	Henry Tate.
Peter Griess, F.R.S.	Philip Worsley.
Dr. D. B. Hewitt.	J. Spiller.
David Howard.	Alex. Chance.

TREASURER

E. Rider Cook.

SECRETARIES

Eustace Carey.	Ludwig Mond.
Thomas Tyrer.	George E. Davis.

Figure 8.2 The first Council of the Society of Chemical Industry.
From the *Journal of the Society of Chemical Industry*, Jubilee Edition (July 1931): 13.

some disagreement over the name, even though the committee had continued to use Society of Chemical Engineers. After a long discussion and not wanting to exclude anyone working in the chemical industry from membership, the meeting finally approved the name as the Society of Chemical Industry.[64] Before adjourning, the first council was elected (see Figure 8.2). The composition of the council clearly reflected the intent of bringing together leading academic and industrial chemists, "dedicated to the promotion of research, development and application in the chemical industries."[65] Roscoe's ability to bring together such a group of leading industrialists and academics in a common endeavor was a remarkable achievement and an indication of his standing in both communities.

The SCI's first General Meeting was held on 28 and 29 June 1881, with Roscoe in the chair. Just three months after the inaugural meeting, Roscoe was able to report that membership already stood at 297, including 59 technical chemists, 54 manufacturing chemists, and 15 chemical engineers. By the first Annual meeting in July 1882, the membership had reached 1,140. Besides agreeing to organize its annual meeting in a different city each year (following the example of the BAAS), the Council had

---

[64] Ibid., 12. The SCI was granted its Royal Charter on 17 June 1907.
[65] Robert Kargon, *Science in Victorian Manchester: Enterprise and Expertise* (Manchester: Manchester University Press, 1977), 204–205.

also decided to organize local sections based in the major cities of Britain; the first, the London Section, had already been formed, and others would follow soon after, including Liverpool (1881), Manchester (1883), and Glasgow (1884). A proposed journal was also being planned by a committee that included Roscoe, Davis, and Mond. Probably at Roscoe's suggestion, the first editor was Watson Smith (demonstrator in technological chemistry at Owens College), who went on to edit the journal for 32 years.[66] The first edition appeared in January 1882 (only six months after the first general meeting) and its editorial outlined its broad remit.[67] Besides including full reports of the Society's affairs and meetings (including those of the sections), the journal would include government information related to chemical industry, abstracts from foreign journals, and a list of British and foreign patents likely to have an interest for members. The journal appeared monthly and the first volume (published in 1882) comprised 518 pages.

With so many leading figures from Manchester preoccupied with the SCI's inception and with the first annual meeting in Manchester in 1882, it is not surprising that the Manchester section was formed soon after, in 1883. Ivan Levinstein, the Manchester dye manufacturer, became the first Chairman. Roscoe served as Chairman for the period 1884–1887 and continued to take an active role after his resignation from Owens College, offering many papers, attending regularly, and chairing meetings whenever other pressures allowed.

The Conflict between Roscoe and Ivan Levinstein

With British industry coming under strong competition and economic pressure during the last 30 years of the nineteenth century, the relationship between academia and industry evolved into a hotly contested debate.[68] Roscoe was not insulated from these debates and accusations that the chemistry department at Owens should provide greater support for industry. He was targeted in particular by the Manchester dye manufacturer Ivan Levinstein, who contrasted Owens College's lack of support with that given to industry by German universities and polytechnic schools. The debates and counter-accusations were recorded in the pages of several journals, including the *Chemical Review* (founded by Levinstein and published between 1871 and 1891), the *Journal of the Society of Dyers and Colourists*, and the *Journal of the Society of Chemical Industry*.[69]

Ivan Levinstein (1845–1916) was born in Germany and trained at the Royal Prussian Gymnasium in Berlin and then at the Gewerbeinstitut (the Royal Trade Academy), studying aniline dyes. With the family's business fortunes fading, Levinstein moved

[66] "Obituary—Watson Smith," *Journal of the Society of Chemical Industry* 39 (1920): 191; and "Obituary—Watson Smith," *Journal of the Chemical Society, Transactions* 117 (1920): 1637–1638.

[67] "Editorial," *Journal of the Society of Chemical Industry* 1 (1882): 1.

[68] Peter Reed, "George E. Davis (1850–1907): Transition from Consultant Chemist to Consultant Chemical Engineer in a Period of Economic Pressure," *Ambix* 67 (2020): 252–270, on 254.

[69] Martin Saltzman, "Academia and Industry: What Should Their Relationship Be? The Levinstein-Roscoe Dialog," *Bulletin for the History of Chemistry* 23 (1999): 34–41.

in 1864 to Britain, then experiencing a surge in interest in dye manufacture following William Henry Perkin's discovery in 1856 of mauve. After an initial works in Salford, Levinstein & Company was established at Blackley, north of Manchester, in 1865 and began manufacturing a range of aniline dyes.[70] By the 1870s the competition from Germany and Switzerland was taking a serious toll on the British dye industry due to their wider range, higher production, and lower prices. Feeling this commercial pressure, Levinstein began to speak out at every opportunity, seeing the issues in straightforward terms: Britain's weak patent laws, and the lack of support of academic institutions. The existing patent laws urgently needed revision, especially with re-gard to an effective working clause, and Levinstein, in conjunction with the powerful Manchester Chamber of Commerce and its sectional committee for the chemical and allied trades, eventually persuaded the President of the Board of Trade, David Lloyd George, in 1908 to revise the regulations.[71]

While Roscoe was not a Member of Parliament when the patent laws were under review and the new British patent law was approved, he was challenged by Levinstein over Owens College's perceived lack of support for industry, and in particular the lack of research collaboration between the College and his company. During his Chairman's Address to the Manchester Section of the SCI, Levinstein took a firm stance when comparing Britain and Germany:

I do not remember during the period referred to [the last 10 years], with perhaps few exceptions, any important original research bearing on this subject [dyestuffs] undertaken by English professors, and I certainly cannot call to mind any work done by them which has been of practical utility to this industry ... and while the Germans are constantly developing this department of chemistry.[72]

Levinstein was incensed about the neglect of organic chemistry, especially the chem-istry of aromatic compounds, which he believed could lead to new classes of dyestuffs.

Roscoe had a different focus based on *his* experiences in Germany; the principle that guides good management:

is the absolute necessity of having high-trained chemists, not only at the head of the works but at the head of every department of the works where a special manufacture is being carried out. In this respect this method of working stands in absolute con-trast to that too often adopted in chemical works in this country, where the control of the processes is left in the hands of men whose only rule is that of the thumb, and whose only knowledge is that bequeathed to them by their fathers.[73]

[70] Anthony S. Travis, *The Rainbow Makers: The Origins of the Synthetic Dyestuffs Industry in Western Europe* (Bethlehem, PA: Lehigh University Press, 1993), 135.

[71] Peter Reed, "The British Chemical Industry and the Indigo Trade," *British Journal for the History of Science* 25 (1992): 113–125, on 117 and 118.

[72] I. Levinstein, "Address by the Chairman," *Journal of the Society of Chemical Industry* 3 (1884): 69–73, on 72.

[73] H. E. Roscoe, "President's Address," *Journal of the Society of Chemical Industry* 1 (1882): 250–252, on 251.

Roscoe stood by his standard position which emphasized the importance of the highly trained chemist to lead every department, including a company's research (or R&D) department, where new dyes were more likely discovered. The technological chemistry course at Owens College only provided the theoretical principles behind industrial processes, and Roscoe remained steadfast that it should not deal with the practical operation of processes.[74] Such statements only hardened positions on both sides. Levinstein later became a leading figure in improving technical education in Manchester and in the Manchester Municipal Technical School (housed in the Sackville Street Building from 1902). As a result of an agreement between Victoria University and Manchester Municipal Corporation in 1905, the professors at the Technical School (renamed Manchester Municipal College of Technology after World War I) would constitute the Faculty of Technology of the Victoria University.[75]

Roscoe and Levinstein crossed paths again in 1883 during a patent infringement court case brought by the German firm Badische Anilin und Soda Fabrik (BASF, but usually referred to as Badische in this period) against Messrs. Levinstein. Badische had taken out a British patent in February 1878 and felt Messrs. Levinstein's "Fast Blackley Red" was an infringement of their patent.[76] Patent infringement court cases meant that the judge and lawyers for both parties had to become acquainted very quickly with the science underpinning the disputed patent; some leading lawyers (Lord Moulton was an outstanding example) became very adept at quickly assimilating very technical details through lengthy interviews with the scientific experts, not only to understand the terms of the patent, but to cross-examine scientific witnesses.[77]

The case started on 8 March, and its 10-day duration included detailed scientific evidence for Badische by James Dewar (Professor of Chemistry, University of Cambridge), Henry Armstrong (Professor of Chemistry, London Institution), and William Odling (Waynflete Professor of Chemistry, University of Oxford). Levinstein's defense was that "it was simply impossible to carry out these processes by the means stated in the specification" and further asserted that using the chemicals in the concentrations specified in the patent "would be a danger to human life."[78] After the plaintiffs' evidence, Justice Pearson declared that he would not decide the case until an experiment had been carried out by an independent chemist.[79] The experiment was duly carried out, but Levinstein did not accept the result, maintaining that his dye was manufactured by a "secret process."

Subsequently, the judge asked Roscoe (as an independent authority agreed by both parties) to carry out Levinstein's "secret process," which he subsequently did outdoors to mitigate any danger. Roscoe's report was in favor of the plaintiffs (the patentees),

---

[74] H. E. Roscoe, "Remarks on the Teaching of Chemical Technology," *Journal of the Society of Chemical Industry* 3 (1884): 592–594, on 594.

[75] History of UMIST, https://www.manchester.ac.uk/discover/history-heritage/history/umist/. See also P. J. Short, "The Municipal School of Technology and the University, 1890–1914," in *Artisan to Graduate*, ed. D.S.L Cardwell (Manchester: Manchester University Press, 1974), 157–164.

[76] "Badische Anilin und Soda Fabrik v Levinstein," *The Times*, 9 March 1883, 3.

[77] Peter Reed, "John Fletcher Moulton and the Transforming Aftermath of the Chemists' War," *The International Journal for the History of Engineering and Technology* 87 (2017): 1–19.

[78] "BASF v Levinstein," *The Times*, 9 March 1883, 3.

[79] Ibid., 4.

and he further reported that "the supposed secret process was only a modified form of the plaintiffs' invention as described in their specification," and Justice Pearson found in favor of Badische Anilin und Soda Fabrik.[80]

Levinstein refused to back down on this claim and took the case to the Court of Appeal, where a majority of the court reversed the earlier judgment by Justice Pearson. An appeal was taken to the House of Lords, where on 8 December 1884 the Lords Justices found in favor of Badische, and Justice Pearson's judgment was restored. The use of an independent advisor by a Court (in this case, Roscoe) was not unprecedented, but it was rare.[81] How Roscoe came to be chosen to act as an independent expert is not known, but both parties knew Roscoe and trusted his expertise.

Company Directorships

Soon after his election as a Member of Parliament, Roscoe's industrial connections took a new direction. In 1887 he was appointed a director of the Aluminium Company Limited and then in 1895 a director of the Castner–Kellner Alkali Company Limited. These appointments drew not only on Roscoe's reputation as an outstanding academic chemist who always welcomed the latest advances, especially when they benefited industry, but also on his standing among business leaders as a founder (and first President) of the SCI and his knighthood in 1885.

Roscoe was always intrigued by aluminum ("the 'wonders' of the silver-white metal," and "the 'crystallised gold' of aluminium bronzes") and he was aware of Robert Bunsen's investigations into the electrolytic production of the metal in 1854.[82] An ingot of aluminum was one of the spectacular attractions at the Paris Exhibition in 1855 after the Frenchman Henri St. Clair Deville had developed the production of aluminum by reacting the double chloride of aluminum and sodium with metallic sodium in the flux of the mineral cryolite. Production of aluminum remained modest (about 10,000 pounds per annum), while retaining its high price. However, by 1887 production levels had reached 100,000 pounds per annum and at a lower cost of 20s per pound. This dramatic change was made possible by the American chemist Hamilton Young Castner, who had reduced the cost of sodium production by 80%.[83] In 1887 the Aluminium Company Limited was formed to work the Castner patents at Oldbury (in the west Midlands), with Castner as managing director; and the company acquired the Aluminium Crown Metal Company of James Webster at Oldbury. During the development phase, Roscoe acted as director of the metallurgical department, and when the company's prospectus appeared in *The Times* on 27 June 1887, Roscoe was listed as a director, alongside Gerald Balfour MP, brother of the future prime minister Arthur Balfour and a fellow Manchester MP with Roscoe. Appended

[80] Viscount Alverstone, *Recollections of Bar and Bench* (London: Edward Arnold, 1914), 189–191. Viscount Alverstone (as Richard Webster QC before he was raised to the peerage in 1900) was one of the plaintiffs' counsels.

[81] Alverstone, *Recollections of Bar and Bench*, 190.

[82] H. E. Roscoe, "Aluminium," *Proceedings of the Royal Institution* 12 (1889): 451–464, on 451 and 462.

[83] Aluminium Company, *Grace's Guide*, https://www.gracesguide.co.uk/Aluminium_Co.

to the prospectus was a report by Roscoe reviewing the patents, the site operation at Oldbury, and the finances underpinning the company:

> It is obvious that the perfection attained in the respective processes is the result of long and careful research and experiment, and I concur in the praise given to the skill and labour spent upon the works, and their fitness for the purpose for which they were designed.[84]

Roscoe's report concluded with a projection for the future, pointing out that the process using sodium was also applicable to the production of magnesium. However, by 1891 the cheaper electrolytic manufacture of aluminum (making use of hydroelectric power) had taken over, and the Aluminium Company's production of aluminum was gradually abandoned, leaving the manufacture of sodium as the company's main business.[85]

In an effort to improve the manufacture of sodium, Castner was investigating how to make purer caustic soda when he indirectly developed a radically new process for producing caustic soda and chlorine (which could be used to make bleaching powder). This process involved electrolyzing brine in an ingenious electrolytic cell using mercury; the cell became known as the "rocking cell."[86] Investigations had showed that the cost of producing caustic soda undercut the Solvay process of Brunner Mond at Winnington (Cheshire). But other mercury cells were under development, including a cell by Carl Kellner, an Austrian chemist whose patent was owned by the Belgian company, Messrs Solvay et Cie (except for Austria). With the increasing demand for cheap and high-quality caustic soda and chlorine, these electrolytic cells attracted considerable interest. The United Alkali Company (UAC) showed keen interest in these electrolytic processes as a replacement for their antiquated, uneconomic, and heavily polluting Leblanc process. However, their judgment was influenced by the projected cost of electricity and whether equipment for generating electricity on an almost continuous basis was possible. Having inspected the Oldbury operations, the head of UAC's Central Laboratory, Ferdinand Hurter, advised the UAC Board to reject the approach of the Aluminium Company Limited, and a short time later UAC's interest officially ended.[87]

In 1895 the Castner–Kellner Alkali Company was formed to work the patents of Hamilton Young Castner (owned by the Aluminium Company) and of Carl Kellner (owned by Messrs Solvay et Cie) at Weston Point (Cheshire), close to the chemical manufacturing centers of Widnes and Runcorn.[88] The move from Oldbury to Weston Point gave access to the brine pipeline (constructed by the Salt Union) connecting the Cheshire salt fields to the centers of the Lancashire alkali trade. Directors of the

---

[84] "The Aluminium Company Limited," *The Times*, 27 June 1887, 4.

[85] Aluminium Company, *Grace's Guide*.

[86] D.W.F. Hardie, *History of the Chemical Industry in Widnes*, 185.

[87] This decision is often alluded to as a major error of judgment, but generating electricity cheaply and continuously remained unresolved. See Hardie, *History of the Chemical Industry in Widnes*, 189.

[88] Anonymous, *Fifty Years of Progress: The Story of the Castner-Kellner Alkali Company Told to Celebrate the Fiftieth Anniversary of Its Formation, 1895–1945* (Birmingham: Kynoch Press, 1947).

new company included Castner, Kellner, William Mather (Chairman of Mather & Platt), Gerald Balfour MP, and Roscoe. It was Mather who played the pivotal role in developing and building the electrical generating equipment upon which the cell operation depended.[89] He readily accepted the invitation of the other directors to become Chairman, even though he was also Chairman of Mather and Platt, Engineers in Manchester, and when he resigned at the 1906 Annual General Meeting (AGM), Gerald Balfour became Chairman.[90]

The company prospered from the first year of operation, steadily expanding its operations, and has continued to do so (with only minor technical changes) into the twenty-first century.[91] Drawing on his experiences with the Leblanc soda process and with the Alkali Inspectorate, Roscoe at the third AGM commented on how "their works might be considered as model works in every respect. Chemical works were generally a nuisance to the neighbourhood, but with their process there were no noxious fumes whatever."[92]

The business was also extremely profitable and in most years the shareholders agreed to pay a bonus (of up to £1,000) to the directors.[93] His association with the Castner–Kellner Alkali Company proved interesting and remunerative for Roscoe.

[89] "The Castner-Kellner Alkali Company Limited," *Chemical Trade Journal*, 3 April 1897, 238. See also *Fifty Years of Progress*, 33.
[90] *Fifty Years of Progress*, 46.
[91] The works at Weston Point is now part of INEOS.
[92] "Castner-Kellner Alkali Co.," *Chemical Trade Journal*, 3 June 1899, 367.
[93] "Castner-Kellner Alkali Co.," *Chemical Trade Journal*, 4 June 1898, 381.

9

Securing Britain's Economic Future

Sounding the Alarm

Academics, educationalists, and MPs were concerned following the International Exhibition in Paris in 1867 that Britain had lost, or was beginning to lose, its position as the world's most advanced industrial nation—a process which left unchecked could adversely affect the country's economic position. As we have already discussed in Chapter 1, this perception was not wholly accurate in the late 1860s. Nonetheless, concern was now in the air, and Lyon Playfair, then Professor of Chemistry at Edinburgh University who was responsible for organizing the juries judging the exhibits in Paris, wrote to Lord Taunton, chairman of the Schools Inquiry Commission about the concerns shared by many (though not all) visiting the Paris exhibition and lamenting Britain's system of technical education compared to those on the continent. Playfair concluded his letter with a call for a government inquiry.[1]

In response to Playfair's letter and the issues it raised, the Society of Arts convened a conference on technical education in London on 23 January 1868. Invitations had been extended to:

> the mayors of large towns, presidents of chambers of commerce, presidents of learned and professional societies and the City Companies (it was felt the City Guilds could help), university teachers, school and factory inspectors and, in short, all concerned in one way or another with technical development.[2]

Several jurors at the Paris Exhibition also attended. Among the 247 notables present were many who would take up "the baton" for technical education and play a prominent role in a variety of initiatives across Britain during the remainder of the nineteenth century. These included Playfair himself, Bernard Samuelson, James Bryce, and Thomas Huxley; Roscoe did not attend, but Joseph Greenwood (Principal of Owens College) did.[3] Many of those attending were supporters of educational reform, and many wanted the government to take a leadership role to avoid relying on ad hoc initiatives across the country, however effective. A committee was set up to develop the proposals discussed at the conference; both Huxley and Samuelson were members of the committee. When the committee presented its report in July 1868, it offered a comprehensive definition of technical instruction: "general instruction in those

---

[1] Letter from Lyon Playfair to The Right Hon. Lord Taunton, dated 15 May 1867, reproduced in *Journal of the Society of Arts* 15, no. 759 (7 June 1867): 477–478.
[2] D.S.L. Cardwell, *The Organisation of Science in England* (London: Heinemann, 1972), 112–113.
[3] "Conference on Technical Education," *Journal of the Society of Arts*, 14, no. 793 (31 January 1868): 183–209.

Henry Enfield Roscoe. Peter J.T. Morris and Peter Reed, Oxford University Press. © Oxford University Press 2024.
DOI: 10.1093/oso/9780190844257.003.0009

sciences, the principles of which are applicable to the various employments of life."[4] Meanwhile, Gladstone's government had acted promptly after receiving Playfair's letter, according to Donald Cardwell: "they [the government] instructed ambassadors and consuls abroad to report on technical education in the various countries, as the Schools Commission [Taunton Commission] had recommended that they should."[5]

Samuelson Committee

The Liberal government set up a select committee on scientific instruction with Samuelson as chairman on 24 March 1868. This select committee followed three earlier Royal Commissions on education. The Royal Commission on the State of Popular Education in England was established in 1858 to consider whether the current elementary education was working, under the chairmanship of Henry Pelham-Clinton, 5th Duke of Newcastle (himself a product of Eton and Christ Church, Oxford). He was also the Colonial Secretary (under Lord Palmerston) for much of the period of this Commission. The Commission decided in 1861 that on the whole the system was working, but recommended that part of the state funding for elementary schools should be paid on the basis of the results of an annual test of the pupils, the so-called payment by results. It ultimately led to the Elementary Education Act of 1870. This report was immediately followed by the Royal Commission on the Public Schools, chaired by the former Foreign Secretary George Villiers, 4th Earl of Clarendon (who was educated by a private tutor before entering St John's College, Cambridge, at the age of 16), which examined the nine leading public schools following concerns raised about the running of Eton College, with its focus on classics to the exclusion of science. The Commission's report produced in 1864 waxed lyrical about the value of these schools to the nation, describing them as "the chief nurseries of our statesmen," but recommended changes to the curriculum (including more science) and their governance. This report led to the Public Schools Act of 1868, which only covered boarding schools. Set up immediately after the Clarendon Commission, the Schools Inquiry Commission had a broader remit to examine the 782 endowed grammar schools in England and Wales (in fact it was asked to examine all schools not covered by the previous two commissions), under the chairmanship of Henry Labouchère, Baron Taunton (Winchester and Christ Church, Oxford), who had been Newcastle's predecessor as Colonial Secretary. The problems with the endowed schools were that they had moved away from their original aim of educating the poor, and were unevenly distributed across the country, as they tended to be in areas which had been wealthy in the Middle Ages or the Tudor period. Girls' schools were also underrepresented. When it reported in 1868, the commission recommended that the endowments be used for modern purposes rather than their original intentions under the aegis of an Endowed Schools Commission. Thus schools currently offering free classical education were converted into modern fee-paying grammar schools. Its recommendations were given effect by the Endowed Schools Act of 1869.

[4] Cardwell, *Organisation of Science*, 113–114.
[5] Ibid., 115.

Hence all of elementary education and much of secondary education had now been examined; what was lacking was a similar investigation of scientific education, especially for those going into industry. That was the remit of the Samuelson Committee, although it lacked the authority of a Royal Commission. The other members of the committee included Lord Frederick Cavendish (MP for the West Riding), who was famously assassinated as Lord Lieutenant of Ireland in 1882; Lord Robert Montagu (MP for Huntingdonshire), the Conservative Vice-President of the Committee on Education; Henry Austin Bruce (MP for Merthyr Tydfil, later Lord Aberdare), the Liberal Vice-President of the Committee on Education; Edmund Potter (MP for Manchester), a calico printer (and Roscoe's father-law); George Dixon (MP for Birmingham), an educational reformer; Thomas Bazley (MP for Manchester), a cotton spinner; and Thomas Acland (MP for Devonshire North), also an educational reformer. The committee's remit was "to inquire into the Provisions for giving instruction in Theoretical and Applied Science to the Industrial Classes" under two headings: the state of scientific instruction and the relation of industrial education in industrial progress. Evidence was taken from witnesses representing a wide range of organizations: Department of Science and Art (South Kensington), the Committee of Council for Education (the government body responsible for education), universities, secondary schools and mechanics institutes, and those engaged in major industries in the principal manufacturing areas.[6]

Roscoe had been Professor of Chemistry at Owens College for almost 11 years when he gave evidence to the Select Committee on Scientific Instruction in June 1868. Responding to questions from the Committee's members, Roscoe's evidence (provided by oral testimony and submitted papers) concentrated on Owens College, its comparison with other universities in Britain and Germany, the funding of higher education in Britain, recruitment of well-qualified students, engaging working men in scientific subjects, the role of Working Men's Colleges, and the training of science teachers. The published report shows the rather ad hoc and repetitive questioning in responding to the particular interests of individual Committee members.[7] Nevertheless, in his responses Roscoe was able to draw attention to the limited accommodations at Owens College for both lecturing and laboratory work, and the current plans for an extension, his low pay as professor compared with professors at other universities (though he did also receive a portion of the students' fees), from which materials and equipment had to be provided. He sharply contrasted this provision with those found in Germany, which were expansive to cater to a large student population (with a strict examination system) and were funded by the state. On several occasions Roscoe drew attention to the need for government financial support from "the national exchequer" since institutions such as Owens College, acting as the University of the North and serving the needs of regional industry and commerce, could not rely on voluntary contributions or municipal funding.[8] Success at higher education necessitated well-qualified students, placing a responsibility on schools to broaden their curriculum and include science; Roscoe was able to point out that more recently the standard of

---

[6] *Report from the Select Committee on Scientific Instruction*, P.P. 1868 (432), iii.
[7] Ibid., 276–290.
[8] Ibid., 285.

students applying to study at Owens College had improved because of better schools provision, and used this point to emphasize the urgent need to increase the number of good teachers and the provision for teacher training.

Asked about the best way of interesting the working class in scientific subjects, Roscoe spoke about the successful series of Science Lectures for the People in Manchester which were published at a penny each:

> By the help of a few friends who assisted me, I arranged a series of Science Lectures for the People ... and which were published at a penny each ... the number of persons attending the 13 lectures was upward of 4,000; I had over 400 people at my four lectures on elementary chemistry, which I gave to begin with; I charged an admission fee of one penny.[9]

Roscoe was then pressed by Edward Potter on the funding required to sustain such series, and he reiterated that voluntary funding was not a way forward and that "we have found throughout the world that high-class education does not pay its own way."[10] There was a more fundamental issue for Roscoe when comparing Britain with Germany that was not readily nor quickly resolved: "the love of scientific education, and that the intellectual culture of the people is probably greater throughout the masses in Germany than it is in Britain."[11]

The committee reported very quickly in July 1868 and without any dissent. It made a large number of recommendations, and arguably too many. The key recommendations were that every child should have an elementary education, which should teach drawing, physical geography, and "the phenomenon of nature." Hence elementary school teachers should be taught more science. Furthermore, secondary education should be reformed and include more science, with some schools even being science schools. Colleges of science and schools of scientific education should be given some state funding, as they cannot be run on the basis of fees alone. However, in contrast to the earlier Royal Commissions, the report of the Samuelson Committee did not directly result in any parliamentary legislation. Nonetheless, it set the framework for later Royal Commissions on scientific and technical education.

Campaign for Science Funding

Soon after the Samuelson Committee had reported, a quite different campaign started up. At the meeting of the British Association for the Advancement of Science in Norwich in August, a retired Indian Army officer, Lieutenant-Colonel Alexander Strange, gave a paper on the "Necessity of State Intervention to Secure the Progress of Physical Science." While it might seem strange that a retired army officer would be concerned about such matters, Lt-Col. Strange was no ordinary officer. While serving in the Indian Army between 1834 and 1859, he took part in the Great Trigonometrical Survey

[9] Ibid., 283.
[10] Ibid., 283.
[11] Ibid., 285.

of India. He was also an amateur astronomer. After he retired, Strange proposed to the Indian government that it set up a department to inspect scientific instruments, and he duly became the inspector of instruments. So he was already familiar with the relationship of the state (albeit in this case a colony) and science when he spoke at the Norwich meeting. He argued that there was a crisis in science, in part because science graduates could not afford to follow a scientific career; hence why should anyone study science? He argued that this shortfall in funding could only be solved by state intervention.

The British Association for the Advancement of Science set up a committee in November to study whether there was sufficient provision for physical science in Britain. This committee included many of the leading scientific reformers (and the members of the X-Club), including Tyndall, Huxley, Frankland, Playfair, and Lockyer. The scientific community was asked if there was a course of action, either by government or private initiative, which would improve their lot. Perhaps unsurprisingly, the committee then concluded that there was inadequate provision for science in Britain; the nation had benefited, but the individual scientist had not. While many scientists were keen to see state funding of science, others feared the effect of patronage, so it would become a matter of "who you know" rather than "what you know" (Lockyer had gained his sinecure at the War Office through patronage). Alfred Russel Wallace argued that public competition was "a greater stimulus to true and healthy progress than any Government patronage."[12]

For academic scientists such as Roscoe, there was a danger that this campaign would divert government money from the civic universities to the funding of institutions such as a national science museum, scientific research outside the universities, and the support of individual scientists in particular. However, the pressure for an inquiry into the relations of the state to science was growing, not least in the pages of *Nature*, the new journal founded by Lockyer. The other Royal Commissions on education having completed their task, Gladstone agreed in February 1870 to set up a Royal Commission to consider scientific instruction, and it was established under the chairmanship of William Cavendish, 7th Duke of Devonshire, who was the Chancellor of Cambridge University and the President of Owens College, in May 1870.

Devonshire Commission

The formal remit of the Royal Commission on Scientific Instruction and the Advancement of Science was "to make Inquiry with regard to Scientific Instruction and the Advancement of Science and to Inquire what aid thereto is derived from Grants voted by Parliament or from Endowments belonging to the several Universities in Great Britain and Ireland and the Colleges thereof and whether such aid could be rendered in a manner more effectual for the purpose." Besides the Duke of Devonshire, the members of the Royal Commission on Scientific Instruction and the Advancement of Science including several outstanding scientific, educational, and political figures. They included the Marquess of Lansdowne, John Lubbock (MP, banker, and scientific writer),

---

[12] Alfred Russel Wallace, "Government Aid to Science," *Nature* 1 (13 January 1870): 288–289.

Figure 9.1 Sir Joseph Norman Lockyer.
Photograph by Walery. Wellcome Collection.

Sir James Kay-Shuttleworth (civil servant and educationalist), Bernhard Samuelson (MP), William Sharpey (Secretary of the Royal Society), Thomas Huxley (Professor of Natural History, Royal School of Mines), William Miller (Professor of Chemistry, Kings College, London), and George Stokes (Lucasian Professor of Mathematics, University of Cambridge). Following the death of Miller in September 1870, Henry J.S. Smith (Savilian Professor of Geometry, University of Oxford) was appointed as a Commissioner.[13] Norman Lockyer, although he was busy editing his new journal, was appointed Secretary for the Commission as he was still a civil servant in the War Office and it was standard practice for the Secretary to be a civil servant (Figure 9.1).[14]

---

[13] *Report of Royal Commission on Scientific Instruction and the Advancement of Science, Volume 1, First, Supplement, and Second Reports, with Minutes of Evidence and Appendices*, P.P. 1872 (C. 536), iii–iv.
[14] A. J. Meadows, *Science and Controversy: A Biography of Sir Norman Lockyer, Founder of Nature*, 2nd ed. (London: Macmillan, 2008), 82.

It would be impossible to cover the work of the Devonshire Commission, which produced eight reports over a five-year period, in detail.[15] Clearly the most important report for Roscoe and Owens College was the fifth report on civic universities, which appeared in August 1874. However, the other reports sometimes touched on matters of importance or interest to Roscoe. The first volume, on state-funded institutions in London such as the Science and Art Department, the Royal College of Chemistry, and the School of Mines, was of interest to Roscoe for two reasons. He was heavily involved with the examinations set by the Science and Art Department, and the potential expansion of colleges such as the Royal College of Chemistry was a threat to civic universities such as Owens College. It was even more of a threat to nearby University College, and Williamson spoke against increased state funding for teaching science. The commission recommended the move of the Royal College of Chemistry to South Kensington and as a result of this report, the building in South Kensington intended to house the Royal School of Naval Architecture became the Normal School of Science with the aim of training science teachers, which was headed by Huxley. The second report was on scientific and technical instruction in elementary schools, an issue which was considered again by the later Samuelson Commission of which Roscoe was a member. The third Report on Oxford and Cambridge was of less interest to Roscoe, except insofar as it examined the issue of where research should be carried out, as well as a topic dear to Roscoe's heart: the importance of a professor being a researcher as well as a teacher.

Museums were also close to Roscoe's heart and this was the topic of the fourth report. The commission supported the separation of the natural history collections of the British Museum from the main museum and its relocation to South Kensington. It also proposed that the collections of the Patent Office Museum should be merged with the science collections of the South Kensington Museum. Although this merger did not occur for almost another decade, the proposal that the science collections be expanded led to the scientific loan exhibition of 1876. Leaving aside the fifth report (and the related seventh report) for now, the sixth report dealt with the teaching of science in public and endowed schools. While not directly connected with Roscoe, it did encourage science teaching in public schools, most notably Rugby and Eton.

The final report, published in June 1875, was in many respects the most contentious. The early reports had dealt only with scientific and technical education, but

---

[15] *Report of Royal Commission on Scientific Instruction and the Advancement of Science, Volume 1, First, Supplement, and Second Reports, with Minutes of Evidence and Appendices,* P.P. 1872 (C. 536); *Report of Royal Commission on Scientific Instruction and the Advancement of Science, Volume 2, Minutes of Evidence, Appendices, and Analyses of Evidence,* P.P. 1874 (C. 958); *Report of Royal Commission on Scientific Instruction and the Advancement of Science, Volume 3, Minutes of Evidence and Appendices, Analyses of Evidence, Index to the Eight Reports (with their Appendices) and the General Index to the Evidence; and to the Appendices to the Evidence Given in Volumes I–III,* P.P. 1875 (C. 1363); *Third Report of Royal Commission on Scientific Instruction and the Advancement of Science,* P.P. 1873 (C. 868); *Fourth Report of Royal Commission on Scientific Instruction and the Advancement of Science,* P.P. 1874 (C. 884); *Fifth Report of Royal Commission on Scientific Instruction and the Advancement of Science,* P.P. 1874 (C. 1087); *Sixth Report of Royal Commission on Scientific Instruction and the Advancement of Science,* P.P. 1875 (C. 1297); *Seventh Report of Royal Commission on Scientific Instruction and the Advancement of Science,* P.P. 1875 (C. 1297); and *Eighth Report of Royal Commission on Scientific Instruction and the Advancement of Science,* P.P. 1875 (C. 1298).

this report covered the second part of its remit: the advancement of science. How indeed was science to be advanced, and how should the state assist its advancement? The report surveyed the current support given by the government to science, and it proved to be both extensive and piecemeal. The various institutions and less formal arrangements that linked the state to science had grown up organically over the previous 200 years and hence had become a haphazard jumble. There was strong support for new scientific institutions, especially an astrophysical observatory which was close to the heart of Lockyer, the Commission's Secretary. There was also a demand that the state should fund individual scientists. There was a call for a government ministry of science to be established. The report supported the idea of the astrophysical observatory and proposed that the Royal Society's current grant for research of £1,000 to be increased to £5,000 (£600,000 in today's terms; tiny compared with modern science budgets). In the end, unlike the earlier commissions, the Devonshire Commission did not result in a specific act of Parliament or even a series of concrete actions. The main result of the commission's deliberations over and above what might have happened anyway (such as the move of the Royal College of Chemistry and the British Museum's natural history collection to South Kensington) was the increase in the grants disbursed through the Royal Society. These grants were eagerly sought (although the number seeking them soon decreased) and doubtlessly much valuable research was supported in this way, as leading scientists such as William Thomson received grants from the scheme. However, the impact of the Devonshire Commission after five years of work did not transform British science. There are two reasons why this was the case. First the commission had been set up by the Liberals, and the Conservatives (under Benjamin Disraeli) came to power in February 1874 while the commission was still at work. While this may seem a setback, and perhaps was in some ways, the leading Conservative peers, the 15th Earl of Derby and the 3rd Marquess of Salisbury, were more enthusiastic supporters of science than Gladstone. However, what the Conservative cabinet lacked were the strong advocates of reform, such as Playfair and Anthony Mundella. Furthermore, Derby was Foreign Secretary and Salisbury was Secretary of State for India. Hence, they were not in a position to directly influence matters. The Permanent Secretary to the Treasury, Ralph Lingen (needless to say, a Classics graduate from Oxford), was strongly opposed to additional expenditure on science (perhaps influenced by his experience of the "payment by results" system at the Education Office while he worked there), which led to a stalemate that had to be broken by Financial Secretary William Henry Smith (of the eponymous news-agent firm). Perhaps for this reason he was elected a Fellow of the Royal Society in 1878. The other reason for the Devonshire Commission's lack of success was the impossibility of it fulfilling all the hopes placed on it. A major science-funding program would have required government spending on goals not hitherto seen as the business of the state on an unimaginably large scale, even if British governments were already moving rapidly away from the purest form of Gladstonian prudence.

Let us now turn to the involvement of Roscoe with the Devonshire Commission and its impact on his activities. Roscoe and Greenwood gave evidence to the Royal Commission on 31 March 1871 as part of its review of the metropolitan universities (University College London and King's College), Owens College, the College of

Physical Science in Newcastle-upon-Tyne, and the Catholic University of Ireland. Their evidence came during a period of intense activity for Owens College; namely, drafting documents for the two parliamentary bills and planning the move to the Oxford Road site. Finance was a key factor in both bills and thus figured prominently in their evidence. Greenwood focused on the wider challenges facing the college and the changes to its governance and administration. The Royal Commission's first report contained not only Greenwood's oral evidence, but also important documents drafted by the college and used during their deliberations on the constitution of the College and the "College extension," as well as detailed financial information.[16] Roscoe gave a detailed account of the chemistry department and compared the provision at Owens with universities in Germany. He benefited from the visit he had made to Germany with Greenwood in the summer of 1868. His evidence contrasted the "hand to mouth" existence of Owens College with the expansive laboratory facilities, the numerous well-paid staff, and the focus on research in many German universities, all funded by the government of the various states.[17] His published evidence includes a list of students working in the chemistry laboratory, their age, duration of their studies, and purpose of their study or proposed occupation. Roscoe was questioned about his role in spreading scientific instruction among the local population and was able to draw attention to the series of Science Lectures organized in Manchester with the assistance of several other scientists.[18]

When the fifth report was published in 1874, the Commissioners were clear that Owens College had made strenuous efforts and had established a claim to aid from the state, and therefore recommended:

> that the Owens College should receive assistance from Government, both in the form of a capital sum, to be regarded as a contribution towards its Building Fund, and also in the form of an Annual Grant, in aid of its working expenses, with the especial view of enabling it to complete the curriculum of studies by the establishment of New Chairs.[19]

Unfortunately, the recommendation had little immediate effect because it was not until 1890 that Owens College (as Victoria University) received its first government grant of £2,250.[20] Nevertheless, the evidence by Greenwood and Roscoe helped raise the profile of Owens College and provided details (and challenges) of its important higher education role in the North of England during a period when the College was trying to advance its university status.

---

[16] This is probably the best source of information on the proposed changes to Owens College that informed the parliamentary bills in 1870 and 1871. See *Report of Royal Commission on Scientific Instruction and the Advancement of Science, Volume 1, First, Supplement, and Second Reports, with Minutes of Evidence and Appendices*, P.P. 1872 (C. 536), 475–497.

[17] *First, Supplement, and Second Reports*, 501–508.

[18] Ibid., 513.

[19] *Fifth Report of Royal Commission on Scientific Instruction and the Advancement of Science*, P.P. 1874 (C. 1087), 21.

[20] H. B. Charlton, *Portrait of a University 1851–1951* (Manchester: Manchester University Press, 1951), 149.

Growth of Technical Education

During the period of these parliamentary inquiries, attention was beginning to focus more closely on technical education. This was a subject of growing interest to Roscoe because of his increasing involvement with industry and industrialists through his work in the chemistry department at Owens. This interest was further encouraged by his association with several leading figures in an emerging technical education (or instruction) movement, including Bernard Samuelson, Philip Magnus, and Thomas Huxley.

The Society of Arts removed science subjects from their examinations in 1870 since they were replicating examinations of the Department of Science and Art at South Kensington. Having just joined the council of the Society of Arts, Major Donnelly, the inspector for science in the Science and Art Department, then proposed that the society "could continue its useful work by instituting examinations in technology, with papers designed specifically to supplement the science papers set by the Department," thus restoring the emphasis on technical education.[21] He went on to suggest that the Livery Companies of the City of London should support technical education because one of the primary purposes of their wealthy Guilds was to support craft apprentices. This proposal was approved by the Society of Arts.[22] After protracted discussions, facilitated by the Prince of Wales and the Lord Mayor of London, a conference agreed in July 1873 to build a teaching institute in the City of London. In the following year, a working party chaired by the Liberal politician Roundell Palmer (recently ennobled as Baron Selborne) recommended allocating £10,000 to the institute and £10,000 toward scholarships and support for existing metropolitan and provincial technical colleges. Selborne's working party also drew up the constitution of the new college.

The City and Guilds of London Institute for the Advancement of Technical Education (CGLI) was formally established in 1878 and the following year took over the technological examinations of the Society of Arts. The original plan was to have a junior technical college in the City and an advanced university-level institute in South Kensington. In May 1880 Philip Magnus (Figure 9.2) was appointed Secretary and Director of CGLI, which was then still without a building. This was a remarkable appointment given his position as a rabbi at a Reform synagogue, although he had lectured at Stockwell Training College (in the theory and practice of education) and had published *Lessons in Elementary Mechanics*.[23] Progress on the advanced institute was very slow, and at the second annual meeting of the governors of the City and Guilds Institute in 1880 it was agreed that the Commissioners would provide £50,000 toward the building and £50,000 annually to maintain the university. The laying of the foundation stones of the two institutions followed in 1881: Finsbury Technical College on 10 May and the advanced institute at South Kensington on 18 July. Finsbury Technical College opened in February 1883 and the Central Institution (as it was now called)

---

[21] Cardwell, *Organisation of Science*, 126–127. For Sir John Donnelly, see *ODNB*.
[22] Cardwell, *Organisation of Science*, 128.
[23] Bill Baily, "Sir Philip Magnus (1842–1933)," *ODNB*.

Figure 9.2 Sir Philip Magnus.
Image courtesy of the Royal Society of Chemistry Library.

opened in 1884, alongside the Royal School of Mines (RSM) and the Royal College of Chemistry (RCC) in South Kensington.[24]

In the period after the report of the Samuelson select committee and while the Devonshire Commission was still gathering evidence, two powerful pleas for the importance of academic chemistry were made. In October 1870, at the inauguration of the Science Faculty of University College London, Alexander Williamson, Roscoe's former professor and employer, gave a speech entitled simply "Plea for Pure Science." This was not, as the title might suggest, a justification for pure science rather than

[24] In 1907 the RSM and the RCC were incorporated into Imperial College, and the Central Institution became the City and Guilds College.

applied science in academia (a matter with which Roscoe was much concerned), nor was it simply "an outline of the benefits which science confers on mankind," as Williamson claimed at the beginning of his lecture. He began by stating a need for opposing points of view: "that every worker know and acknowledge the point of view at which he is placed and from which he sees." He then postulated that there were two schools of thought in education. One side emphasizes efficiency and the acquisition of skills. The other side puts the stress on general principles and the development of one's mental powers. Hence the first party wishes a young man to have technical instruction, by which Williamson meant professional education, an apprenticeship or pupilage. By contrast, the second party (of which Williamson was clearly a member) wishes a young man, having acquired a knowledge of general principles "accompanied by practice in the methods of their application to simple cases," to be able to learn a more specific aspect of their business, and "improving its details" in later life. This was precisely the model of technical education advocated by Roscoe at Owens College. Williamson then goes on to show how the science teaching at University College London was an excellent example of the latter, more enlightened method of training the country's young men. Hence his lecture was a head-on attack on narrow professional training and especially apprenticeships, even presumably when this specialized training was obtained at technical colleges.

Roscoe's approach was similar to Williamson's, but at the same time his argument was subtly different in that he put the emphasis on scientific research as the best way of training young men for a technical career. Since his appointment as Professor of Chemistry at Owens College in 1857, Roscoe had held strong opinions on the relationship between the courses at Owens College and industry. These opinions were informed through his experiences studying with Robert Bunsen at Heidelberg, his visits and interactions with other German scientists, and his knowledge of the German education system, with its universities and polytechnics. The universities were focused primarily on research and advancing knowledge and understanding in science, while the polytechnics were institutions concerned with technical education and training for industry. Roscoe had taken numerous opportunities to emphasize the importance and benefits of studying the physical sciences (but not to the exclusion of other subjects) and then later to promote original research as a means of education; again, not to the exclusion of research in other subjects, including the arts. His Introductory Lecture following his appointment at Owens in 1857 was titled *The Development of Physical Science*, and Roscoe not only provided a historical survey and the benefits of studying the physical science for civilization and Britain, but also drew attention to the important role of laboratory instruction alongside lectures.[25]

By contrast, in his important lecture "Original Research as a Means of Education" at the formal opening of the extension of Owens College in October 1873, Roscoe stressed the importance of original research:

> The subject of the value of original scientific investigation may be considered from many points of view. Of these, that of the national importance of original research is

---

[25] Henry Enfield Roscoe, *Introductory Lecture on the Development of Physical Science* (London: Longman, Brown, Green, Longmans, and Roberts, 1857).

the one which naturally first engages attention; and it does not take long to convince us that almost every great material advance in modern civilisation is due, not to the occurrence of haphazard or fortuitous circumstances, but the long-continued and disinterested efforts of some man of science.[26]

He continued on the theme of national importance:

> As soon as English people see clearly the imperious necessity for encouraging, stimulating and upholding original research as contained in the seeds of our future national position, they will not be behindhand in securing the free growth of those seeds. It is, therefore, the bounden duty of all those whose employment or disposition has led them to feel the truth of this great principle, to leave no stone unturned to make widely known and keenly felt the importance of the national encouragement of original investigation.[27]

He briefly surveyed the important role that original research was playing in France, Germany, and United States, before pointing out how little is known about the chemical elements and the opportunities that lie ahead:

> Among the sixty-three different elements of which the earth, so far as we know, is made up, there are many which have been found only in the minutest quantity. A few only of these rare substances are employed in the arts and manufactures, or are known to play any part in the economy of nature; the rest are substances of interest at present only to the scientific chemist. It would, however, be presumptuous on our part were we to assume that the existence of these bodies is a matter of no moment, for we are constantly learning that substances hitherto supposed to be useless are of the most vital importance. Hence it is obviously our duty to get to know all we can about the properties of each, even the rarest, of these elementary bodies, and especially about their relation to, and mode of action on, the other elements.[28]

Again, Roscoe emphasized the importance of research in all subjects, not just in the physical sciences, but noted the particular benefits that research in chemistry could bring to industry and the prosperity of the Britain.

The Samuelson Commission

The second Gladstone Liberal government replaced the Conservative government of Disraeli in April 1880. Gladstone appointed the ardent educational reformer Anthony Mundella as Vice-President of the Committee of the Council and hence the de facto education minister.[29] Between 1880 and 1886, Mundella "set about a series of inquiries

---

[26] H. E. Roscoe, "Original Research as a Means of Education," in *Essays and Addresses by Professors and Lecturers of Owens College, Manchester* (London: Macmillan, 1874), 21–57, on 21.

[27] Ibid., 23.

[28] Ibid., 54.

[29] Cardwell, *Organisation of Science*, 132.

and reorganizations covering the whole sphere of educational activity," especially technical education. As a manufacturer in Leicester, he had been struck by the differences with Leicester and the similar town of Chemnitz in Saxony, where he also had a factory. Keen to alert his fellow politicians and other policymakers about the superior nature of German education (as he saw it), he induced his agent in Chemnitz, Henry Felkin, to gather information about the educational situation in Saxony. This was published as *Technical Education in a Saxon Town* in May 1881 for the Council of CGLI, thereby showing the close links between that institution, Magnus, and Mundella.[30] The preface by an anonymous author associated with Gresham College remarks that the aim of the pamphlet was to show:

> what kind of provision is being made in Germany to meet these new wants, may help those who are endeavouring to promote technical education in England to avoid errors resulting from inexperience, to appreciate the educational requirements of manufacturers and artisans, and to understand the best means of providing for these requirements.

He admitted, however, "What will strike every one who carefully examines the figures contained in the following pages, is the great cost of a well-organised system of technical instruction."[31] Felkin gave a thorough survey of the education provided in Chemnitz from elementary schools upward, but played particular attention to the Royal Technical Institution, with floor plans and details of its courses. This institution had replaced the "old apprenticeship system [which had] become obsolete and gone overboard for ever";[32] a provocative statement given the importance attached to the apprenticeship by British manufacturers and trade unions. Felkin's little book fueled public concern about the English technical education system falling behind its German counterpart (as it had been intended to do). This anxiety enabled Mundella to persuade Gladstone to establish the Royal Commission on Technical Instruction in August 1881, under the chairmanship of Samuelson, who had chaired the earlier select committee on the same topic. Given the existence of this committee and the Devonshire Commission, one might wonder why this royal commission was needed. However, given the growing anxiety (partly fomented by Mundella) about English technical education falling behind its rivals, the remit of the commission was:

> to inquire into the Instruction of the Industrial Classes of certain Foreign Countries in technical and other subjects, for the purpose of comparison with that of the corresponding classes in this Country; and into the influence of such Instruction on manufacturing and other Industries at home and abroad.[33]

The members of the commission were a distinguished group from politics, academia, business, and education, including Roscoe himself, seasoned by his experience

---

[30] Henry M. Felkin, *Technical Education in a Saxon Town* (London: Kegan Paul, 1881).
[31] Ibid., 3.
[32] Ibid., 30.
[33] *First Report of the Royal Commissioners on Technical Instruction*, P.P. 1882 (C. 3171), 3.

of serving on the Royal Commission on Noxious Vapours between 1876 and 1878. John Slagg was a Manchester businessman and a Liberal MP for Manchester (1880–1885) who became President of the Manchester Chamber of Commerce.[34] Swire Smith had served an apprenticeship in the textile industry in Keighley before starting his own business, and later was appointed Secretary of a new trade school council in Keighley and the local Mechanics' Institute; in 1873 he published a widely circulated book, *Educational Comparisons*, that reviewed education provision in France, Germany, and Switzerland.[35] William Woodall was Liberal MP for Stoke on Trent (1880–1886) and senior partner in a china-manufacturing business at Burslem.[36] The architect and art historian Gilbert Redgrave was Secretary of the Commission; he later became a schools inspector and Assistant Secretary to the Board of Education in 1900. Roscoe later grumbled that unusually the Commissioners had to pay their own expenses because the government was only covering the secretarial and printing costs.[37]

The Commissioners divided themselves into small groups to undertake visits to major cities in France, Switzerland, Germany, Austria, Belgium, Holland, and Italy to see how schools and other educational institutions approached "instruction of the industrial classes in technical and other subjects."[38] Besides taking evidence from a number of British witnesses, the Commissioners also visited educational establishments in London, Oxford, Cambridge, Manchester, Liverpool, Oldham, Barrow, Birmingham, Leeds, Sheffield, Bradford, Keighley, Saltaire, Macclesfield, Burslem, Nottingham, Bristol, Bedford, Kendal, Edinburgh, Glasgow, Dublin, Belfast, and Cork. Besides the Commissioners, several other respected and knowledgeable experts were asked by the chairman to undertake investigations on behalf of the commission and submit reports on their findings.

With his knowledge and experience of Germany, Roscoe took responsibility for investigating "science-teaching in that country and generally on the Continent." The Commissioners' travels through Britain and abroad "were of a most interesting and successful character":

> our object being to ascertain how far the systematic education given to the workmen, to the overseers or foremen, and to the masters abroad, and the comparative lack of a system at home, told in favour of the industrial progress of continental nations and against that progress with us. The evidence we gathered from all quarters abroad, from manufacturers and workmen alike showed us that the beneficial effect of technical-school training upon industry was universally admitted; and our visits proved that the sums of money voted by both State and municipality were far in excess of what were then applied to British education.[39]

---

[34] "Obituary," *The Times*, 8 May 1889, 7.

[35] For Sir Swire Smith, see *ODNB*.

[36] For William Woodall, see *ODNB*.

[37] Henry Enfield Roscoe, *The Life and Experiences of Sir Henry Enfield Roscoe DCL, LLD, FRS, Written by Himself* (London: Macmillan, 1906), 190.

[38] *Second Report of the Royal Commission on Technical Instruction, Volume 1*, P.P. 1884 (C. 3981), 15.

[39] *Second Report of the Royal Commission on Technical Instruction*, 191.

While the definition of technical instruction proved contentious, Roscoe for his part was very clear on the nature of technical instruction under investigation by the commission:

> No one whose opinion is of value pretends that the technical school can supplant the workshop or the factory. To be an adept in any handicraft the experience of long continuous work in the shop or at the loom is, of course, essential. The school teaches the principles, scientific, mechanical, or artistic, upon which the industry is based. The workshop puts those principles into practice. The school does not consider economy of production; at any rate it cannot carry out economy in detail; whilst in practice economy, that is, the proper relations between production and expenditure and between quality and quantity, is an essential condition of successful work.[40]

The commission published two reports. The first report in 1882 was only 30 pages long and was just a preliminary survey. The second report, published two years later, was much more substantial. In addition to the main report, there were four other volumes. The second volume contained reports by Herbert Jenkins, Secretary of the Royal Agricultural Society of England, on agricultural education in North Germany, France, Denmark, Belgium, Holland, and the United Kingdom, and by William Mather, on technical education in the United States and Canada. The third volume comprised Mather's report on technical education in Russia; a report by Thomas Wardle, a silk dyer who collaborated with William Morris, on the English silk industry; and a report on technical education in Ireland by William Sullivan, a chemist who became President of Queen's College, Cork (now University College Cork). Evidence relating to Ireland was also the subject of the fourth volume, and the fifth volume contained reports from other countries. It is perhaps a little curious that a Commission whose origins lay in a book about German education should devote so much space to Irish education and seemingly little to Germany.

Completing the final draft of this report and its companion volumes must have been a lengthy task that cut into vacation periods, as Roscoe relates in his autobiography:

> With regard to the preparation of the Report—a most difficult and tedious task—had I not invited several members of the Commission and our indefatigable secretary to visit me during my summer holidays, on more than one occasion at Graythwaite, on Windermere, it would, I believe, never have been licked into shape. Here we worked steadily at it, and at the finish Mr. Swire Smith, Mr. Magnus and myself gave the last touches to our recommendations at Sir Bernard Samuelson's country house in Devonshire.[41]

The commission's recommendations included:

> Rudimentary drawing be incorporated with writing as a single elementary subject
>   Advocated more teaching of agriculture and craft work

[40] Roscoe, *Life and Experiences*, 188.
[41] Roscoe, *Life and Experiences*, 202.

Advocated more teaching of science and art in training colleges

More support for [City and Guilds of London Institute]

Greater powers for local authorities to establish more technical and secondary schools

Advocated less part-time employment for children

Recommended more systematic training for young workers in work schools and that employers and trade organizations should make financial contributions to help realise this recommendation.[42]

Broadly, as Margaret Gowing concluded, while "their reports noted great educational advance at home, their evidence showed still greater progress abroad, and they urged, above all, local authority provision of secondary and technical schools."[43]

These recommendations attracted broad support, and *The Times* was sympathetic:

whatever may be the progress of other nations in technical education and in manufactures, our own industries also are full of vigour; that we already possess considerable opportunities for theoretical instruction in the technical sciences and in art applied to industry; that these opportunities are capable of increase on their present lines; that the value of such instruction and the necessity for its further development are felt by those most directly interested; and that this development is making sure, though gradual and perhaps somewhat tardy, progress.[44]

But with so much detailed information covering so many different aspects, it was not surprising that various criticisms were expressed: pointing out where good practice was already undertaken; the lack of adequate financial support, especially from government; and the lack of a coordinating and regulating body. Among workers there was strong opposition to the principles of technical education, as the historian, Bernard Cronin has pointed out:

The industrial or craft apprenticeship system was a form of technical education in relation to which new forms of technical education were to be developed. Technical education, as perceived by those for whom it was intended, the workers, was not merely a system of theoretical training instituted as an adjunct to practical work carried on in the workshops. In the separation of theoretical and practical elements in a definition of technical education, unquestioned importance is given to the former.[45]

After all, since a trade or industry had developed in the past under the apprenticeship system, it was better to retain this tradition and rely "on the skill and energy and

---

[42] Richard Evans, *A Short History of Technical Education*, chapter 8, "The Developments at the End of the 19th Century" (June 2009), https://web.archive.org/web/20230426153622/https://technicaleducationmatt ers.org/2009/06/17/chapter-8-the-developments-at-the-end-of-the-19th-century/.

[43] Margaret Gowing, "Science, Technology and Education: England in 1870: The Wilkins Lecture, 1976," *Notes and Records of the Royal Society of London* 32 (July 1977): 71–90, on 76.

[44] "The Technical Education Commission," *The Times*, 16 May 1884, 4.

[45] Bernard Cronin, *Technology, Industrial Conflict and the Development of Technical Education in 19th-Century England* (Aldershot, Hants: Ashgate, 2001), 4.

industry" of the worker.[46] But through the nineteenth century the mode of production became more technological, necessitating a transition from an artisan workforce to a technically trained one.

Many manufacturers were also resistant to the changes embodied in technical education. They were concerned about outside interference in the workings of industry and "were not going to encourage something which would bring all the workmen from the different works together to discuss matters in which trade secrets were involved."[47] But on the other hand, there were enterprising manufacturers such as William Mather of Messrs. Mather and Platt who organized a tour of the company's private technical evening school for a group of the Technical Instruction Commissioners. The school had 68 scholars and provided science teaching (with drawing exercises) for the apprentices employed in the works. Echoing Roscoe's position on the relationship between workshop and school (separate but related):

> In Mr Mather's opinion, you must bring the school to the workshop; you cannot bring the workshop to the school. Bringing the school to the workshop is simple and inexpensive. The teachers here are draughtsmen in the works ... he [the teacher] knows what each person is working at each day, and has the opportunity of pointing out something connected with work he is doing. The teaching has an actual bearing on his every day work. The students are rewarded not only for proficiency in drawing, but for regular attendance, and actual proficiency in their manual work. It is also a condition of employment that they should be regular in their attendance here.[48]

The Mather and Platt school was an enlightened example but unique to Lancashire, and confirmed the voluntary approach supported by most manufacturers. This stance was consistent with the laissez-faire approach of governments toward industry, as Cronin concluded:

> Commitment to laissez-faire doctrines was a prominent feature of Victorian Britain's economic affairs and was reflected in attitudes to technical education ... and [the Commission of 1884] revealed the government's reliance upon the workings of the market with respect to technical education, demand for technical education would call forth appropriate supply; if demand was not forthcoming supply would respond accordingly.[49]

Adherence to the philosophy of market forces on the part of government and industrial leaders would play a major role in the shaping of legislation on technical education for the remainder of the nineteenth century and would thwart the ambitions of the leading figures of the technical education movement such as Samuelson, Magnus and Roscoe.

---

[46] Stephen Cotgrove, *Technical Education and Social Change* (London: George Allen & Unwin, 1958), 24.
[47] Evidence by Col. John F.D. Donnelly (Science and Art Department), *Second Report of the Royal Commission on Technical Instruction, Volume 3*, P.P. 1884 (C. 3981-II), Minute 2862, 286.
[48] *Second Report of the Royal Commission on Technical Instruction, Volume 1*, P.P. 1884 (C. 3981), 429.
[49] Cronin, *Technology, Industrial Conflict*, 163–164.

However, for Roscoe himself there was one consolation. Samuelson sent Roscoe an appreciative letter commenting on his outstanding contribution to the Commission, remarking that during the period that the Commission was active, he had "learnt that your loyal support as a colleague and your advice on all matters of business have been almost of equal value as your ability as a man of science." Given his sterling contribution to the Samuelson Commission, Mundella put Roscoe's name forward to Gladstone for a knighthood, not only for his "great and valuable services on the Technical Commission" but also in recognition of "your distinguished service to science and education," which was duly communicated to Roscoe by Gladstone in June 1884. In due course, he would probably have been given a knighthood for his presidency of the British Association for the Advancement of Science in 1887, something that Roscoe implicitly acknowledges by putting his knighthood on the facing page to the one about his presidency of the British Association in his autobiography.[50] Nevertheless, his knighthood does show his massive contribution to the Samuelson Commission and to the development of technical education more generally. Thomas Huxley's letter of congratulation had an amusing P.S.: "Shall I tell you what your great affliction henceforward will be? It will be to hear yourself called "S'enery Roscoe" by the flunkies who announce you."[51]

The publication of the Commission's reports and his knighthood did not end Roscoe's commitment to technical education. Like several of his fellow Commissioners, Roscoe undertook many speaking engagements throughout Britain between 1882 and 1885. While acknowledging that he "was not behind my colleagues in their missionary work ... to make the conclusions and recommendations arrived at widely known," Roscoe nevertheless challenged more fundamentally the psyche of the English nation:

> English characteristics have been, and still are, eminently practical. We prided ourselves, and still do, I presume, on being a practical nation, and we have been rather in the habit of looking upon professors and schoolmasters as theoretical kind of people, who are not up to very much good in the battle of life. Now, I think we are beginning to acknowledge that theory is only a systematic practice—and practice without theory is very often poor practice—and we recognise that science is but ordered knowledge, and that if we are to succeed in the great endeavours which we as a nation have to make, if we are to keep abreast of the progress other countries are making, and are to preserve the position of superiority for our manufactures and trade we have enjoyed in the past, every effort must be made to put our educational house in order and to see that the sciences upon which the manufactures and industries of this country are based are taught to the people of every class from top to bottom.[52]

Roscoe's mission to improve technical education in Britain was a major achievement, which stands on equal terms with his important contributions to Owens College. Severing his ties with Owens College and the Victoria University in 1886

---

[50] Roscoe, *Life and Experiences*, 228.
[51] Ibid., 229.
[52] Ibid., 203–204.

did not diminish his determination and commitment to bring about educational reform. Having been a strong and effective advocate on behalf of Owens College and the chemistry department, Roscoe could contemplate now a broader canvas where changes could bring national benefits. During his time as an MP (1885–1895), when legislation on technical instruction was being promoted more forcibly, Roscoe continued to take every opportunity within Parliament and beyond to promote the major reforms that he (and others) felt were so urgently needed to improve science and technical education. His contributions to further Royal Commissions and Select Committees, his commitment to secure technical education legislation, and his position as Secretary of the National Association for the Promotion of Technical and Secondary Education are all testament to his determination to see Britain retain its position as a major trading nation, able to compete on an equal footing with Germany and the United States.

10

A Chemist in the House

The General Election of 1885

When Roscoe was selected as the Liberal candidate for South Manchester in 1885, he claimed that he had not hitherto been involved in politics.[1] However, this statement belies his family's strong involvement with the Liberal Party and its predecessors, at least in part because of their Unitarian beliefs. His grandfather William Roscoe had been the MP for Liverpool in 1806–1807, and his father-in-law Edmund Potter was a Liberal MP for Carlisle between 1861 and 1874. He was a close friend of many leading Manchester Liberals, especially C. P. Scott, the editor of the *Manchester Guardian*, and he was a member of the Manchester Reform Club, the social center of Manchester Liberalism.[2] Indeed, he must have had some connection with Liberal politics to be at the meeting where he intended to support the candidature of a friend and was unexpectedly adopted as the candidate. However, before considering his candidature, we have to explore how this election came about.

The Liberal Party had been in government under William Gladstone since 1880.[3] The government's main focus had been on parliamentary reform in the form of the Corrupt and Illegal Practices Prevention Act of 1883 and the Representation of the People Bill of 1884 (known as the Third Reform Bill). The bill easily passed the Commons where the Liberals had a large majority, but was rejected by the Conservative-controlled House of Lords because of fears about the impact of the extension of the suffrage to all male house dwellers in rural seats. This defeat increased the antipathy of many Liberals (including Roscoe) to the upper chamber. Under pressure from Queen Victoria, Gladstone, and the Conservative leader, the Marquess of Salisbury, the Conservatives agreed to pass a reform bill, but insisted that the existing parliamentary seats had to be replaced for the most part by single-member

---

[1] Henry Enfield Roscoe, *The Life and Experiences of Sir Henry Enfield Roscoe DCL, LLD, FRS, Written by Himself* (London: Macmillan, 1906), 257–259.

[2] Roscoe, *Life and Experiences*, 270 (for Scott) and 309 (for his chairmanship of the Manchester Reform Club).

[3] The political history of this period and the history of the Liberal Party has been taken from G. R. Searle, *A New England? Peace and War, 1886–1918* (Oxford: Oxford University Press, 2005), which is the only general history of England in this period that mentions Roscoe; the first three chapters of Searle, *The Liberal Party: Triumph and Disintegration, 1886–1929* (Basingstoke: Macmillan, 1992); H.C.G. Matthew, *Gladstone, 1809–1898* (Oxford: Oxford University Press, 1999); Richard Shannon, *Gladstone: Heroic Minister, 1865–1898* (Harmondsworth, Middx: Allen Lane, 1999); and Ian St John, *Gladstone and the Logic of Victorian Politics* (London: Anthem Press, 2010). D. A. Hamer, *Liberal Politics in the Age of Gladstone and Rosebery: A Study in Leadership and Policy* (Oxford: Clarendon Press, 1972), covers the period when Roscoe was in the House of Commons, but typically does not mention him; however, it discusses the issues that Roscoe adopted, for example temperance and Welsh disestablishment. Older works should not be overlooked as they concentrated more on political history, notably R.C.K. Ensor, *England, 1870–1914* (Oxford: Clarendon Press, 1968) and Philip Magnus, *Gladstone: A Biography* (London: John Murray, 1970).

Henry Enfield Roscoe. Peter J.T. Morris and Peter Reed, Oxford University Press. © Oxford University Press 2024.
DOI: 10.1093/oso/9780190844257.003.0010

constituencies. The Conservatives hoped that these new suburban seats would largely vote Conservative (so-called Villa Toryism). Hence the Representation of the People Act of 1884 was followed by the Redistribution of Seats Act of 1885. At this point, the Liberal Party was divided over the Irish Parliamentary Party's demand for Home Rule. Charles Stewart Parnell, the leader of the Irish Parliamentary Party (IPP), which represented the supporters of Home Rule in Ireland, transferred his support to the Conservatives, and the Liberal government was defeated over its budget. The Conservatives now took office under Salisbury with the support of the IPP. A General Election was inevitable but was delayed until the new electoral register created by the 1884 Reform Act could be implemented. It was eventually held between 24 November and 18 December 1885.

The Redistribution of Seats Act of 1885 largely (but not completely) replaced multi-member constituencies with single-member constituencies. This changed the nature of election campaigns, as it removed the basis for pre-election deals whereby the seats were parceled out beforehand, either by the Whigs and Radicals in the Liberal Party or by the Conservative and Liberal Parties. This meant that elections in many constituencies now became more competitive. Furthermore, constituencies which had covered towns and cities as a whole were replaced by constituencies which represented an inner city and its suburbs. It was these new suburban seats that the Conservatives were eager to capture, and thereby replace rural seats that might be lost because of the extension of the male suffrage to better-off agricultural workers.

The introduction of single-member constituencies also meant a large increase in the number of constituencies. In Manchester, there had been a single three-member constituency, but for the 1885 election, there were six single-member constituencies. The boundaries of the Manchester constituencies were expanded, mostly in South Manchester, which saw the transfer of Moss Side, Rusholme, and part of Gorton from Lancashire. After Richard Pankhurst had stood as an independent Radical against the wishes of the Manchester Liberal Association in 1883, the Radicals had become more influential in the local party.[4] Furthermore, the older, generally more moderate, leading figures in the Manchester party had moved away from Manchester—in one case (Sir Thomas Bazley) as far as Stow-on-the-Wold in the Cotswolds, which was over a hundred miles away!

Candidates had to be found for these new constituencies, and they were now selected by entirely new constituency associations. In May 1885, Roscoe was invited to propose Charles Schwann at an adoption meeting for South Manchester.[5] Schwann was a Manchester merchant of German ancestry who was born in London. To his astonishment, however, Roscoe was put forward as the candidate, rather than Schwann. At this point, Roscoe could have said he was not interested or that he was unworthy of the honor. However, he accepted this sudden change of plan without needing to be persuaded, although he did think that someone with a commercial background would

---

[4] James R. Moore, *The Transformation of Urban Liberalism: Party Politics and Urban Government in Late Nineteenth-Century England* (Aldershot, Hants: Ashgate, 2006), chapter 1, "The Rise of Manchester Radicalism," 25–52. Moore (b. 1972) should not be confused with the historian of science James R. Moore (b. 1947).

[5] Ibid., 48.

be more appropriate.[6] As the constituency association's letter of thanks to Roscoe in 1895 noted, the seat was effectively a university seat as well as a center of trade and manufactures, so the nomination of a leading academic from the local university was not inappropriate.[7] Roscoe apologized to Schwann for taking his place, but Schwann was happy at this turn of events (and probably even connived in the "ambush").[8] In the event, Schwann stood for the Liberal Party in North Manchester for which he had already been tapped. He lost that election, but then held the seat between 1886 and 1918.

South Manchester was a good constituency for Roscoe, as he lived there for many years. It comprised the area around Oxford Road, where Owens College was now situated, Moss Side, Victoria Park, Rusholme, Fallowfield, and Kirkmanshulme.[9] The area had a strong nonconformist character and inclined toward support for the social reforms favored by Roscoe.[10] It also had a large Welsh population, who largely supported the Liberals. Furthermore, leading Liberals such as Scott and the former MP John Slagg were his neighbors. However, James Moore sees it as the typical well-off suburban constituency that the Conservatives hoped to gain as a result of this reorganization of constituencies.[11]

Roscoe also received offers of nomination from Eccles (which was won by the Conservatives with a narrow majority) and the university seat of Edinburgh and St. Andrews.[12] This latter constituency might have been more congenial for Roscoe, who eventually became irritated by the need to visit his constituents, but he lacked any connection with the Scottish universities. In theory he could have stood in more than one constituency, but clearly decided not to do so. Roscoe was probably proposed by the current MP Lyon Playfair. The distinguished surgeon John Eric Erichsen of University College Hospital stood for the Liberals, but was narrowly defeated by the Conservative lawyer John Macdonald, who was a local man, and the seat remained in Conservative or Unionist hands until it was abolished in 1918.

A key factor in his acceptance of the nomination (and here the timing was very important) was the unexpected death of his son Edmund at Oxford at the beginning of 1885 as a result of a ruptured appendix, which caused fatal peronitis.[13] His death left Roscoe and his wife completely bereft. In a letter to Thomas Huxley dated 18 January 1885, Roscoe said, "I feared that this death of her only son—part of her very soul would have nearly killed my wife. But I am thankful to say she bears it bravely contenting herself as 'blessed amongst women' in having had such a son."[14] He now wished to leave Manchester and find new pastures in which to both live and work.[15] Becoming

---

[6] *Manchester Guardian*, 14 May 1885, 5; Roscoe, *Life and Experiences*, 259 and 310.

[7] Roscoe, *Life and Experiences*, 311–313.

[8] Ibid., 258.

[9] For a contemporary description of the constituency, see *Manchester Guardian*, 18 November 1885, 5.

[10] Roscoe, *Life and Experiences*, 260; Moore, *Transformation of Urban Liberalism*, 172.

[11] Moore, *Transformation of Urban Liberalism*, 171.

[12] Roscoe, *Life and Experiences*, 259.

[13] Beatrix Potter recorded his death in her journal in the entry for 2 January 1885 and gives a very candid description of him, calling him "a queer one" (in the old-fashioned sense of the word) and said that his heaviest sin was conceit; Leslie Linder, ed., *The Journal of Beatrix Potter, 1881–1897* (London: Frederick Warne, 1989), 124–125.

[14] Huxley papers 25.285, Imperial College Archives and Records Unit. By permission of the Archives, Imperial College London.

[15] Roscoe, *Life and Experiences*, 260 and 371.

an MP, with a new outlet for his energies and a new home in London, was the perfect solution in this sad situation.

The Conservatives went into the election in a loose alliance with the Irish Parliamentary Party, and Irish votes may have made a difference in some English seats. Indeed, such was the strength of the Irish vote in the Scotland constituency in Liverpool (named after a local road rather than the country) that an IPP candidate T. P. O'Connor won the seat and held it (as the last member of the IPP) until his death in 1929. The Irish vote may partly explain why the Liberals lost North Manchester, as it contained the strongly Irish ward of St. Michael's, which was strongly Liberal from 1886 onward.[16] Roscoe's South Manchester constituency, which did not have many Irish voters, was the only Manchester seat won by the Liberals in this election, despite having two out of the three members for Manchester in the previous election of 1880.[17] Beatrix Potter remarked in her journal entry for 27 November that:

> Uncle Harry's position is most extraordinary, the only Liberal member for Manchester, Liverpool, Bolton, Preston and Stockport. Never since Manchester had a member has such a thing happened. I fancy he has got rather more of a position than he bargained for.[18]

Roscoe only managed to defeat his Conservative opponent Peter Royle with a fairly narrow majority of 670 (out of 6,912 votes cast). There were attempts to link Royle with the defense of (Anglican) religion and Roscoe with atheism, which greatly annoyed him. He always tried to avoid any kind of religious sectarianism, and the Ven. George Anson, the Archdeacon of Manchester and Rector of St. James's, Rusholme, was a friend (and a Liberal).[19] He was also a friend of the Rev. Alfred Steinthal, the Unitarian minister in Manchester.[20] Several other clerics of different denominations took part in his electoral campaigns. At a hustings, when someone shouted out that the Liberal program included atheism, Roscoe replied that the heckler who shouted "'atheism' wished to help the cause of Toryism by endeavoring to make people believe that everybody who was not a Tory was an atheist."[21]

In contrast to his view of his three later opponents, Roscoe was dismissive of Royle in his autobiography, clearly considering him both a makeshift candidate and a lightweight.[22] He scoffed that Royle, an eminent physician and former President of the British Medical Association, was "a club doctor" (a physician who serves a club of subscribers, similar to a modern Health Maintenance Organization in the U.S.), and

---

[16] For the Irish community in Manchester, see Mervyn Busteed, *The Irish in Manchester, c. 1750– 1921: Resistance, Adaptation and Identity* (Manchester: Manchester University Press, 2016), especially 190–193.

[17] For the low number of Irish voters in South Manchester, see Roscoe, *Life and Experiences*, 260; Moore, *Transformation of Urban Liberalism*, 174; and the figure of 300–400 for the Irish vote in South Manchester (against 1,200 in North Manchester), albeit for 1906, in P. F. Clarke, *Lancashire and the New Liberalism* (Cambridge: Cambridge University Press, 1971), Appendix C, 431.

[18] Linder, *Journal of Beatrix Potter*, 164.

[19] Roscoe, *Life and Experiences*, 260–261.

[20] *Manchester Guardian*, 17 November 1885, 5.

[21] Roscoe, *Life and Experiences*, 262.

[22] Ibid., 261.

a member of the Order of Oddfellows (a prominent health insurance society) and a Freemason. He poked fun at the grandiose titles of Freemasonry. "It will, perhaps, be wondered why, when a gentleman possessing such titles and uniting the offices of a local club doctor with that of the Grand Standard-bearer [actually just the Standard-bearer] of England came forward, I did not at once retire in his favour."[23] Royle was a leading member of the Manchester Conservative Party, so clearly, he had an interest in politics. One wonders if Royle rattled Roscoe more than he cared to admit.

On the eve of the election, via his son, Herbert Gladstone, William Gladstone declared his support for Irish Home Rule, the so-called Hawarden kite, named after Gladstone's country residence at Hawarden in north Wales. This meant that the Parliament thus elected was hung (319 for the Liberals, 247 for the Conservatives, and 86 for the IPP). The IPP combined forces with the Liberals to defeat the short-lived Conservative government when Parliament met in January. But by declaring for Home Rule—as Salisbury hoped—Gladstone split his own party. Most of the Whigs, led by the Marquess of Hartington, and some Radicals, led by Joseph Chamberlain, left the Liberals and formed a new Liberal Unionist grouping. In Scotland, the physicist Sir William Thomson (soon to become Baron Kelvin), who had been born in Belfast, was a leading force in the formation of the Scottish Liberal Unionist Party. Roscoe himself remained with the Liberals. He admitted in his autobiography that he had initially been antagonistic toward the Irish members, but had been won over by their wit and intellect.[24] They had a certain light-hearted bonhomie which appealed to him. More importantly, he was intensely loyal to Gladstone as the respected leader of the Liberals, although he found the Grand Old Man's High Church views uncongenial. Roscoe had no innate sympathy for Irish separatism, but he accepted Gladstone's argument that it was essential to give the Irish people Home Rule in order to prevent a more fundamental split further down the road, thereby preserving the unity of the British Empire.[25] Roscoe was a strong believer in the importance of the British Empire, and this was a clinching argument for him. His decision was made easier by the fact that religious sectarianism was not a major feature of Manchester society, in strong contrast to Liverpool.

There was a belief among the Gladstonian Liberals that by giving the Irish this measure of devolution—an assembly which would deal with purely local matters but would reserve Imperial affairs (trade, defense, and foreign policy) to the Imperial Parliament—the British government would be able to head off demands for full Irish independence.[26] By contrast, some of the Irish nationalists saw Home Rule merely as a stepping stone to full Irish independence. There are clear parallels here with the more recent devolution of government in Scotland, but the Liberals in the 1880s did not have any such historical parallels to draw upon. The Conservatives and the Liberal Unionists, on the other hand, saw Home Rule as being "Rome rule" and a betrayal of

---

[23] For a biography of Royle and his various titles, see *Manchester Guardian*, 18 November 1885, 5, and his obituary, *Manchester Guardian*, 13 November 1891, 8.

[24] Roscoe, *Life and Experiences*, 279.

[25] Ibid., 268; *Manchester Guardian*, 21 October 1885, 8.

[26] For the history of Irish Home Rule in this period, see Nicholas Mansergh, *The Irish Question, 1840–1921*, 3rd ed. (London: George Allen and Unwin, 1975); and Alan O'Day, *Irish Home Rule, 1867–1921* (Manchester: Manchester University Press, 1998).

the Irish unionists, especially, but not exclusively, in the northeast of Ireland. The parliamentary majority Gladstone had gained with his alliance with the IPP by espousing Home Rule was canceled out by the Liberal Unionists joining forces with the Conservatives. The government lost the vote on the Home Rule bill (Government of Ireland Bill 1886) in June and a new general election was called. The third Gladstone administration had lasted just 20 weeks, one of the shortest UK governments in history.

General Election of 1886

Roscoe now faced a new election. His position was precarious, perhaps more than he liked to admit, as South Manchester was a local hotbed of Liberal Unionism.[27] Furthermore, he did not have a bloc of Irish votes to rely on, unlike his friend Charles Schwann in North Manchester. If a Liberal Unionist were to stand, his position would have been dire, although he was well-regarded personally, even by the Liberal Unionists. He did in fact face a challenge from Captain North Dalrymple,[28] but despite receiving support from Chamberlain and Hartington, Dalrymple withdrew to avoid splitting the unionist vote, after holding a hustings "characterised by great disorder" (because of heckling by Roscoe's supporters) in late June.[29] The strength of the Liberal Unionists in South Manchester may have been a weakness in elections. They felt entitled to put up the Unionist candidate in this constituency, but the Conservatives refused to give way.[30] Annoyed at this rebuff, the Liberal Unionists seemed to have been reluctant to vote Conservative, thus weakening the Unionist vote. We will see that when the local Unionists did eventually put up a Liberal Unionist, they won the election.

In contrast to his disdain for Peter Royle, Roscoe had a high regard for his opponent in this election, Thomas Sowler, proprietor of the Conservative-leaning *Manchester Courier* and a director of the Manchester Royal Exchange. He was a strong supporter of the Church of England and Honorary Colonel of the 19th Lancashire Volunteer Artillery, having served as its Lieutenant-Colonel. The outcome was very close, Roscoe won by just 335 votes; but all the Liberal wins in Manchester were tight. Despite the assistance of the Irish vote, Charles Schwann managed to gain North Manchester by just 96 votes and Jacob Bright won South-West Manchester by 111. C. P. Scott lost in North-East Manchester by 327 votes. Arthur Balfour, a rising star in the Conservative Party and the nephew of Salisbury, held East Manchester for the Conservatives with a reduced majority of 644. He was soon to become Chief Secretary for Ireland, a key post in a period of increasing Irish agitation. Nationally the result was a crushing

---

[27] Moore, *Transformation of Urban Liberalism*, 89; and *Manchester Guardian*, 27 June 1892, 9.

[28] The Hon. Captain North de Coigny Dalrymple was the son of the 10th Earl of Stair, and although connected with Ayrshire, he lived in Windsor and Bruton Street, London (where Queen Elizabeth II was born almost 40 years later). He had just fought in the Egyptian and Sudan campaigns; see *The Times*, 30 June 1886, 5 (which, however, erroneously states he was the son of the 9th Earl of Stair).

[29] Roscoe, *Life and Experiences*, 265–266; *Manchester Guardian*, 29 June 1886, 6; 1 July 1886, 7 (Dalrymple's withdrawal); and 3 July 1886, 8.

[30] Moore, *Transformation of Urban Liberalism*, 96–97 and 175; and *Manchester Guardian*, 1 July 1892, 8.

Figure 10.1 Roscoe—South Manchester. An election campaign poster that Roscoe sent to Bunsen.
Courtesy of Heidelberg University Library (ref: Heid. Hs. 2741 IV B-1).

defeat for the Liberals, who lost 127 seats, and Salisbury headed a new administration supported by the Conservative and Liberal Unionist alliance which lasted six years. This meant that Roscoe was now in opposition, which inevitably reduced his impact as an MP and removed any possibility of holding government office.

General Election of 1892

When the next general election was held in July 1892, two important political events had occurred. Charles Stewart Parnell, the charismatic leader of the Irish nationalists, had been named as the co-respondent in the divorce of his fellow Irish Nationalist

William O'Shea from his wife Katherine at the end of 1889. With his usual self-confidence, Parnell was convinced that he could carry on, but he lost the support of the Roman Catholic hierarchy in Ireland and of many nonconformist MPs in the Liberal Party. The IPP then split into Parnellite and anti-Parnellite wings. In the event, the Irish nationalists did not suffer a major electoral setback in Ireland; they got 81 seats in 1892, compared with 85 six years earlier. Roscoe did not know Parnell well, and with his usual broad-mindedness, he was probably untroubled by this turn of events. As Roscoe remarked in the context of the Sunday opening of museums, the Liberal nonconformist MPs often claimed not to be puritanical themselves, but were in thrall to their more strait-laced constituents.[31] The second event was the adoption of the so-called Newcastle program by the National Liberal Federation at its annual meeting in Newcastle-upon-Tyne in 1891.[32] This was a radical platform which put Irish Home Rule first, but also included reform of the House of Lords, shorter parliaments, land reform, and the disestablishment of the Welsh and Scottish churches. Although not intended as such, it was effectively the first modern election manifesto in British politics. The immediate effect of this program was to torpedo any hope of reconciliation between the two Liberal factions. It is striking that Roscoe does not mention the Newcastle program in his autobiography, although he agreed with its proposals.[33] Gladstone was willing to go along with the Newcastle program, although he wanted to concentrate on Irish Home Rule, but it was not an electoral success. The Newcastle program was abandoned as a vote-loser by the more right-wing Lord Rosebery when he succeeded Gladstone as premier in 1894. Meanwhile, the Conservative-dominated House of Lords blocked any of the measures based on the program which managed to get through the House of Commons.

In South Manchester, Roscoe again faced a Conservative (rather than a Liberal Unionist) candidate, namely Viscount Emlyn, the son and heir of the 2nd Earl Cawdor.[34] Emlyn had been the Conservative MP for Carmarthenshire between 1874 and 1885, but had lost the election for the new seat of Carmarthenshire West in 1885. Fighting accusations that he was a carpet-bagger, Emlyn's supporter Captain Frederick Pratt tartly noted that Roscoe claimed to be a "Manchester man," but "he had been born in London and educated in Liverpool and on the Continent."[35] Recognizing the high regard for Roscoe in the constituency, whom he was careful to praise as a person, Emlyn largely based his campaign on the single issue of Home Rule.[36] A major local issue in the election was the (temporary) muzzling of dogs, which Roscoe had supported to prevent the spread of rabies, and the Conservatives hoped to gain the votes of dog owners.[37] At the very end of the campaign, a row broke out as Roscoe's agent

[31] Roscoe, *Life and Experiences*, 296.

[32] The Newcastle program is a thread which runs through Hamer, *Liberal Politics*; see in particular chapter 7, "The End of the Irish Preoccupation," 154–184, and 211–217; Naomi Lloyd-Jones, "The 1892 General Election in England: Home Rule, the Newcastle Programme and Positive Unionism," *Historical Research* 93 (2020): 73–104.

[33] *Manchester Guardian*, 20 June 1892, 6.

[34] Emlyn claimed to be called by the Unionists in South Manchester, but he was registered as a Conservative candidate; *Manchester Guardian*, 14 November 1891, 8.

[35] *Manchester Guardian*, 5 July 1892, 5.

[36] *Manchester Guardian*, 14 November 1891, 8; and 30 June 1892, 9.

[37] *Manchester Guardian*, 30 June 1892, 9; 1 July 1892, 8; and 6 July 1892, 11.

insisted that Emlyn was legally required to be listed on the ballot paper under the name by which he had been registered, namely Frederick Campbell, rather than his courtesy title. The Conservatives blamed Roscoe for this, but his agent pointed out that it was the fault of his supporters for registering him under that name.[38] Given the narrowness of the vote, the absence of his usual moniker on the ballot paper may have played a role in his defeat. Curiously, Roscoe has nothing to say about his opponent in this election in his autobiography, but he tersely noted that his majority was reduced.[39] It was now only 181 votes, in contrast to his fellow Manchester Liberals, Schwann and Bright, who both increased their majorities.

Nationally, the Liberals gained 80 seats, but failed to make the major electoral breakthrough they had hoped for. The Conservatives and Liberal Unionists had a majority of 42 over the Liberals, but the Liberals could rely on the support of the Irish nationalists. Salisbury refused to resign, but the Conservatives lost the vote of confidence on 11 August by 40 votes. Gladstone became Prime Minister for the fourth and last time. As a minority administration, Liberal MPs had to be continuously in attendance for the frequent votes that took place at short notice, and being an MP became rather wearisome. At the same time, because of the government's dependence on the Irish vote, little of note was achieved apart from Irish Home Rule, and that was eventually blocked by the House of Lords, leading to Gladstone's resignation in 1894. He was replaced by Rosebery, with whom Roscoe did not enjoy a close relationship, in contrast to Gladstone, and he was perhaps unhappy that Rosebery rather than Roscoe's close friend John Morley had become Prime Minister. For some reason, Roscoe also began to find meeting his constituents in Manchester irksome:

> My supporters were anxious, of course, to see me frequently in their midst and were desirous that I should not lose touch; but not infrequently one went down, perhaps at considerable inconvenience, to a tea-party or bazaar or local gathering and found the room half-full.[40]

His thoughts now turned to retirement, and he hoped to find a replacement of "high repute ... a man of advanced Liberal views and a keen politician."[41] From the outset of his political career, Roscoe believed that South Manchester needed someone with a commercial background. However, the Rosebery administration then fell more quickly than expected, as a result of a political ambush by the Conservatives. In a committee debate about the supplies of cordite bullets for the army on 21 June 1895, the Conservatives realized that they had a majority in the thinly attended house, committee debates not being considered important, and called a vote on the Secretary of War's salary—in effect a vote of no confidence in the Secretary of War, the future Prime Minister Henry Campbell-Bannerman. They won by seven votes (132 versus 125). Rosebery could have considered it simply a vote of no confidence in the Secretary of War, but decided to regard it as a vote of no confidence in his administration. He

---

[38] *Manchester Guardian*, 4 July 1892, 9; 5 July 1892, 5; and 6 July 1892, 11.
[39] Roscoe, *Life and Experiences*, 310.
[40] Ibid., 310.
[41] Ibid., 310.

tendered his resignation to the Queen, who then called on Salisbury to form a govern-ment. As Salisbury lacked a majority, he called a general election for 13 July.

General Election of 1895

Having failed to find a successor, Roscoe fought his fourth election with little enthu-siasm. Matters were complicated by the fact that the marriage of his daughter Margaret to Charles Mallet (later Liberal MP for Plymouth between 1906 and 1910) had been arranged for 11 July. For the first time, he faced a candidate who was a Liberal Unionist and well-connected: the Marquess of Lorne, the son and heir of the 8th Duke of Argyll. Lorne married Princess Louise, the fourth daughter of Queen Victoria, in 1871, the first non-royal person to marry a member of the Royal Family since 1515. He then served as the Governor-General of Canada between 1878 and 1883. According to Roscoe, he had "assiduously nursed the constituency" for two years and had con-sidered standing there in 1886.[42] Uncharacteristically, Roscoe obsequiously refers to his "engaging manners and his high position."[43] It would now be considered improper for a member of the Royal Family, even if only connected by marriage, to stand for election to Parliament. Even at the time, some of Roscoe's supporters expressed dis-quiet, on these grounds, about Lorne standing.[44] As early as March 1894, Roscoe was campaigning for the forthcoming election on the issue of the House of Lords versus the people.[45] This was on the grounds that the House of Lords had blocked several Liberal bills which had passed through the Commons, including the latest Home Rule bill, but it was a particularly apposite campaign in South Manchester given that his opponent was the heir to a duke. He rhetorically asked in an election leaflet, "Are the People or Peers to Prevail?"[46]

The *Manchester Guardian* was confident of victory, saying that a majority of 181 (Roscoe's majority in 1892) represented the "irreducible Liberal minimum."[47] On the day, having rushed back to Manchester after his daughter's wedding, Roscoe lost by 78 votes. An attempt has been made to explain Roscoe's defeat in terms of the Liberal vote as a whole, and his defeat has been regarded as surprising (as indeed it presumably was to the *Manchester Guardian*), but to us the reasons appear to be more personal.[48] Roscoe was facing a Liberal Unionist and had lost his enthusiasm for parliamentary politics. Truth be told, Roscoe was glad to be shot of the constituency, and if he lost to a major figure in society, all the better.[49] As it turned out, Lorne never faced another election. He succeeded his father as Duke of Argyll in April 1900 and had to resign his seat. In the following by-election, the seat was won by another Liberal Unionist,

---

[42] For attempts to tap him for this seat in 1886, see *Manchester Guardian*, 30 June 1886, 6.

[43] Roscoe, *Life and Experiences*, 311.

[44] *Manchester Guardian*, 31 July 1893, 6.

[45] *Manchester Guardian*, 1 March 1894, 6; 29 March 1894, 8.

[46] Election leaflet of 1895 in the Manchester local history library, quoted in Clarke, *Lancashire and the New Liberalism*, 167.

[47] *Manchester Guardian*, 12 July 1895, 4.

[48] Moore, *Transformation of Urban Liberalism*, 177–178.

[49] Roscoe, *Life and Experiences*, 311; Sir Edward Thorpe, *The Right Honourable Sir Henry Enfield Roscoe, PC, DCL, FRS: A Biographical Sketch* (London: Longman, Green, 1916), 156.

William Peel, the grandson of the Tory Prime Minister Sir Robert Peel, with a large majority of 2,039 votes, and he held the seat with a reduced majority in the general election a few months later. In the Liberal landslide of 1906, as Roscoe had predicted after his defeat, South Manchester was won back for the Liberals by Arthur Haworth, a leading Congregationalist and a Manchester businessman, exactly the type of person that Roscoe believed should represent the constituency.

Roscoe's Political Views

What were Roscoe's political views? As he was a loyal member of the party, it is difficult to distinguish between the party line and Roscoe's personal opinions, but the effort can be made on the basis of his autobiography (although it is not very revealing in this respect) and his speeches to his constituents as reported in the *Manchester Guardian*. It is striking that he never described himself as a Radical, although he called his constituency a Radical one in his autobiography.[50] A Radical in this context means the wing of the Liberal party which campaigned for major reforms, in particular the disestablishment of the Church and the abolition of the House of Lords. Or, as it has been succinctly expressed, "against the Peerage, the beerage and war."[51] Roscoe's caution is perhaps a consequence of the upheaval in the Manchester party caused by Richard Pankhurst, who was also associated with Owens College.[52] Nonetheless, he held "advanced Liberal" or progressive views and supported the radical Newcastle program put forward in 1892.[53] James Moore, the only political historian to discuss his political career in any depth, considers him to have been a Radical.[54] The nearest one finds to a self-definition is Roscoe's desiderata for his successor as the Liberal candidate for South Manchester, which described the ideal person as "a man of advanced Liberal views."[55] Leaving labels aside, he was a member of the modern progressive wing of the Liberal Party which sought to retain the support of the working man by espousing social policies. In this respect he was typical of the Manchester Liberal Party with its industrial base and urban Liberalism more generally.[56]

It is fair to say that Roscoe believed in Liberal principles rather than a specific program, regarding reform and progress as self-evidently logical and a moral imperative. He said that the material improvement of recent times had to be matched with moral

---

[50] Roscoe, *Life and Experiences*, 311.

[51] For a discussion of the nature of Radicalism in this period, see H. V. Emy, *Liberals, Radicals and Social Politics, 1892–1914* (Cambridge: Cambridge University Press, 1973), 47–54. This wonderful phrase is a summary of the views of the Radical Liberal baronet Wilfred Lawson by Emy (on 48), based on the description by Norman Longmate, *The Waterdrinkers: A History of Temperance* (London: Hamish Hamilton, 1968), 216–227. Although the phrase is in quote marks, it cannot be found in *The Waterdrinkers*. As a temperance reformer, Lawson coined the term "beerage" because many brewers were created peers or were MPs; an alternative version is the brewocracy (a portmanteau of brewer and aristocracy).

[52] Moore, *Transformation of Urban Liberalism*, 26–34. Pankhurst is conspicuously absent from Roscoe, *Life and Experiences*.

[53] *Manchester Guardian*, 20 June 1892, 6.

[54] Moore, *Transformation of Urban Liberalism*, 46 and 174–175.

[55] Roscoe, *Life and Experiences*, 310.

[56] This is a major thesis of Moore, *Transformation of Urban Liberalism*; in particular, see chapter 9, "Manchester and the Rise of Progressivism," 213–235.

improvement, which he equated with prisons being replaced by schools. Clearly his idea of moral improvement was education which made citizens more law-abiding, rather than the promotion of religious morality.[57] He believed that anyone who opposed such obviously desirable reforms probably had a vested interest, for example, members of the House of Lords, Anglican parsons, or publicans. He also thought that such reforms were inevitable and that the Conservatives would oppose them in vain.[58] His bafflement to any opposition to progressive reforms not based on vested interests is well illustrated by his puzzlement at the continuing obstruction of compulsory metrication (see below).

His major preoccupation was education and in particular technical education, as both his Liberal friends and his political opponents recognized. His tireless work on the promotion of technical education has been covered in Chapter 9, and it suffices here to say that he saw the worth of a technical education well beyond the obvious value for industry, as he also saw it as the answer for the depressed state of Britain's agriculture in this period.[59] Roscoe also campaigned for free elementary and secondary education and university education for women.[60] While he believed in non-sectarian education, he thought that the good voluntary schools (i.e., church schools) would survive and the bad ones would die out.[61] Closely allied to Roscoe's educational agitation was his promotion of educational activities for adults on Sunday, in particular the opening of museums and the holding of public lectures, in the face of opposition from nonconformist Sabbatarian organizations.[62] He was President of the Sunday Society, set up in 1875 to promote the positive use of that day.[63]

Roscoe was also concerned with political reform. First and foremost, there was the introduction of universal male suffrage.[64] It might be assumed that Britain was close to achieving this after the passage of the Third Reform Act, but this was far from being the case. In South Manchester, there were just 8,534 voters in 1886 out of a total population (including women and children) of 67,346.[65] In this context, he was also anxious in 1895 to have the faggot vote abolished.[66] The faggot (also spelt fagot) vote was an additional vote held by an individual as a result of holding property in another constituency.[67] Apart from the issue of multiple votes, it was open to abuse as a landowner could give titles to property to other men who would then be expected to vote

[57] *Manchester Guardian*, 13 October 1894, 9.

[58] *Manchester Guardian*, 29 October 1889, 12.

[59] Roscoe, *Life and Experiences*, 205; HC Deb 22 May 1895, *Hansard*, vol. 34, cc7–62.

[60] *Manchester Guardian*, 21 October 1885, 8; 29 October 1889, 12; and 27 July 1891, 5 (free education); *Manchester Guardian*, 5 October 1894, 7.

[61] *Manchester Guardian*, 21 October 1885, 8, and 2 November 1885, 3. He laid the memorial stone of St. Joseph's Catholic Day School in Longsight in 1891, *Manchester Guardian*, 19 October 1891, 6.

[62] Roscoe, *Life and Experiences*, 295–297; for the importance of Sabbatarianism in the Liberal Party, see D. W. Bebbington, The *Nonconformist Conscience: Chapel and Politics, 1870–1914* (London: George Allen & Unwin, 1982), 57–59.

[63] *Manchester Guardian*, 24 May 1886, 8.

[64] *Manchester Guardian*, 20 November 1894, 12.

[65] *Manchester Guardian*, 3 July 1886, 8.

[66] *Manchester Guardian*, 15 January 1895, 12. He said that someone with property in one constituency would have one vote, but the same property distributed over 20 constituencies would give him 20 votes.

[67] Neil Johnston, "The History of the Parliamentary Franchise," *Research Papers of the House of Commons Library* 13, no. 14 (2013): 8.

as he dictated. Sometimes the property would be transferred only for the duration of the election. The name ultimately comes from the term faggot for a stick added to a bundle. The faggot vote had been abolished for borough constituencies in the Second Reform Act of 1867 and for county constituencies in the Third Reform Act of 1884, but a current faggot voter retained his vote for life. Eventually the remaining faggot votes were abolished in 1918. Roscoe was also in favor of women having the vote.[68] He wanted shorter Parliaments and reforms of the voting procedure; for example, that general elections should be held on the same day throughout the country and to be held on a Saturday.[69] In 1895, he was concerned with the registration of voters, demanding that they should all be registered and allowed to vote after a shorter period of residence in a constituency.[70]

Roscoe was surprisingly cautious about the House of Lords and at least in public, he did not share the view of many Liberals, especially in Manchester, that the House of Lords should be abolished.[71] Until 1894, he only made a few suggestions, including the removal of Anglican bishops from the House of Lords and the removal of the hereditary element from the making of legislation (neither of which has so far been achieved).[72] However after the House of Lords vetoed the second Irish Home Rule Bill in September 1893, his position hardened and he now wondered if the best way forward might be the abolition of the Upper House's veto as proposed by John Bright in 1883. He argued that an emasculated House of Lords able to review (but not veto) legislation was preferable to a democratic second chamber or no second chamber.[73] Accepting the logic of the Irish Home Bill, Roscoe proposed a measure of devolution for Scotland and Wales.[74] He sought to increase the number of labour members of parliament[75] and believed this could only be done if MPs were given a salary, which was introduced in 1911.[76] Roscoe was also keen on the reform of parliamentary procedure which he felt was both too slow and inefficient.[77]

Roscoe supported the disestablishment of the Church of Wales, and while he did not call for the immediate disestablishment of the Church of England, he said he would support it when the public wanted it.[78] He was also in favor of civil parishes which took over certain civic functions from the parish vestries of the Church of England in 1894 and thus formed the lowest level of local government.[79]

---

[68] *Manchester Guardian*, 21 October 1885, 8.

[69] *Manchester Guardian*, 29 October 1889, 12; Roscoe, *Life and Experiences*, 294.

[70] *Manchester Guardian*, 15 January 1895, 12.

[71] Roscoe rather obliquely supported the abolition of the House of Lords in his campaign in 1886, see *Manchester Guardian*, 29 June 1886, 6.

[72] *Manchester Guardian*, 21 October 1885, 8 (removal of bishops); *Manchester Guardian*, 29 October 1889, 12 (hereditary legislators).

[73] *Manchester Guardian*, 1 March 1894, 6; 20 November 1894, 12; and 15 January 1895, 12.

[74] *Manchester Guardian*, 16 August 1885, 5.

[75] By this he meant working-class MPs, the Labour Party did not exist at this time.

[76] *Manchester Guardian*, 18 November 1891, 5.

[77] *Manchester Guardian*, 16 November 1885, 7; 11 December 1885, 4; and 22 March 1886, 8.

[78] *Manchester Guardian*, 21 October 1885, 8; 1 March 1894, 6; 20 November 1894, 12; and 18 January 1895, 8 (Church of Wales); *Manchester Guardian*, 29 October 1889, 12 (Church of England).

[79] *Manchester Guardian*, 12 October 1894, 6.

It might surprise some readers to learn that his third area of political activity was temperance.[80] He rejoiced that his neighborhood of Victoria Park did not have a public house ("beerhouse").[81] This was the key to his approach: he did not actively seek national prohibition (attempts by others to secure national prohibition had failed in the 1870s), but argued that each district should be able to decide whether to allow the sale of alcohol or not.[82] Such an approach has been adopted in America, but did not catch on in England, although it existed until the late twentieth century in some areas of Scotland. In common with many Liberals he was opposed to publicans being compensated for the loss of their alcohol license.[83] Roscoe introduced the Manchester Licensing Bill into the House of Commons in March 1894.[84] It provided for the reduction of the number of licenses to one per thousand people in the urban boroughs at a time when there were almost 3,000 licensed houses for a population of roughly 600,000.[85] There would be referenda of local ratepayers on the issues of hours of sale and Sunday opening, although a high hurdle of a 75% majority would be required. This was a modest proposal produced by a coalition of moderate forces, including the United Kingdom Alliance for the Suppression of the Traffic in all Intoxicating Liquors, the Honorary Treasurer of which was William Crossley of the Manchester engineering firm of Crossley Brothers, and the Church of England Temperance Society along with the local Anglican and Roman Catholic bishops. The bill was scorned by the more hardline temperance campaigners and it failed to reach the statute book. Roscoe's support of temperance reflected the views of several non-conformist churches, including the Unitarians.[86]

As a Manchester Liberal he was strongly in favor of Free Trade, which he described as "a blessing to this country."[87] However, there was one area in which he was perhaps less progressive than one would have expected, namely the regulation of industry. While he was far from being an exponent of laissez-faire policies, Roscoe was concerned about impact of regulation on industry. Manchester was one of the important industrial conurbations in the world and Roscoe was a consultant to the chemical industry. While he recognized that pollution had to be tackled and working conditions ameliorated, he sought to deal with these problems by persuasion rather than by legal means. Early in his career as an MP, he successfully opposed a bill to control river pollution, arguing that it would place too heavy a burden on industry.[88] When he was working with the Mersey and Irwell Joint Committee, he encouraged manufacturers polluting the waterways to install equipment to prevent their harmful effluents as far

---

[80] For example, he sought a reduction in the number of public houses (bars) in 1894, *Manchester Guardian*, 20 November 1894, 12. For a general history of temperance in Britain see Longmate, *The Water Drinkers* and for the importance of temperance in the Liberal Party in this period, see Searle, *Liberal Party*, 40–41; Hamer, *Liberal Politics*; Emy, *Liberals, Radicals*, 44–45; and especially David M. Fahy, *The Politics of Drink in England, from Gladstone to Lloyd George* (Newcastle-upon-Tyne: Cambridge Scholars, 2022).

[81] *Manchester Guardian*, 21 October 1885, 8.

[82] *Manchester Guardian*, 12 April 1888, 7; and 29 October 1889, 12.

[83] *Manchester Guardian*, 12 April 1888, 7.

[84] Fahy, *Politics of Drink*, 122–124; *Manchester Guardian*, 7 April 1894, 8.

[85] Longmate, *The Water Drinkers*, 238, gives the number of licensed houses in Manchester as 2,992 in 1897.

[86] Bebbington, *Nonconformist Conscience*, 46–51.

[87] Roscoe, *Life and Experiences*, 207.

[88] HC Deb 16 March 1886, *Hansard*, vol. 303, cc1049–1056; *Manchester Guardian*, 22 March 1886, 8.

as reasonably practicable rather than threaten them with legal action.[89] On the other hand, he spoke strongly in favor of the control of air pollution by the *users* of chemicals being brought under the control of the Alkali Acts which already regulated the *manufacturers* of chemicals.[90] This was not inconsistent with his general approach, as the Alkali Inspectorate employed the cooperative approach he favored.[91] In general his economic views appear to be similar to those of his cousin, the economist Stanley Jevons. Of Jevons it has been said that:

> In his view, there was an important but limited view for government intervention on a range of social issues, such as sanitation, housing and recreation. He specifically advocated more public education and public libraries and various forms of labor legislation … [but] [w]ages and working hours were to be determined, quite simply, by the marketplace.[92]

The question of whether Jevons influenced Roscoe's economic views is difficult to determine, but they probably held similar views because of their shared upbringing, education, and employment at Owens College.

There is no doubt that Roscoe later looked back on his political career with a certain degree of exasperation. As a man of action, he looked forward to promoting a program of moderate progressive reform that would modernize Britain. Furthermore, he had hoped to promote British industry, especially in Manchester. In the event, despite being an MP for nearly 10 years, little progress was made on political reform, social reform, or the promotion of industry. This was partly because the Liberals were only in power for three and a half years during his time as an MP. Furthermore, the business of Parliament was dominated for the entire decade by the Irish question, either by Home Rule if the Liberals were in power, or the pacification of Ireland if the Conservatives were in charge. As an English MP and especially as a Manchester MP, Roscoe must have regarded this preoccupation with the Emerald Isle with some frustration. On the one matter of reform he promoted assiduously, metrication, he met passive resistance which he found unfathomable, and made little headway. On other more limited matters, where his expertise was acknowledged, he had more success, as we will see.

Scientific Matters

As a chemist, Roscoe was sometimes called to speak about scientific matters. The most prominent example and one he makes much of in his autobiography was the use

---

[89] Roscoe, *Life and Experiences*, 304–305; *Manchester Guardian*, 8 December 1891, 7; and 3 September 1895, 8.

[90] HC Deb 10 May 1889, *Hansard*, vol. 33, cc1734–1735.

[91] For the early history of the Alkali Inspectorate and the historical background to its formation, see Peter Reed, *Acid Rain and the Rise of the Environmental Chemist in Nineteenth-Century Britain: The Life and Work of Robert Angus Smith* (Farnham, Surrey: Ashgate, 2014).

[92] Margaret Schabas, *A World Ruled by Number: William Stanley Jevons and the Rise of Mathematical Economics* (Princeton, NJ: Princeton University Press, 1990), 29.

of the then newly marketed artificial sweetener saccharin (then spelled "saccharine") in beer in 1888.[93] For the Treasury, the use of saccharin was not a matter of public health, but a restraint of revenue, as sugar was taxed and saccharin was not. Roscoe had spoken about the manufacture of saccharin at the Royal Institution and had an interest in its use. It was argued that saccharin acted as a mild antiseptic and gave the beer "body." The Customs and Inland Revenue Bill gave the Treasury (with the assistance of its laboratory, soon to become the Laboratory of the Government Chemist) the power to prohibit the use in the manufacture of articles of excise (i.e., goods being taxed) of "any substance or liquor of a noxious or detrimental nature" or which "being a chemical or an artificial extract or product, may affect prejudicially the interests of the Revenue."[94] It was proposed by Lyon Playfair that the second grounds for objection, "the interests of the Revenue," be altered to "regulate" rather than "prohibit" as it was clearly aimed at saccharin; he emphasized the value of saccharin as an antiseptic in light beers.[95] He suggested that if there was a danger of losing revenue, the Treasury should tax saccharin, rather than banning it. The Chancellor of the Exchequer George Goschen argued that the use of 12 tons of saccharin by British brewers would deprive the Treasury of a million pounds in sugar duty.[96] It should be noted in passing that the excise laboratory had to develop a test to detect saccharin in beer.[97] The difficulty of detecting low levels of saccharin in beer was raised by Roscoe as an objection to banning its use. He also objected to the equating of noxious substances with harmless substances such as saccharin. Furthermore, he felt that the ban would potentially undermine "a very important and interesting English industry."[98] The amendment was lost by 107 votes. Subsequently, in 1901 an import duty was placed on saccharin.

Interestingly, the debate about saccharin also touched on the usefulness of coal tar. Goschen, in defending the ban on saccharin, referred to it as a choice between sugar and coal tar, since saccharin was ultimately made from coal tar. Roscoe congratulated Goschen on his scientific knowledge (perhaps ironically), but objected to his use of the word "tar" to refer to saccharin. Saccharin was "likely to be an important modern discovery … and it was not right to treat it as the introduction of tar into beer."[99] Theodore Fry, the Liberal MP for Darlington, who was a director of the Bearpark Coal and Coke Company and the owner of the Darlington iron manufacturers Fry Janson and Company, then mentioned the use of coal tar to make "beautiful dyes."[100]

Roscoe was preoccupied with the rehousing of two scientific institutions in South Kensington. He was concerned about the state of the Normal School of Science, although it had only opened in 1871, and in particular the physics laboratory which was then inadequately housed on the other side of Exhibition Road in the buildings

[93] Roscoe, Life and Experiences, 270–271.

[94] H. Martin, "Present Safeguards in Great Britain against Pesticide Residues and Hazards," Residue Reviews 4 (1963): 17–32, on 25.

[95] HC Deb 27 April 1888, Hansard, vol. 325, cc764–768.

[96] Ibid., cc768–770.

[97] P. W. Hammond and Harold Egan, Weighed in the Balance: The History of the Laboratory of the Government Chemist (London: HMSO, 1992), 145–147, which, however, errs in believing that saccharin was first introduced into the United Kingdom in 1900; it was clearly in use by 1887.

[98] HC Deb 27 April 1888, Hansard, vol. 325, c773.

[99] HC Deb 27 April 1888, Hansard, vol. 325, c772.

[100] HC Deb 27 April 1888, Hansard, vol. 325, c781.

erected for the 1862 International Exhibition alongside the science collections of South Kensington Museum.[101] Roscoe was a member of the committee that considered the disposal of unwanted artifacts in the collections of the South Kensington Museum, and he was anxious that a new building should be erected for these collections which had just become the Science Museum after the merger of the science collections with the Patent Museum (formerly the Patent Office Museum), also located in South Kensington.[102] Eventually his nagging of government ministers about the Normal School was at least partly responsible for the government agreeing to fund the construction of new science laboratories across the road from the original building in 1890. A year later Roscoe was alarmed that the land might be used to build a Gallery of Modern Art (which became the Tate Gallery), but fortunately this idea did not come to anything.[103] The laboratories were designed by Aston Webb and opened in 1906. The Normal School became the Royal College of Science in 1890 and Imperial College in 1907 following a merger with the Royal School of Mines. This site was also intended to accommodate a new building for the Science Museum, but Roscoe's endeavors for the museum were unsuccessful in this period. We will see that his efforts were eventually crowned with success in the next chapter.

Another persistent concern was the low profile of science in the entrance examinations to the British Army's colleges at Woolwich and Sandhurst.[104] Although Roscoe never mentioned this, the need for scientifically competent entrants for Woolwich, where the Royal Engineers were trained, was particularly important as this regiment provided scientifically trained staff to the Science and Art Department at South Kensington, such as Major-General Sir John Donnelly; to the Science Museum, the first director of which was Major-General Edward Festing; and to Imperial College, notably Sir William de Wiveleslie Abney (although his prior employment had been at the Science and Art Department which became the Board of Education in 1903).[105] The government was never able to satisfy Roscoe on this point, and Edward Stanhope,

---

[101] HC Deb 07 March 1887, *Hansard*, vol. 311, c1418; HC Deb 06 April 1888, *Hansard*, vol. 324, cc611–612; HC Deb 15 May 1888, *Hansard*, vol. 326 cc331–332; HC Deb 26 June 1888, *Hansard*, vol. 327, cc1274–1275; HC Deb 03 March 1893, *Hansard*, vol. 9, cc955–956. *Manchester Guardian*, 7 April 1888, 5. For the historical context, see "Imperial College," *Survey of London*, vol. 38: *South Kensington Museums Area* (London: Athlone Press, 1975), 233–247, https://www.british-history.ac.uk/survey-london/vol38/pp233-247; Hannah Gay, *The History of Imperial College London, 1907–2007* (London: Imperial College Press, 2007), 46–53. For the broader context, see S. Forgan and G. Gooday, "'A Fungoid Assemblage of Buildings': Diversity and Adversity in the Development of College Architecture and Scientific Education in Nineteenth-Century South Kensington," *History of Universities* 13 (1994): 153–192. Also see S. Forgan and G. Gooday, "Constructing South Kensington: The Buildings and Politics of T. H. Huxley's Working Environments," *British Journal for the History of Science* 29 (1996): 435–468.

[102] HC Deb 07 March 1887, *Hansard*, vol. 311, cc1396–1397; HC Deb 06 April 1888, *Hansard*, vol. 324, cc612–614; HC Deb 26 November 1888, *Hansard*, vol. 331, cc155; HC Deb 25 February 1890, *Hansard*, vol. 341, cc1190–1191; HC Deb 06 February 1891, *Hansard*, vol. 350, cc142–143. For the early history of the Science Museum, see Robert F. Bud, "'Infected by the Bacillus of Science': The Explosion of South Kensington," in *Science for the Nation: Perspectives on the History of the Science Museum*, ed. Peter J.T. Morris (Basingstoke, Hants: Palgrave Macmillan, 2010), 11–40.

[103] HC Deb 19 March 1891, *Hansard*, vol. 351, cc1424; HC Deb 20 April 1891, *Hansard*, vol. 352, c925; HC Deb 26 May 1891, *Hansard*, vol. 353, cc1074–1075.

[104] HC Deb 23 March 1888, *Hansard*, vol. 324, cc222–224; HC Deb 18 June 1888, *Hansard*, vol. 327, c422; HC Deb 25 June 1888, *Hansard*, vol. 327, c1112; HC Deb 14 March 1889, *Hansard*, vol. 333, c1643; HC Deb 27 February 1891, vol. 350, cc1835–1856; HC Deb 07 April 1892, vol. 3, cc824–825.

[105] For Donnelly and Abney, see *ODNB*. An entry for Festing is forthcoming.

the Secretary of State for War, finally stated in 1894 with evident weariness that "I have received a paper on the subject from the hon. Member, which is being carefully looked into at the War Office."[106]

Medical Matters

Although he was not medically qualified, as a chemist Roscoe was sometimes involved with medical issues in Parliament. Because of his great respect for Louis Pasteur as a scientist, Roscoe promoted the use of Pasteur's rabies vaccine in Britain. Pasteur and his colleagues at the Institut Pasteur in Paris developed the vaccine in the summer of 1885. As early as March 1886, Roscoe asked Joseph Chamberlain (then still a member of the Liberal Party), the President of the Local Government Board, if he would set up a committee to investigate the reliability of Pasteur's method. Chamberlain responded that he would examine how such an inquiry could be carried out.[107] Two years later, Roscoe asked if the government would give financial support to the Institut Pasteur, as 178 British subjects had been treated for rabies there free of charge, reducing the number of deaths from the expected 28 to 3.[108] Goschen wondered if any other government had given similar support, but proposed that if any of the British patients had been persons of means, they should make a contribution to the institute themselves. In reply, Roscoe asked Goschen to find out if any other government had made a contribution and the Chancellor agreed. The following summer, given the rise in the number of rabies cases, Roscoe asked the government for the muzzling of dogs and the destruction of stray dogs.[109] William H. Smith, the First Lord of the Treasury,[110] replied that dogs were muzzled in London but no general order had been issued by the Privy Council, and only the Metropolitan Police had the power to destroy dogs. As we have seen, Roscoe's support for muzzling dogs was controversial in the general election of July 1892. In June 1890, Roscoe objected to the government's policy of replacing muzzling by the registration of dogs as such a policy had failed in continental cities in which it had been tried.[111] Henry Chaplin, President of the Board of Agriculture, replied that the replacement of the muzzle by a collar only applied to areas where the disease no longer existed. Three years later, Roscoe supported the use of Pasteur's anthrax vaccine in cattle, a policy which has never been introduced by the UK government.[112]

In his autobiography Roscoe mentions that he was entrusted with the task of shepherding the Pharmacy Acts Amendment Bill, a bill to amend the 1868 Pharmacy Act,

---

[106] HC Deb 07 April 1892, *Hansard*, vol. 3, cc824–825.

[107] HC Deb 11 March 1886, *Hansard*, vol. 303, cc435–436; *Manchester Guardian*, 22 March 1886, 8.

[108] HC Deb 17 December 1888, *Hansard*, vol. 332, c441.

[109] HC Deb 15 July 1889, *Hansard*, vol. 338, c405. For a general discussion of the issue of muzzling and destruction of dogs, see John K. Walton, "Mad Dogs and Englishmen: The Conflict over Rabies in Late Victorian England," *Journal of Social History* 13 (1979): 219–239.

[110] The office of First Lord of the Treasury was not tied to the prime ministership until 1905. Smith was acting in the House of Commons for the Prime Minister, the Marquess of Salisbury, who sat in the House of Lords.

[111] HC Deb 09 June 1890, *Hansard*, vol. 345, cc331–332.

[112] HC Deb 01 June 1893, *Hansard*, vol. 12, c1729.

through the parliamentary process.[113] He does not say who sought his help, but he may have been approached by three influential Manchester chemists and druggists, Robert Hampson, Scott Brown, and James Woolley.[114] The need for these amendments arose from the failure of the 1868 Act to prevent unqualified assistants from dispensing dangerous drugs and scheduled poisons. The sponsors of the 1868 act had intended to achieve this, but it was poorly drafted and soon ran into problems.[115] The bill sponsored by Roscoe aimed to bar unqualified assistants from dispensing drugs. In his characteristic manner, Roscoe took the opportunity to relate two striking examples of unregulated dispensing in his autobiography. One case, the 15-year-old son of a pharmacist in Lewisham, who dispensed too much strychnine in a drug, was used by Roscoe's co-sponsor, Dr. Robert Farquharson (1836–1918), a former military physician and Liberal MP for West Aberdeenshire, to ask the government for its support for the bill, which was refused for lack of parliamentary time; the usual answer, as Roscoe remarked.[116] The other example involved the adulteration of sweets by arsenic [trioxide] instead of the usual adulterant "daff" [plaster of Paris] thanks to the peddler of the sweets being given the wrong white powder by a shop boy. The bill passed the House of Lords, but a decision in the Commons was postponed for a lack of a quorum when it reached its second reading on 6 April 1888; it then ran out of time, as most private bills do, as the date for the second reading slipped down the timetable.[117] Roscoe says it was opposed for the fear of creating a monopoly, but there is no evidence of this in *Hansard*. This was one of at least five similar bills that failed in the late 1880s and 1890s, before success was achieved with the Pharmacy Acts Amendment Act of 1898.[118] But just how important was Roscoe's role with this bill? Like its predecessor in 1887, it had originally been introduced in the House of Lords by the 6th Earl of Milltown, an Irish representative peer.[119] He was more successful with the parallel bill for Ireland, which went through Parliament in the same period and became the Pharmacy Act (Ireland) Amendment Act, 1890.[120] Dr. Farquharson seems to have

[113] Roscoe, *Life and Experiences*, 273–275.

[114] Suggestion made by Stuart Anderson, personal communication, 3 November 2022. Also see Sydney W.F. Holloway, *Royal Pharmaceutical Society of Great Britain, 1841–1991: A Political and Social History* (London: Pharmaceutical Press, 1991), 256; the work of Hampson to get women admitted to the Pharmaceutical Society would have gained Roscoe's approval, see Ellen Jordan, "'The Great Principle of English Fair-Play': Male Champions, the English Women's Movement and the Admission of Women to the Pharmaceutical Society in 1879," *Women's History Review* 7 (1998): 381–410.

[115] Holloway, *Royal Pharmaceutical Society of Great Britain, 1841–1991*, 239–241; Chapters 5 and 6 of this book give the legal and historical background to this bill, although the bill itself is not mentioned.

[116] HC Deb 28 June 1888, *Hansard*, vol. 327, c1544.

[117] For the reading on 6 April, see *Journals of the House of Commons*, 9 February 1888 to 24 December 1888, Sess. 1888. 51 & 52 Victoria, vol. 143, 136. The second reading was then rescheduled no less than 27 times until the end of November; the session then ended and it was lost. "Medico-Parliamentary," *British Medical Journal* 2, no. 1439 (July 28, 1888): 208, mentions one of these postponements (mentioned on 391 of the *Journals*). The exchange between Dr. Farquharson and Dr. Tanner, MP for Mid Cork, published here is not recorded in *Hansard*. But why did Roscoe not speak in favor of the bill? He was in the House that day asking a question about the Astrographic Catalogue, HC Deb 31 July 1888, *Hansard*, vol. 329, cc953.

[118] For the Pharmacy Acts Amendment Act of 1898, see Holloway, *Royal Pharmaceutical Society of Great Britain, 1841–1991*, 272–273.

[119] HL Deb 16 February 1888, *Hansard*, vol. 322, c540.

[120] HL Deb 14 May 1888, *Hansard*, vol. 326, c120.

Figure 10.2 Harry Roscoe in middle age.
Image courtesy of the Royal Society of Chemistry Library.

been the only MP to have spoken in favor of the bill in the Commons, although he always spoke on behalf of Roscoe.

Between the General Elections of 1892 and 1895, Roscoe spoke only 23 times in Parliament. On 11 of these occasions, he raised the issue of cholera, especially outbreaks of cholera among pilgrims at the Islamic holy city of Mecca. In July 1893, he asked if the government would request that the Ottoman government (the Porte[121]) take measures against an outbreak of cholera in Mecca.[122] Sir Edward Grey, the Under-Secretary of State for Foreign Affairs, dodged the issue by saying that the outbreak would be dealt with by the Ottoman board of health. The following day Roscoe asked when the Ottoman board of health's report would be made public.[123] Grey waffled that "no effort will be spared to secure the adoption of suitable sanitary measures by the Ottoman Authorities." In his question, Roscoe referred to the danger presented to "the lives of thousands of Her Majesty's Mahomedan subjects." At this time there were few Muslim people living in Britain, so he is clearly referring to the Muslim subjects of the

[121] The Porte here is the Sublime Porte, the Bāb-ı Ālī in Arabic, the ceremonial entrance to the Topkapi Palace in Constantinople (as it was then called), which acted as a synecdoche for the Ottoman imperial authorities.

[122] HC Deb 24 July 1893, *Hansard*, vol. 15, cc307–308. Edward Frankland had been concerned with the insanitary state of the water supplies at Mecca since 1881 and especially since 1891; see Colin A. Russell, *Edward Frankland: Chemistry, Controversy and Conspiracy in Victorian England* (Cambridge: Cambridge University Press, 1996), 393–396.

[123] HC Deb 25 July 1893, *Hansard*, vol. 15, c483.

British Empire, above all in British India, which then included what is now Pakistan and Bangladesh. As this Muslim population was recorded as being 57 million in the 1891 Indian census, the situation in Mecca posed a major health risk in India, as well as the danger of cholera reaching Europe. Clearly skeptical about the Ottoman board of health, Roscoe asked when the board was set up, and of course Grey did not know. The following month, Roscoe asked what measures the Indian government was taking to suppress cholera both among pilgrims to Mecca and at Indian religious festivals.[124] In response, George Russell, the Under-Secretary of State for India, claimed that the Indian government and the Porte were making efforts to combat the disease. At the end of August, Roscoe raised the question of British subjects (presumably from India) being illegally detained by the Governor of Mecca despite the presence of cholera, with Grey, who said that successful representations had been made to the Porte.[125] A week later, Roscoe asked if the Isolation Hospitals Bill could be passed in view of the outbreak of cholera.[126] A mere five days later, he raised a case of cholera in Westminster and mentioned the unsatisfactory drains there, an issue he had broached in its own right a month earlier.[127] In June 1894, in his final intervention on this issue, Roscoe pressed the Foreign Office once more, asking if the Porte accepted the recommendations of the recent Sanitary Congress. Grey responded that the health situation in Mecca was reported to be good and that the Porte had been represented at the congress, but its delegates did not sign the recommendations.

Metrication

Roscoe was the chairman of the select committee on possible changes to the system of weights and measures which had been set up in February 1895 as a result of difficulties encountered in international trade.[128] In July 1895, the committee recommended that the metric system should be legalized, made compulsory after a space of two years, and taught in schools. The advantages of the metric system for industry and trade seemed obvious to Roscoe and he argued that "what the negroes and Arabs in Tunis can easily learn, the British public should be able to comprehend."[129] Balfour, on behalf of the incoming Conservative government, introduced a bill which made the use of metric system legal (as the Weights and Measures (Metric System) Act, 1897) and it was taught in schools as a result of the Educational Code of 1900. However, the House of Commons refused to make use of the metric system compulsory. Roscoe blamed what he called "party spirit" for this setback, but the situation appears to have been more complex.[130] He became one of the vice-presidents of the Decimal Association,

---

[124] HC Deb 17 August 1893, *Hansard*, vol. 16, cc410–411.

[125] HC Deb 31 August 1893, *Hansard*, vol. 16, c1584.

[126] HC Deb 07 September 1893, *Hansard*, vol. 17, c477. He raised the issue of isolation hospitals again in November; see HC Deb 06 November 1893, *Hansard*, vol. 18, cc331–332.

[127] HC Deb 12 September 1893, *Hansard*, vol. 17, c954; For the issue of sewer gas in Westminster, see HC Deb 11 July 1893, *Hansard*, vol. 14, cc1271–1273.

[128] Roscoe, *Life and Experiences*, 285–288; T. C. Mendenhall, "The Metric System in England," *Science* 2, no. 31 (1895): 119–120; HC Deb 13 February 1895, *Hansard*, vol. 30, c691.

[129] Roscoe, *Life and Experiences*, 286.

[130] Ibid., 286.

founded in 1841, and started a public campaign for metrication.[131] This brought about the Weights and Measures (Metric System) Bill of 1904, which was introduced in the House of Lords by the Scottish representative peer Lord Belhaven and Stenton (who had served in the Royal Engineers) and the physicist Lord Kelvin (appropriately, given that his title is now the SI unit for the absolute temperature scale).[132] In his speech for the second reading, Lord Belhaven referred to the report of Roscoe's select committee and Parliament's failure to implement all its recommendations.[133] The bill was passed but metrication was delayed. A new select committee considered the issue following the Liberal landslide of 1906 and it recommended compulsory metrication by 1910. But any further progress was halted by World War I. Roscoe found the opposition to compulsory metrication baffling, and in his autobiography, he optimistically declared that "we may thus hope that, in spite of opposition, which no doubt will occur in the House of Commons, a measure will ere long be passed to make [use of the metric system] compulsory."[134] In retrospect, it is perhaps surprising that he thought this would be brought about by "the pressure of public opinion."

Ventilation

Roscoe was closely involved with the issue of ventilation. In 1857, he had worked for the departmental committee chaired by John Campbell of Islay and produced a report on ventilation of Wellington barracks in London, schoolrooms, and working-class dwellings.[135] Nowadays we would think of ventilation in terms of the free movement of fresh air and avoidance of stuffiness and mold.[136] For Roscoe, it was more about removing particles from the air, carbon dioxide and the combustion products of interior gas lighting.[137] However, these particles were not the small inorganic particles that are of concern in the twenty-first century, but microorganisms. Ventilation, mainly with the aim of removing chemicals from the air by a forced draught, was also an important aspect of laboratory design in this period when laboratories had relatively few fume cupboards and chemists preferred to work on the open bench.[138]

However, Roscoe's initial involvement with ventilation after he became an MP was closer to home, namely the ventilation of the Commons chamber. Since the new Houses of Parliament were built in 1837, there were problems with foul smells both coming in from the River Thames outside—most famously the Great Stink of

---

[131] Ibid., 287.

[132] William H. Seaman, "The Discussion in the British Parliament on the Metric Bill," *Science* 21, no. 524 (1905): 72–75.

[133] *Hansard*, HL Deb 23 February 1904, *Hansard*, vol. 130, cc674–689.

[134] Roscoe, *Life and Experiences*, 286.

[135] Ibid., 100–101. More recently, the ventilation of schoolrooms has been an important issue because of the COVID pandemic.

[136] Curiously there has been no scholarly history of ventilation, in contrast to the history of air conditioning.

[137] Roscoe, *Life and Experiences*, 278.

[138] Peter J.T. Morris, *The Matter Factory: A History of the Chemistry Laboratory* (London: Reaktion, 2015), 104–106.

1858—and within the Commons chamber itself.[139] For some reason (it does not appear to have been a warm spring), the latter problem became noticeable in the spring of 1886, as shown by this parliamentary remark in the middle of a debate on Irish affairs on 27 May 1886. Edward Rider Cook (Liberal, West Ham North) said:

> I do not rise ... to take part in the discussion upon the Amendment before the Committee, but to call attention to the abominable atmosphere in which we are sitting. It seems to me that the air of this House is not only disagreeable, but that we are really sitting here at the risk of our lives. Unless something can be done, and that immediately, to remedy the evil, we ought, out of respect for ourselves and respect for our wives and children, to report Progress, and adjourn the House until such time as we can have an atmosphere in which it is proper for us to sit.[140]

As the senior partner in the soap manufacturers Edward Cook & Company of Bow, he was on familiar terms with Roscoe, and he was the first Treasurer of the Society of Chemical Industry. Previously, on 6 April 1886, Arthur Duncombe (Conservative, Howdenshire) had asked the Junior Lord of the Treasury George Leveson-Gower (Liberal, North-West Staffordshire) whether he would empower some competent sanitary engineer to investigate the causes of, and, if possible, provide some effectual remedy for, the disagreeable odors that constantly prevailed in the House.[141] As a result of these complaints a Select Committee on the Ventilation of the House was set up, and despite being a new MP, Roscoe was made its chairman. The original plans for the building of the House of Commons by Sir Charles Barry could not be found, and Roscoe was forced to order the cutting into the foundations to access the sewage workings.[142] The sewage created by its inhabitants fell into a pit which had been subsequently linked to London's new sewerage system completed by Sir Joseph Bazalgette in 1865, but the system completely lacked pressure. As a result, the internal sewage lingered in the pit and external sewage could enter it. Clearly the whole setup needed to be changed, and Roscoe decided to use the hydro-pneumatic ejector invented by the Welsh engineer Isaac Shone (1836–1918) who drew on his prior experience of being a mining engineer.[143] The internal sewage and rainwater were collected into large iron spheres underneath the Speaker's garden (rather than in the basement of the building), which were then periodically pneumatically ejected into the main sewer above its water level. Subsequently, Roscoe successfully campaigned for the improvement of

---

[139] The original ventilation system in the Houses of Parliaments had been designed by the chemist and pioneer of ventilation science, David Boswell Reid, and installed despite the misgivings of the architect Charles Barry. See Edward J. Gillin, *The Victorian Palace of Science: Scientific Knowledge and the Building of the Houses of Parliament* (Cambridge: Cambridge University Press, 2017), chapter 4, "Chemistry in the Commons," 121–183. I am indebted to Roland Jackson for this reference. It is unfortunate that Gillin did not seek to cover the developments after 1860, at least briefly. Also see Henrik Schoenefeldt, "The Historic Ventilation System of the House of Commons, 1840–52: Re-visiting David Boswell Reid's Environmental Legacy," *The Antiquaries Journal* 98 (2018): 245–295.

[140] HC Deb 27 May 1886, *Hansard*, vol. 306, c267.

[141] HC Deb 06 April 1886, *Hansard*, vol. 304, c914.

[142] Roscoe, *Life and Experiences*, 277–278.

[143] https://londonist.com/london/history/the-story-of-london-s-other-great-stink and https://www.gracesguide.co.uk/Isaac_Shone.

ventilation in the House of Commons chamber to make it less stuffy. He also sought to replace gas lighting with electric lighting to prevent damage to the internal stonework of the building, though it was not replaced in the House of Commons until 1912.[144] However, he was able to get a glass ceiling put into the chamber so the fumes from the burning gas could not enter the room.

Roscoe also had an impact on another matter of ventilation, namely the use of steam in cotton-weaving sheds in Lancashire to enable the weaving of heavily sized cloth.[145] The cloth weavers were complaining that the moisture-laden air was injuring their health, but the manufacturers claimed that they were unable to measure the amount of moisture present. The issue was considered by a committee chaired by Lord Cranborne, the son of the Prime Minister, and Roscoe was a member of the committee. Roscoe showed that it was a simple matter to use hygrometers to measure the moisture in the air and thereby to set a limit. Roscoe then produced a table of these limits, which were added to the Cotton Cloth Factories Act of 1889 in 1893. To the dismay of the Amalgamated Weavers' Association the act did not forbid the use of steaming above 70°F, but rather the heating of the shed when that temperature was reached. They were further discomfited that the use of steaming spread from the weaving of heavily sized cotton to finer yarns following the passage of the act. However, it was agreed on all sides that increased ventilation proved beneficial, and in 1896 Roscoe chaired a committee to review the working of the act and, in particular, the issue of ventilation. The committee decided that there should be an upper limit for the amount of carbon dioxide in the air. Under pressure from Dr. James Wheatley, the Medical Officer of Health for Blackburn, the Roscoe committee also looked at the purity of the water used in steaming, another topic close to Roscoe's heart, as we will see.[146] The committee wanted to enforce the use of drinking water for steaming, but as many weaving sheds had no access to tap water the final recommendation was simply that the water should be as pure as possible. The Roscoe committee reported in 1897, and its recommendations were implemented a year later. The controversy over steaming continued for many years; there was an ongoing debate about the importance of carbon dioxide levels versus the cooling effect of ventilation.

Sewage

If the condition of the air was a concern, the situation regarding the rivers going through large cities and industrial areas was also causing problems. In 1887[147] Roscoe was invited to become chemical adviser to the Metropolitan Board of Works. This was responsible for the infrastructure of London in the absence of a local authority which covered the whole city—it replaced by the new London County Council two years later. The problem of smells from the River Thames had come up in 1855 and

---

[144] Roscoe, *Life and Experiences*, 275; HC Deb 9 July 1888, *Hansard*, vol. 328, c731.

[145] Roscoe, *Life and Experiences*, 300–301. An excellent history of this issue is provided by Alan Fowler, *Lancashire Cotton Operatives and Work, 1900–1950* (Aldershot, Hants: Ashgate, 2003), 148–166.

[146] James Wheatley later became Medical Officer of Health for Shropshire and died in 1928.

[147] Roscoe (*Life and Experiences*, 303) says 1889, but this cannot be correct as the Metropolitan Board of Works was abolished that year; Thorpe (*Roscoe*, 157) says 1887.

1858 and Michael Faraday had been called in to solve the issue. Despite the creation of a sewerage system by Bazalgette in the early 1860s, the problem arose again in 1887 thanks to the hot weather that summer.[148] Bazalgette's sewers were part of the problem, as they efficiently delivered untreated sewage to the River Thames with the concomitant smell. The Metropolitan Board of Works employed the chemist William Dibdin (1850–1925), who tried to remove the foul odors by applying expensive sodium permanganate to the dried-up Thames.[149] At this point E. Rider Cook, who was on the board, lost patience with Didbin and brought in his fellow Liberal MP, Roscoe. He stopped the use of chemicals and created a temporary solution by improving the ventilation of the sewers. Roscoe then sought to address the problem of untreated sewage pouring into the Thames from sewerage outlets on both sides of the river. He soon realized that the sewage had to be treated before entering the river, a task which was beyond his scope. The newly established London County Council took over the sewage system in 1889 and reinstated Dibdin. Within a few years Dibdin pioneered the use of bacterial treatment of sewage, thus avoiding the excessive use of chemicals.[150]

Subsequently, Roscoe was approached by Sir John Hibbert (1824–1908), Chairman of the Lancashire County Council (and also Liberal MP for Oldham) to examine the sewerage situation for the Mersey and Irwell in south Lancashire.[151] He assisted the passage of the Mersey and Irwell (Prevention of Pollution) Act through Parliament in 1892, although he did not speak in the main debate. Roscoe was then appointed the chemical advisor to the Joint Committee in April 1893 for a year at a salary of £700, worth about £70,000 in 2023 and much more than the salary of £400 paid to MPs for the first time in 1911.[152] Some members of the committee objected to this, but they were overruled.[153] He was also given the use of a laboratory in Manchester for chemical analyses.[154] He was paid £725 to serve for a year and additional expense allowances (such as traveling costs) in March 1896.[155]

He sought to control industrial pollution, clearly a major problem in heavily industrialized south Lancashire, by collaborating with the manufacturers. As a result of his patient negotiation, 306 of the 444 works in the area covered by the Joint Committee set up their own purification works without the need for legal enforcement.[156] This

---

[148] There was a major drought in 1887 but not in 1889, which confirms that Thorpe was correct. Roscoe was keen in 1888, presumably following the drought of 1887, to avoid a water famine in dry weather and unsuccessfully asked that a Royal Commission be set up to consider the issue. As he was prompted to raise the matter as a result of the views of Sir John Lawes, his concern was presumably for the needs of agriculture rather than domestic consumers. HC Deb 20 February 1888, *Hansard*, vol. 322, cc877–878.

[149] For William Dibdin, see Grace's Guide, https://www.gracesguide.co.uk/William_Joseph_Dibdin; and J. H. Coste, "William Joseph Dibdin," *The Analyst* 50 (1925): 369–371.

[150] Christopher Hamlin, "William Dibdin and the Idea of Biological Sewage Treatment," *Technology and Culture* 29 (1988): 189–218. Roscoe gives a more jaundiced view of Dibdin's earlier work in *Life and Experiences*, 303–304, and does not mention Dibdin, giving the credit for the introduction of bacterial treatment to Dibdin's successor Frank Clowes (1848–1923). For Clowes, see H. B. Baker, "Frank Clowes," *Journal of the Chemical Society, Transactions* 125 (1924): 985–987.

[151] *Manchester Guardian*, 8 December 1891, 7. For Hibbert, see *ODNB*.

[152] *Manchester Guardian*, 7 March 1893, 12.

[153] *Manchester Guardian*, 11 April 1893, 9.

[154] Thorpe, *Roscoe*, 160.

[155] *Manchester Guardian*, 3 March 1896, 3; and 31 March 1896, 9.

[156] Roscoe, *Life and Experiences*, 304.

was a triumph of the voluntary principle which governed most of Britain's pollution and pesticide management until the 1980s, when the control of pesticides was taken over by the European Community. In order to control urban sewerage which was largely composed of human waste, Roscoe proposed the use of artificial filtration combined with chemical treatment. The area of these artificial filters soared from 6.5 acres (2.6 hectares) in 1898 to 105 acres (42.5 hectares) only six years later.[157] Roscoe hoped that a National Water Board would be set up to control sewerage across the United Kingdom; this was eventually achieved, in part, with the creation of the regional water authorities in 1974.[158]

Technical Education

As the Honorary Secretary of the National Association for the Promotion of Technical Education—formed after the Samuelson Commission had delivered its report in 1884—Roscoe was tasked with the job of getting legislation for secondary technical education through Parliament. However, while the Conservatives and their Liberal Unionist allies were in power, the chances of a Liberal MP getting a bill through Parliament were slim. Roscoe introduced a technical education bill in February 1888 and again exactly a year later.[159] William Hart Dyke introduced a technical instruction bill in May 1888.[160] Arthur Acland, the Liberal MP for Rotherham and the Secretary of the association, introduced the Technical Schools (Local Authorities) Bill in March 1889.[161] None of these bills reached the statute book. The government then presented its own bill, which became the Technical Instruction Act of 1889 in August.[162] This act permitted local authorities to support and aid technical instruction. Roscoe commented, "Although this Act was not altogether in accord with the views I had put forward, I gladly accepted the measure as a satisfactory instalment."[163] A year later, he tried to introduce a bill to remove any possible doubt about the legality of the provision of technical instruction in public elementary schools, but was not successful.[164]

While the Technical Instruction Act of 1889 put technical education on a firm footing, it did not provide any government funding for it. Under the act, local councils were permitted to raise one penny on the rates to fund technical education, but few

---

[157] Ibid., 305.

[158] Ibid., 306. For a general history of water supply, see John Hassan, *A History of Water in Modern England and Wales* (Manchester: Manchester University Press, 1998).

[159] Roscoe, *Life and Experiences*, 209–210. Introduced: *Journals of the House of Commons*, 1888, vol. 143, 16, and vol. 144, 16. Also see Alan W. Jones, *Lyulph Stanley: A Study in Educational Politics* (Waterloo, Ontario: Wilfrid Laurier Press, 1979) 71–75; W.H.G. Armytage, *A. J. Mundella 1825–1897: The Liberal Background to the Labour Movement* (London: Ernest Benn, 1951), 269–275; and P. R. Sharp, "The Entry of County Councils into English Educational Administration, 1889," *Journal of Educational Administration and History* 1 (1968): 14–22.

[160] Introduced: *Journals of the House of Commons*, 1888, vol. 143, 235. For Roscoe's views of this bill, see *Manchester Guardian*, 21 June 1888, 8.

[161] Introduced: *Journals of the House of Commons*, 1889, vol. 144, 162.

[162] Introduced: *Journals of the House of Commons*, 1889, vol. 144, 370; Second Reading: HC Deb 01 August 1889, *Hansard*, vol. 339, cc157–160; received Royal Assent: *Journals*, vol. 144, 475.

[163] Roscoe, *Life and Experiences*, 209.

[164] HC Deb 20 February 1890, *Hansard*, vol. 341, cc748–749.

showed any inclination to do so. Almost immediately, however, a major source of possible funding for secondary and further education arose by chance. This windfall was a result of the debates over temperance and the licensing of publicans in the late 1880s and which went on into the 1890s. In 1888, the Conservative-Unionist government reformed local government and as part of that reform introduced local elected licensing boards instead of the magistrates. The main problem, apart from the possible corruption of local voters by the alcohol trade, was the issue of compensation: money paid to publicans for losing their licenses. The temperance lobby (including Roscoe) was strongly opposed to such an idea, not least because the remaining licenses would become more valuable. Goschen and his ally Charles Ritchie (who was MP for Tower Hamlets, although he came from Dundee) proposed that a scheme be set up to reduce the number of licenses; in effect, new licenses would only be permitted if an area's population was growing. There would be money for compensation without there being a right to it. The temperance reformers saw this for the fudge that it was and objected to it. Goschen and Ritchie could ignore the Liberal temperance lobby, but a Liberal Unionist William Caine took up the fight. Although his amendment failed, this row put the alliance between the Conservatives and Liberal Unionists in danger. In June 1890, Smith, as the head of the government in the Commons, threw in the towel and the scheme was withdrawn.[165] However, a fund for this compensation clause had already been set up in the budget using a beer levy introduced in 1889 and a new tax on whisky, hence its popular name, the "whisky money." Almost uniquely, a government had an income stream without any clear purpose to spend it on. The maverick Irish nationalist Tim Healy, the MP for North Longford, argued that not having an immediate purpose for the tax as laid down in the budget was against the procedure of the House and the Speaker agreed with him.[166]

Acland then proposed that the money raised in this way be used to fund technical education.[167] On 21 July, Goschen, the Chancellor of the Exchequer, rose in the House to say that the county and county borough councils in England could use the money to fund "intermediate, technical, or agricultural education." The amount of this "whisky money" depended on how much beer and whisky was drunk in a given year (which may have troubled temperance reformers, including Roscoe). In the financial year 1890–1891, it was just under £750,000 (worth around £76 million in 2023) and it was never less than this in subsequent years.[168] It thus exceeded the total expenditure of the Science and Art Department in every year during the 1890s. This unexpected bounty was distributed to the counties and county boroughs in the same way as probate duty under the 1888 Local Government Act, which in turn reflected the various government grants the local authorities had received in 1877–1878. Hence it bore no relationship to their actual need for technical education funding or even their population. The largely rural county of Herefordshire received as much whisky money per

---

[165] For the history of this ill-fated scheme, see Fahy, *The Politics of Drink*, 62–67.

[166] HC Deb 24 June 1890, *Hansard*, vol. 345, cc1799–1805. Healey was later the first Governor-General of the Irish Free State.

[167] HC Deb 10 June 1890, *Hansard*, vol. 345, cc567–568. For the history of the whisky money, see P. R. Sharp, "'Whiskey Money' and the Development of Technical and Secondary Education," *Journal of Educational Administration and History* 4 (1971): 31–36.

[168] For a discussion of this bounty by Roscoe, see *Manchester Guardian*, 11 October 1890, 7.

head as Gateshead, Middlesbrough, South Shields, and Wigan combined. Some councils, most notably the new London County Council, decided to give the money back to their rate-payers, which was the other option available to them.[169] Nevertheless, the whisky money was a boon for technical education and was mostly spent on technical education in evening classes and for science education in schools.

In Roscoe's view, there were deficiencies in the Technical Instruction Act. No scholarships (then a major element of education funding) could be awarded, and there were limitations on how the funds could be spent.[170] Private member's bills, introduced by an individual MP rather than the government or the opposition, rarely become law unless they are accepted by the government, as the Pharmacy Acts Amendment Bill had shown. Nonetheless, Roscoe was able to get his bill on technical education passed (as the Technical Instruction Act of 1891) to remove these problems and allow the whisky money to be used more widely.[171] Both these acts were later superseded by the Conservative government's Education Act of 1902, which retained the funding of technical education with the whisky money. The organization and funding of secondary education then remained practically unchanged until the Education Act of 1944.

Could Roscoe Have Flourished as a Liberal Unionist?

Historians are often warned not to indulge in counterfactual history, but there is an interesting counterfactual point to be made here. Roscoe strongly supported Home Rule, but it was not an issue that resonated with him.[172] He supported it because it was proposed by his political hero Gladstone, and he saw it as giving justice to the Irish people. As it was the only fair solution to the Irish problem, he believed even the Conservatives would eventually support Irish Home Rule.[173] By contrast, John Tyndall was a firm unionist, like Kelvin, and he said to Charles Grant in November 1887: "At present science is represented in the House of Commons by that silly creature Stewart [sic, James Stuart, Liberal MP for Hoxton and a Cambridge physicist], and by good natured Roscoe who knows hardly nothing about the Irish question. Playfair barely counts."[174] As his speeches to his constituents, as recorded in the *Manchester Guardian*, show, this was hardly a fair assessment of Roscoe's knowledge of Home Rule after 1887, but it is perhaps a good indication of his concern with the issue despite

---

[169] For the London County Council and the whisky money, see E.J.T. Brennan, "Sidney Webb and the London Technical Education Board," *The Vocational Aspect of Secondary and Further Education* 11, no. 23 (1959): 85–96.

[170] For Roscoe's description of these events, see Roscoe, *Life and Experiences*, 209–211.

[171] HC Deb 26 March 1891, *Hansard*, vol. 351, c1812. It seems that Roscoe never spoke in the House on the bill and the only lengthy discussion on the bill was in the House of Lords, HL Deb 19 March 1891, *Hansard*, vol. 351, cc1358–1377.

[172] The strength of his support for Home Rule increased over the years, e.g., his stirring call for Home Rule in *Manchester Guardian*, 18 November 1891, 5.

[173] *Manchester Guardian*, 30 June 1886, 6; and 31 July 1893, 6.

[174] Roland Jackson, *The Ascent of John Tyndall: Victorian Scientist, Mountaineer and Public Intellectual* (Oxford: Oxford University Press, 2018), 435–436, citing a letter to Charles Grant, dated 6 November 1887, Royal Institution MS JT/1/T/449.

his protestations to the contrary. As Nicholas Mansergh pointed out many years ago, most British Home Rulers (including Roscoe) labored under a fundamental misconception.[175] They believed that limited Irish Home Rule would strengthen the Imperial Parliament at Westminster and the British Empire, not weaken them.[176] Roscoe specifically stated that Home Rule would not affect the supremacy of Parliament.[177] They assumed that the majority of Irish Home Rulers shared this view, but many of them saw Home Rule as a stepping stone to full independence. It is telling that Roscoe avoided friendship with Charles Stewart Parnell, claiming that "his character was not one to encourage acquaintanceship on the part of the English Members," even when a chemical injury Parnell suffered gave Roscoe an opportunity to talk to him.[178] Had Roscoe been aware of this contradiction, it is possible that as an Empire Liberal he might have become a Unionist.

If he had joined his fellow Unitarian Joseph Chamberlain in leaving the Liberal Party over Home Rule, his political career could have been completely different. He would have been completely secure in South Manchester. It was the most Unionist of all the Liberal seats in Manchester, and he would have not been under threat from the suburban villa Tories as the Conservatives would not have put up a candidate against a sitting Liberal Unionist. Indeed, the Conservatives played tribute to Roscoe's record as the local MP and to his academic standing in the elections he fought.[179] They argued that their reason for opposing him, if not perhaps the sole reason, was his support for Home Rule. Liberal voters would have been reluctant to abandon such a well-qualified Liberal, even if he was a Unionist. Hence Roscoe would have won all his elections easily and would have been returned in 1895. We can reasonably speculate that he would have stood again in 1900 and won. Let us further assume that Roscoe would have stood down in 1905 after two decades in the House of Commons rather than being swept away in the Liberal landslide of 1906. Surely, he would have been ennobled for his service and sent to the House of Lords to join his fellow Liberal Unionist scientist Kelvin just before the latter's death. The Parliament Bill of 1911 would have brought his Unionism into conflict with his democratic principles. He would have doubtlessly been one of the Conservative peers counseling moderation.

Now let us consider his putative political career. The Conservatives would have welcomed the support of a leading scientist and academic. Both the Prime Minister Lord Salisbury and his nephew Arthur Balfour (a fellow Manchester MP) had active scientific interests, in stark contrast to Gladstone's outdated belief in Naturphilosophie.[180]

---

[175] Mansergh, *The Irish Question*, 319–320 (this book was first published under a different title in 1940).
[176] Roscoe held this view very strongly; see *Manchester Guardian*, 2 July 1886, 7.
[177] *Manchester Guardian*, 20 January 1886, 8.
[178] Roscoe, *Life and Experiences*, 279.
[179] For example, by Lord Emlyn, *Manchester Guardian*, 14 November 1891, 8.
[180] For Gladstone's outdated views about science, see Roscoe, *Life and Experiences*, 281–283. Sadly, there has been no extended treatment of Salisbury's interest in chemistry, who was assisted by Herbert McLeod, but see his obituary in *Nature* 68 (1903): 392–393; Hannah Gay, "Science, Scientific Careers and Social Exchange in London: The Diary of Herbert McLeod, 1885–1900," *History of Science* 46 (2008): 457–496; and Andrew Roberts, *Salisbury: Victorian Titan* (London: Weidenfeld and Nicolson, 1999), 111–112 and 593–596. Balfour, like his uncle, was a president of the British Association and gave a lecture at its annual meeting in Cambridge in 1904 on the new ideas concerning matter; see Arthur James Balfour, "Reflections Suggested by the New Theory of Matter," *Science* 20 (1904): 257–266.

Salisbury gave a speech in favor of chemistry at the jubilee dinner to celebrate the fiftieth anniversary of the founding of the Chemical Society in 1891.[181] Roscoe would have doubtlessly been instrumental in bringing forward a bill for technical instruction perhaps sooner than 1889 and he would have advised the government about the best way of distributing the so-called whisky money. As a Liberal Unionist, he would then have been out of power between 1892 and 1895, but he might have been in the Cabinet in the Conservative government between 1895 and 1905. Hence, he may well have been one of the architects of the Education Act of 1902, working closely with Balfour and his fellow chemist William de Wiveleslie Abney at the Board of Education.[182] If this had happened, he would have in effect been the Liberal Unionist counterpart of Playfair and Richard Haldane. We would now be celebrating Harry Roscoe as a leading politician and educational reformer, rather than as a chemist, academic, and educational campaigner. This is entirely speculation, of course, but I would argue that it is a reasonable conjecture. The major argument against this is that Roscoe had no time for the upper classes on the whole and avoided being friends with Conservatives, but no one in 1885 would have expected Chamberlain to join forces with the Conservatives.

Conclusions

The career of Roscoe as an MP is a classic example of a leading figure in another field who fails to become an important politician after entering the House of Commons. It is striking that Beatrix Potter is mentioned in Colin Matthew's magisterial biography of Gladstone, but not Roscoe.[183] He was unfortunate that his party was out of office for seven of the 10 years he was an MP and a minority administration even when it was in office. Furthermore, politics during this time was dominated by an issue (Irish Home Rule) to which he could bring neither particular expertise nor any great enthusiasm. The paramount importance of Home Rule meant that relatively little attention was given by Parliament to the issues that mattered to Roscoe, such as education, metrication, social reforms, constitutional reforms, and the promotion of British industry. Only in the field of educational reform did he have any significant impact, and he already had made his mark in this area even before he became an MP. Perhaps Roscoe would have been much more successful and more prominent in the reforming Liberal government of 1906 which enjoyed a large majority—if he had been 20 years younger. There was a counterpart to Roscoe in that administration, namely Richard Haldane, the Secretary of State for War, who also had a German education and a scientific background as a philosopher. Like Roscoe, he was a strong reformer and a Germanophile.[184]

[181] *The Jubilee Celebration [of the Chemical Society of London]* (London: Chemical Society, 1896), 89–92.

[182] For Abney, see *ODNB*.

[183] Matthew, *Gladstone*. The reference is on 520 where Rupert Potter, Beatrix's father and Harry Roscoe's brother-in-law, assisted the artist J. E. Millais with a portrait of Gladstone by giving him a photograph of Gladstone to work from, thus reducing the number of sittings required.

[184] For Haldane, see John Campbell, *Haldane: The Forgotten Statesman Who Shaped Modern Britain* (London: C. Hurst, 2020).

Roscoe was nominally a dissenter MP at a time when nonconformists were at their peak in the House of Commons and were a major influence on the policies of the Liberal Party.[185] This should have made him influential given his intellectual powers and energy. However, he had little influence on his fellow nonconformist MPs. It is significant that he was a Unitarian rather than a mainstream nonconformist. Yet he did not engage much with his fellow Unitarians and scorned the mainstream non-conformists as slavish followers of their constituents' puritanical views. One has to conclude that in a period when nonconformists dominated the Liberal Party, Roscoe was not, in a political sense, a nonconformist, even if he shared some of their views, for example, temperance and disestablishmentarianism.

Roscoe's parliamentary career was successful insofar as any backbench MP can be successful. He was more active than most backbenchers and enjoyed a few legislative successes. For his part, he did not seek high office, nor would he have enjoyed being a minister. From the viewpoint of his party leaders, holding on to a seat which was under severe threat from the Conservatives and Liberal Unionists for 10 years was a major achievement, almost certainly because of the great respect his constituents held for his academic achievements.[186] There are two lessons to be drawn from this. Even in the late nineteenth century there was no real scope for half-hearted politicians. Furthermore, to become a successful politician in the British system, it is clear that one has to have a broader experience beyond being an academic scientist. Playfair had experience as a senior civil servant, and two other successful politicians with a scientific background—Haldane and Margaret Thatcher—trained as barristers.[187] If Roscoe had turned down the chance to become an MP in 1885, he would have probably still left Manchester after the traumatic death of his son Edmund and assisted London-based organizations as one of the "Great and Good." This is in fact what he did after losing his seat in 1895, and to this stage of his life we will now turn in the next chapter.

---

[185] Searle, *Liberal Party*, 38–45; Bebbington, *Nonconformist Conscience*, chapter 1, "Non-conformists and Their Politics," 1–17.

[186] The thesis of Clarke, *Lancashire and the New Liberalism*, is that Lancashire as a whole was a key target for the Conservatives in this period, and this would have been particularly true of a relatively affluent and unionist constituency such as South Manchester.

[187] Thérèse Coffey, the Deputy Prime Minister in the short-lived Conservative government formed in September 2022, has a PhD from University College London, Roscoe's alma mater, on the topic of structural and reactivity studies of bis(imido) complexes of molybdenum(VI), which would have been of interest to Roscoe, but she then went into business management on the financial side.

11

An Active Retirement

Move to London

Harry Roscoe began to shift his extra-parliamentary activities from Manchester to London after the British Association for the Advancement of Science meeting in Manchester in 1887. This was partly because Manchester reminded him and his family of his son Edmund, who died at the age of 20 in January 1885.[1] After living briefly in Queen's Gate, South Kensington, in a house lent to him by his mother-in-law, Roscoe bought 10 Bramham Gardens, Kensington, as his London residence.[2]

Roscoe's London-based activities can be considered as part of a broad progressive program of institutional reform which began in the late 1880s and reached its peak in the period leading up to World War I.[3] Although this movement sprang at least partly from long-standing liberal principles of social reform, it was given a sense of urgency by an increasing anxiety that Britain was being overtaken in both economic and military terms by Germany. Thus, it included the reform of the armed forces and the modernization of the navy. A strong element of this program was the belief that Britain's salvation lay in the improvement of education, specifically technical education, and the funding of scientific research. As such, it encompassed the setting up of the Laboratory of the Government Chemist in 1894, headed by his former assistant Thomas Edward Thorpe, and the National Physical Laboratory in 1900. Roscoe, as a leading scientist and a Liberal MP, had played an important role in this drive to promote science.

Vice-Chancellor of the University of London

The establishment of Victoria University in 1880 sparked off a reform movement that led to the University of London Act of 1898.[4] The University of London became a purely examining body in 1858, and anyone could take its examinations.[5] Roscoe

---

[1] Henry Enfield Roscoe, *The Life and Experiences of Sir Henry Enfield Roscoe DCL, LLD, FRS, Written by Himself*. (London: Macmillan, 1906), 371.

[2] Ibid., 371. He does not appear in the Census of 1891.

[3] Robert Bud, "'Infected by the Bacillus of Science': The Explosion of South Kensington," in *Science for the Nation: Perspectives on the History of the Science Museum*, ed. Peter J.T. Morris (Basingstoke, Hants: Palgrave Macmillan, 2010), 11–40, on 12–14.

[4] For the history of the reform of the University of London, see Negley Harte, *The University of London, 1836–1986: An Illustrated History* (London: Athlone Press, 1986), 140–180 and F.M.G. Wilson, *The University of London, 1858–1900: The Politics of Senate and Convocation* (Woodbridge, Suffolk: Boydell Press, 2004), Part VI, 221–461. For the significance of Victoria University, see Harte, *University of London*, 140, and Wilson, *University of London*, 223.

[5] For the lack of a salary, see Harte, *University of London*, 175, and Wilson, *University of London*, 37.

Henry Enfield Roscoe. Peter J.T. Morris and Peter Reed, Oxford University Press. © Oxford University Press 2024.
DOI: 10.1093/oso/9780190844257.003.0011

was an examiner in chemistry for the university between 1874 and 1878.[6] University College and King's College wanted to see the establishment of a teaching (rather than an examining) university, provisionally called Albert University, along the lines of Victoria University. Opponents of this scheme proposed that the University of London could combine the teaching and examining functions. Not surprisingly, Roscoe himself was a supporter of the teaching university model.[7] Opposition to reform was led by the Convocation of existing graduates, which feared the loss of its powers, and by provincial universities that worried about academics marking their own students. Two Royal Commissions were set up to consider the issue. The first Commission, set up in May 1888 and chaired by Roundell Palmer, 1st Earl of Selborne, could not reach a unanimous decision.[8] The second Commission, chaired by Francis Cowper, 7th Earl Cowper, was created in April 1892 to break the deadlock.[9] Roscoe was proposed as a member by the 5th Earl Spencer, a fellow Liberal politician, but he was not appointed. When the Cowper Commission reported its conclusions in January 1894, it recommended that the examining and teaching functions of the University of London should be combined. Somewhat surprisingly, Convocation, consisting of graduates of the university and hitherto opposed to reform, accepted the report. In the same month, Roscoe become a member of the Senate of the University of London as a Crown appointee.[10]

The eminent surgeon Sir James Paget, having served as Vice-Chancellor of the University of London since 1884, decided to resign in mid-1895, perhaps because of the death of his wife in January. Remarkably, the post was unsalaried.[11] Roscoe was considered as a possible successor, but was discounted by an influential member of the Senate, Joshua Fitch, as he had only been in the Senate for a short time and because he was considered to be a cheerleader for science rather than learning as a whole.[12] He had also identified himself as a supporter of the teaching university model as the Vice-President of the Association for the Promotion of a Professorial (Teaching) University for London and thus could not act as a neutral umpire.[13] Sir Julian Goldsmid, a Liberal Unionist MP and a businessman, was elected by the Senate in July 1895.[14] Roscoe got 9 votes in the second round, while Goldsmid had 13. The runner-up was Grant Duff, an eminent politician and colonial administrator who only lost the third round by one vote. Goldsmid may have benefited from being a Liberal Unionist. However,

---

[6] Roscoe, *Life and Experiences*, 337.

[7] Ibid., 340.

[8] For the Selborne Commission, see Harte, *University of London*, 146–148; Wilson, *University of London*, 273–284; and "Report of the Royal Commission on a University for London," *Nature* 40 (6 June 1889): 121–122.

[9] Cowper is pronounced Cooper, and the Commission is sometimes called the Gresham Commission, as it was proposed that the ancient but moribund Gresham's College could be the basis for a new teaching university, an idea rejected by the Commission. For the Cowper (or Gresham) Commission, see Harte, *University of London*, 150–156; Wilson, *University of London*, 332–347; and "The Report of the Gresham University Commission," *Nature* 49 (1 March 1894): 405–409.

[10] Wilson, *University of London*, 341. This contradicts Roscoe's statement in *Life and Experiences*, 338, that he joined the Senate after retiring from Parliament.

[11] Harte, *University of London*, 175; and Wilson, *University of London*, 37.

[12] Wilson, *University of London*, 392.

[13] Roscoe, *Life and Experiences*, 340–341.

[14] Wilson, *University of London*, 393.

he died unexpectedly soon afterward, at the age of only 57. Roscoe was elected in his stead, seemingly unopposed.[15] He was now free from parliamentary duties, and was instrumental in setting up Victoria University, which was seen by many as a model for the reformed University of London. By this time, a bill to establish the new University of London had been drawn up and presented to Parliament, but it failed to pass three times: in 1895, 1896, and 1897. The University of London bill was finally piloted through the Commons by the Liberal MP Richard Haldane with the help of Arthur Balfour, the Conservative First Lord of the Treasury, and in the House of Lords by the 8th Duke of Devonshire, the Liberal Unionist Lord President of Council (in effect the Education Secretary) and the son of the 7th Duke who had chaired the Devonshire Commission in 1898.[16] Roscoe was unable to assist directly, no longer being an MP, but he led a large deputation to the Duke of Devonshire in January 1898.[17] The new statutes were ready by February 1900 and were approved by Parliament.[18]

During this period, while the old University of London waited to be replaced by the reformed version, Roscoe's hands as Vice-Chancellor were largely tied. He concentrated on steering the university through these upheavals with the help of the chancellor of the university, Farrer Herschell, 1st Baron Herschell, who had been Lord Chancellor during the Liberal Government of 1892–1895.[19] Unfortunately, Lord Herschell slipped and broke his pelvis during a trip to Washington, D.C., and died there in March 1899. Under the new statutes, the university had to appoint a chief officer to run the university, called the Principal. Believing in the importance of having the right person in any given post, Roscoe persuaded the physicist Sir Arthur Rücker (1848–1915) to stand, and then campaigned for his election.[20] Roscoe knew him from when he was Professor of Physics at Yorkshire College between 1874 and 1885, and he then became Professor of Physics at the Royal College of Science.[21] He was knighted in the Coronation honors of 1902 and died just six weeks before Roscoe. Roscoe regarded his appointment as "the best day's work I ever did for the University."[22]

The central administration of the university was located at 6 Burlington Gardens, to the north of the Royal Academy, where laboratory-based examinations were carried out. This accommodation was too small to house the enlarged university and was also rather dilapidated. Roscoe strongly supported the move of the administration to the new but underutilized Imperial Institute building in South Kensington in 1900 with the blessing of the Prince of Wales, who was shortly to become King Edward VII.[23] Even this accommodation was soon found to be too small, and Roscoe (by then

[15] Ibid., 394.

[16] For a detailed account of this legislative battle, see Wilson, *University of London*, 374–441. For the crucial role of Haldane, see Harte, *University of London*, 156–157; and John Campbell, *Haldane: The Forgotten Statesman Who Shaped Modern Britain* (London: C. Hurst, 2020), 207–212. Roscoe's own account of the passage of the bill through Parliament (*Life and Experiences*, 341–344) makes clear his detestation of religious tests for academics at King's College.

[17] Roscoe, *Life and Experiences*, 343; *The Times*, 25 January 1898, 8.

[18] Wilson, *University of London*, 458.

[19] Roscoe, *Life and Experiences*, 339–340.

[20] Rücker had unsuccessfully stood as the Liberal Candidate in the Leeds North constituency in the 1885 election that saw Roscoe elected to Parliament, but then became a Liberal Unionist.

[21] Remarkably, there appears to be no obituary of Rücker.

[22] Roscoe, *Life and Experiences*, 348.

[23] Ibid., 346–347; Harte, *University of London*, 159–160; Campbell, *Haldane*, 214–215.

retired) promoted a plan whereby the university would take over the whole Imperial Institute. However, this plan had powerful opponents who wanted the university to have its own building, proposed initially to be sited next to the British Museum. The Art Deco Senate House building in Russell Square was brought into use (but not actually completed) in 1937.[24] As the new university statutes had now been approved and the new university Senate was in place, Roscoe stepped down in 1902. He also insisted that future vice-chancellors were to be limited to a two-year term rather than holding the office for life.[25]

Roscoe was also involved with the scheme to set up a new Imperial Technical College in London. A departmental committee was set up in 1904 and was initially chaired by Francis Mowatt, who had been Permanent Secretary at the Treasury, and then by Haldane, although he was now in the Cabinet. A royal charter was granted in 1907 establishing the Imperial College of Science and Technology, which merged the Royal College of Science and the Royal School of Mines and allowed for the accession of Central Technical College, which took place in 1910 when it was renamed the City and Guilds Institute.[26] Its name reflected Roscoe's desire that the University of London should become the Imperial University.[27]

Lister Institute

The history of the Institute of Preventive Medicine is complex and will only be covered here in terms of Roscoe's association with it.[28] It might be wondered why a chemist (and an inorganic chemist at that) became involved with a medical research institute. The answer lies in Roscoe's admiration of Louis Pasteur (Figure 11.1), whom he had first met in Paris in the 1860s. Roscoe got to know him better in 1882 when he visited Paris on behalf of the Samuelson Commission.[29] Pasteur's work fulfilled an important role for Roscoe by showing the value of chemistry for the public good. While Pasteur is widely considered to be a microbiologist nowadays, to Roscoe he was and remained a chemist; he had, after all, done important work on optical activity in his early research.[30] Roscoe found his work on fermentation, with its value for the brewing industry, impressive; but with the development of the rabies vaccine, his admiration for the French savant knew no bounds. At the Paris exposition in 1889, Roscoe

---

[24] Harte, *University of London*, 186–192 and 225–226.

[25] Ibid., 175.

[26] Hannah Gay, *The History of Imperial College, 1907–2007: Higher Education and Research in Science, Technology and Medicine* (London: Imperial College Press, 2007), chapter 3, "The Founding of Imperial College," especially 53–58; Campbell, *Haldane*, 218–223.

[27] Roscoe, *Life and Experiences*, 353.

[28] There are two histories of the institute: Harriette Chick, Margaret Hume, and Marjorie Macfarlane, *War on Disease: A History of the Lister Institute* (London: André Deutsch, 1971); and Leslie Collier, *The Lister Institute of Preventive Medicine: A Concise History* (Bushey Heath: Lister Institute of Preventive Medicine, 2000). The latter is not generally useful for this early period.

[29] For Roscoe's account of meeting Pasteur and seeing his treatment of rabies, see Roscoe, *Life and Experiences*, 314–319.

[30] For Pasteur, see Gerald L. Geison, *The Private Science of Louis Pasteur* (Princeton, NJ: Princeton University Press, 1995); and Patrice Debré, *Louis Pasteur*, trans. Elborg Forster (Baltimore, MD: Johns Hopkins University Press, 1998).

Figure 11.1. Drawing of Louis Pasteur.

introduced the Prince of Wales and his wife (later Queen Alexandra) to Pasteur and his assistant Armand Ruffer at the Pasteur Institute. A few months later, Roscoe gave a speech to the students of the Midland Institute in Birmingham on "The Life Work of a Chemist."[31] He spoke approvingly of the secret of Pasteur's success being "the application of exact methods of physical chemical research to problems which had hitherto been attacked by less precise and less systematic methods." While he mentioned Pasteur's earlier work, the climax of the speech was Pasteur's conquest of rabies.

Before 1885, rabies (spread by the bite of an infected dog) caused universal dread, as the disease was almost invariably fatal. In the early 1880s there were around 30 deaths a year from rabies in England and Wales, but in 1884 there was a spike of some 60 deaths. Pasteur and Émile Roux introduced the first effective treatment of rabies in 1885 by developing a vaccine based on an attenuated version of the rabies virus. In a lengthy and complex procedure involving numerous inoculations, the physician Joseph Grancher, with the help of Pasteur, was able to save a young boy with the disease. By 1886, the Pasteur Institute had saved the lives of 350 people from all over the world. Learning of Pasteur's success in curing rabies, Roscoe asked Joseph Chamberlain as President of the Local Government Board on 11 March 1886 if he was willing to set up a committee to explore the reliability of Pasteur's treatment, and

[31] Roscoe, *Life and Experiences*, 328; *Manchester Guardian*, 9 October 1889, 6; "The Life-Work of a Chemist," *Nature* 40 (1889): 578–583 (full transcript).

Chamberlain agreed to his request.[32] Roscoe then became a member of a committee chaired by the legendary Sir Joseph Lister and otherwise composed of medical men. In the subsequent development of the institute, Roscoe was a moving force behind the scenes, but was careful to let the supporters of the institute with medical credentials take the lead in public, mostly notably Lister himself. As Roscoe remarked in a letter to Balfour in July 1898, "Lord Lister and I are doing what in us lies to help the good cause in establishing the 'Jenner Institute of Preventive Medicine,' which bids fair to do excellent work in medical research."[33]

When this committee reported favorably on Pasteur's rabies vaccine, Sir James Whitehead, the Lord Mayor of London, decided in 1889 to call a meeting to raise funds for this work, ultimately raising £3,200 (£325,000 in 2023 terms).[34] The Royal Society sent its officers and three other representatives to the meeting, along with Sir James Paget, the zoologist Ray Lankester, and Roscoe, to support the cause.[35] There were three ways that this money could have been spent: it could be sent to Paris to support the work of the Pasteur Institute; to send rabies victims to Paris for treatment; or to establish a rabies treatment station in England. While the third objective seemed admirable, it aroused the fury of anti-vivisectionists, as Pasteur's methods involved the use of live animals. Because of the vehemence of the opposition to the idea of a British treatment station, the funds raised by the Lord Mayor's campaign were used to support the work of the Pasteur Institute and to send people bitten by rabid dogs to Paris. Pasteur himself suggested that Britain did not need such an institute, as it was an island and could control rabies by introducing quarantine for animals entering Britain and by muzzling dogs. However, the idea of establishing a counterpart to the Pasteur Institute as a research institute for the treatment of infectious diseases endured. Charles Roy, Professor of Pathology at Cambridge and Sidney Turner, a well-known dog fancier, strongly supported this proposal and won the support of the Lister committee, including Roscoe himself.

Cambridge University offered a home to the new institute, but funds were slow to come in. The chemical industrial magnate Ludwig Mond (who was a friend of Roscoe) donated £2,000 and Roscoe gave £100. The institute only became a going proposition when it attracted the support of the Berridge trustees, who offered £20,000 if the institute were able to raise £40,000 to pay for the land and buildings.[36] In spite of the opposition of anti-vivisectionists (but with the tacit support of the Royal Family), the British Institute of Preventive Medicine was incorporated in July 1891 to undertake research on infectious diseases; to treat people with these diseases; provide cures for these diseases to medical personnel; and deliver instruction in preventive medicine to public health officials and students. However, the crucial donation from the Berridge trustees was in jeopardy unless a suitable building could be secured. Hugh Grosvenor,

---

[32] HC Deb 11 March 1886, *Hansard*, vol. 303, cc435–436; *Manchester Guardian*, 22 March 1886, 8.

[33] Roscoe, *Life and Experiences*, 345.

[34] For Roscoe's account of the events leading up to the establishment of the Lister Institute and its early years, see Roscoe, *Life and Experiences*, 319–333.

[35] Ibid., 319–320; *The Times*, 2 July 1889, 10.

[36] The Irish brewer and landowner Richard Berridge (née MacCarthy) had left £200,000 (£20.8 million in 2023 terms) in 1887 for the advancement of the sanitary and economic sciences. See https://discovery.natio nalarchives.gov.uk/details/r/b0b73eb8-782d-448e-9d97-5fac98c726cd.

the first Duke of Westminster, offered a building on the Chelsea Embankment. The Berridge trustees then offered another £25,000 if the institute would study the Hermite process, an electrochemical process to sterilize sewage. Finally, the private College of State Medicine, which had also been supported by the Berridge bequest, was amalgamated with the institute in 1893. The institute then moved into the college's premises at 101 Great Russell Street until the Chelsea building was completed.

Installed in its temporary home, the institute was directed by Ruffer, a Swiss bacteriologist of private means who had been suggested by his previous employer, Pasteur. He was assisted by Joseph Lunt, a graduate of Owens College who had been at the College of State Medicine, probably because Roscoe was a member of the college's board. He worked on the Hermite process with Ruffer and Roscoe; there was a trial of the process at Worthing, Sussex, in June 1894 with rather mixed results. Of more long-lasting importance was the setting up of the serum department to produce anti-toxins for infectious diseases in the wake of a diphtheria outbreak in 1893 in which over 3,000 people died. Horses or ponies were needed to produce the anti-diphtheria serum, and they were stabled in various places, including William Henry Perkin's home, the Poplars, in Sudbury. Ruffer contracted diphtheria himself in 1895, and while he was cured by the institute's own anti-toxin, he left the institute a year later and moved to Egypt, apparently on the grounds of his health.[37] He was succeeded by the Scottish bacteriologist Allan Macfadyen. Lunt left in 1897 and was replaced by another Owens graduate and Roscoe protégé, Arthur Harden.

The institute moved into its striking red brick building (it was described perhaps unfairly as an eyesore) next to Chelsea bridge. A fundraising campaign to commemorate the centenary of Edward Jenner's first inoculation in 1796 was organized to raise £100,000, but in the end only £5,770 was donated, £5,000 of which had been given by Edward Guinness, first Baron Iveagh (later Earl of Iveagh) of the famous Irish brewing family. The institute duly changed its name to the Jenner Institute, but then discovered that there was a private firm in Battersea called the Jenner Institute for Calf Lymph Limited, which made smallpox vaccine from calf lymph.[38] This clash forced the institute to change its name to the Lister Institute of Preventive Medicine in honor of the now ennobled Joseph Lister in 1903. Lord Iveagh gave the institute an endowment of £250,000 (£24 million in 2023 terms) in gratitude for his groom Jim Jackson being cured of rabies by the Pasteur Institute two years earlier. He insisted that a new Governing Body should take over the financial responsibility for the institute from its Council. One member would be appointed by the Royal Society (Lister), three by the Council (Roscoe, Burdon Sanderson, and John Rose Bradford), and three by Lord Iveagh himself (Lord Rayleigh, Jacob Pattisson, and himself). Charles Martin took over as the Director of the institute in 1903 and remained there until 1930. Lord Lister stepped down as Chairman of the Governing Body (but was given the honorific

---

[37] Ruffer became a pioneer of palaeopathology, the study of ancient diseases, through his study of mummies. He was knighted in 1916, just before he was lost at sea when his ship was torpedoed by a German U-boat. For his biography, see A. T. Sandison, "Sir Marc Armand Ruffer (1859–1917) Pioneer of Palaeopathology," *Medical History* 11 (1967): 150–156.

[38] For a brief overview of this firm, see the authority file at the Science Museum, https://collection.science museumgroup.org.uk/people/cp111763/jenner-institute-for-calf-lymph-limited. Remarkably, the firm was only dissolved in 1965; see *London Gazette*, 23 November 1965, 11003.

post of President) in 1904 and was replaced by Roscoe, who had hitherto been the Treasurer.

Ironically, one major reason for the establishment of the institute, namely the treatment of rabies, had vanished by this time, as Pasteur had predicted. The government had introduced the muzzling of dogs and quarantine for animals coming into Britain in 1897; within a year, rabies had practically disappeared from the country. In 1913, Roscoe asked to be relieved of his Chairmanship of the Governing Body on the grounds of his health and was replaced by Bradford, who was a professor of medicine at University College. At this point, a possible merger with the new Medical Research Committee (MRC)—set up under the National Insurance Act of 1911—was under consideration. Once again, the institute was in a precarious financial position as it was suffering competition from commercial firms, most notably Burroughs, Wellcome with its extensive research facilities. The idea of merging with the MRC and providing it with a research institute seemed a logical one.

The proposal was warmly supported by Lord Iveagh, as he had seen his endowment as a gift to the nation, and he offered to build a small hospital next to the Lister Institute to become part of the new research institute. The merger was also welcomed by Martin and Roscoe, but was opposed by Bradford. The institute negotiated with John Fletcher, Baron Moulton, the Chairman of the Medical Research Committee, and agreement was reached in July 1914. The merger was put to the Governing Body and it was accepted by five votes to two. Bradford then resigned as Chairman and was replaced by Roscoe, despite his age and poor health. Sir David Bruce, a bacteriologist, had just returned from Africa and vehemently objected to the merger. The complex organization of the institute then began to work against the merger. The Governing Body and the Council were both answerable to the membership of the institute, who were a mixture of the original subscribers, senior staff at the institute, and eminent persons invited by the Council to become members. They elected 12 members of the council. Bruce rallied the members against the merger and a meeting was held on 18 November 1914.[39] Roscoe, as the Chairman, tried to dissipate any concern that a merger with the MRC would lead to political control of the research. The problem was that anything funded by Parliament had to be overseen by MPs, but Roscoe said that any attempt at political interference would be unthinkable. The opposition to the merger thus struck at the heart of all that Roscoe stood for. He had argued for state support for scientific research for decades, but he was now faced with opposition to the very idea of state funding. He was supported by Iveagh, the institute's main sponsor, and by Moulton, who remarked (as a lawyer) that "in scientific minds abstract difficulties loomed too large." Lankester argued against the merger, saying that the members could not agree to the institute being merged "with a committee of which they knew nothing, and the purpose of which was not identical with that of the Lister Institute." The motion to merge was carried on a show of hands, but was lost by the narrow margin of 39 votes to 32 when the proxy votes were counted. Roscoe concluded the meeting by saying that "the governing body must reconsider their plans." However, the Governing Body felt unable to do anything about this reverse

---

[39] "The Lister Institute: Proposal for Government Control Rejected," *The Times*, 19 November 1914, 5. All the quotations are taken from this article.

in the midst of a war. Perhaps Roscoe's increasing ill health also had something to do with this lack of action. The MRC set up the National Institute for Medical Research (NIMR), which went to Hampstead, where premises had already been acquired; it remained there until it moved to Mill Hill in 1950. Many of the staff of the Lister Institute, including Martin, then joined the war effort, leaving the institute with a skeleton staff. After Roscoe's death in 1915, Bruce became Chairman, and the institute remained independent until it was closed in 1978.

Science Museum

Another institution close to Roscoe's heart was the science collections of the South Kensington Museum, which could be used to enhance the cultural status of science. The Science Museum as it exists today had a long and complicated gestation, but it is essentially a child of the Great Exhibition of 1851.[40] Exhilarated by the success of the exhibition, one of the few World Fairs to make a profit, Prince Albert, the Prince Consort, had the idea of using these profits to create a interconnecting network of institutions near the site of the exhibition in South Kensington to promote industrial design and more broadly the sciences and the arts (in the sense of industrial arts, not the fine arts). There would be educational institutions, a museum, and a meeting place (now the Albert Hall). They would all be part of a new government department, the Science and Art Department. The science collections were given a boost in the summer of 1876 by the holding of a special exhibition, across Exhibition Road from the main museum, of historical and modern scientific instruments and apparatus, the so-called Special Loan Collection of Scientific Apparatus exhibition. Roscoe's friend Norman Lockyer was appointed the curator, and Roscoe worked hard to obtain funding for the exhibition.[41]

After the success of the exhibition, visited by almost a quarter of a million people, leading scientists, including Roscoe, attempted to get the government to acquire the loans (as far as possible) for the museum, to become the core of a new science-based collection. This effort petered out when the government refused the offer of £100,000 (£9.2 million in 2023 terms) from the 1851 Commission to pay for this material, although some objects were donated to the museum by the lenders, including demonstration apparatus given by August Wilhelm Hofmann. A committee was set up in 1881 to consider the future of the museum, and in 1882 another committee of engineers was created to consider what to acquire from the neighboring Patent Office Museum, after it was closed as a result of the 1883 Patent Act. The Treasury announcement of this committee used the term "Science Museum" for the first time. Then in 1885 there was a commission on the housing of the South Kensington Museum. If all these commissions were not enough (although they produced little in the way of action as far as the science collections were concerned), there was yet

---

[40] This account is based on Robert Bud, "Responding to Stories: The 1876 Loan Collection of Scientific Apparatus and the Science Museum," *Science Museum Journal* 1 (Spring 2014), http://journal.sciencemus eum.ac.uk/browse/2014/responding-to-stories/; and Bud, "Infected by the Bacillus of Science," 11–40.
[41] Roscoe, *Life and Experiences*, 284–285.

another committee in 1889 and this time, as an MP, Roscoe was a member. According to Roscoe, "an influential Member of Parliament, ... remarked that he thought a great deal of space was wasted by the exhibition of a quantity of old iron and worn-out models that ought to be consigned to the rubbish-heap."[42] The committee, which consisted of Lord Francis Hervey, Sir Bernhard Samuelson, Roscoe, and General Sir John Donnelly of the Science and Art Department, emphasized the historical importance of the collection and urged for the establishment of a small council to supervise the science collections, a recommendation which was not implemented. Roscoe knew the staff of the nascent Science Museum well, and with his fellow members of the British Science Guild, he continued to campaign for the creation of an independent science museum.[43]

Although little progress was made in this respect, the art and science collections continued to grow apart. The art collections were increasingly influenced by the burgeoning Arts and Crafts movement, while the science collections were becoming more about the history of technology, partly as a result of the merger with the Patent Office Museum in 1883. This growing split was symbolized by the appointment of Major-General Edward Festing of the Science and Art Department as the Director of the Science Museum in 1893, although this museum did not exist as a separate institution.

Matters might have continued to drift if it had not been for two major events. The first was the commencement of the construction of the new Victoria and Albert Museum to replace the South Kensington Museum in 1899; the foundation stone was laid by Queen Victoria in one of her last public acts. Could the art and science collections continue to coexist under the same title? Seven years later, the foundation stone of the Deutsches Museum in Munich was laid by Queen Victoria's nephew Kaiser Wilhelm II. This showed how Germany was showcasing the importance of science and technology, while Britain was again lagging behind. Once the new Liberal government was voted in by a landslide in 1906, the intellectual and financial conditions for the creation of a proper Science Museum had been established. But how was it to be brought into being? The Permanent Secretary at the new Board of Education (which replaced the Science and Art Department), Sir Robert Morant, was both a supporter of science and an advocate of the German threat. He was thus a natural supporter of the Science Museum, once the Liberals were able to fund it through increased taxation. There was a need to get the plan for a new science museum in place before the new Victoria & Albert Museum was formally opened at the end of June 1909. Morant saw the plan through with an ingenious multi-pronged attack. He got the 1851 Commission to revive their offer of £100,000, spurned by the government three decades earlier. He then got Roscoe to produce a plea to the government for a new science museum, signed by all the notable scientists of the day in July 1909. Morant then

[42] Ibid., 297.

[43] The British Science Guild was formed in October 1905; "The British Science Guild," Nature (12 October 1905): 585–586. Although Richard Haldane was the first president, Norman Lockyer was the moving force behind the guild; see A. J. Meadows, Science and Controversy: A Biography of Sir Norman Lockyer, Founder of Nature, 2nd ed. (London: Macmillan, 2008), 270–279, and Roy MacLeod, "Science for Imperial Efficiency and Social Change: Reflections on the British Science Guild, 1905–1936," Public Understanding of Science 3 (1994): 155–193.

pulled his masterstroke by arguing to his government masters that Queen Victoria had intended the new title of Victoria and Albert Museum to be used only for the art collections, thereby leaving the science collections without an institutional home. In this way the Science Museum was brought into being. The President of the Board of Trade, Walter Runciman, set up a committee headed by the industrialist Sir Hugh Bell in April 1910 to consider the future development of the Science Museum, which delivered its second and final report in 1912.[44] Roscoe was appointed to the Science Museum advisory council, which replaced the Bell Committee in January 1913, which was also chaired by Bell.[45] However, he was suffering ill-health by this time and his impact on the development of the new Science Museum was limited. As it turned out, the construction of the new building was delayed by World War I, and the museum was not formally opened until March 1928, long after Roscoe's death. Morant himself left the Board of Education under a cloud in 1911 because of the failings of the school inspectorate.[46]

Final Years

Roscoe was keen on foreign travel and in later life went to the south of France, Italy, Egypt, Tunis, Algiers, and even Biskra, 200 miles south of Algiers.[47] This was done partly, no doubt, to alleviate his gout, which was the only major affliction he had until he suffered an attack of pneumonia in the winter of 1902. He recovered after staying in Mürren and Burgenstock in Switzerland during the following summer and the winter in Algiers and Sicily, but it would be fair to say that his health was never quite the same again.[48] Perhaps aware that his time was running out, Roscoe published his autobiography with Macmillan in 1906.

Roscoe celebrated the fiftieth anniversary ("Jubilee") of his graduation at Heidelberg on 22 April 1904 at Whitworth Hall, Manchester, an event organized by his old friend, student, and colleague Sir Edward Thorpe.[49] Roscoe was presented with a large number of addresses from universities (including of course Heidelberg) and messages from chemists, most notably Marcellin Berthelot, Henri Moissan, Dimitri Mendeleev, Konrad Beilstein, and Stanislao Cannizzaro. For whatever reason, there were few foreign chemists present, but Wilhelm Ostwald and two Japanese chemists gave addresses. Roscoe was very moved by this celebration and reproduced some of the addresses as an appendix to his autobiography.

Another highlight of this period was the Seventh International Congress of Applied Chemistry in May and June 1909 in South Kensington, with the plenary lectures held

---

[44] *The Times*, 1 April 1910, 14.

[45] *The Times*, January 4, 1913, 9.

[46] Morant was then given the task of implementing the National Insurance Act of 1911, including setting up the Medical Research Committee, which we have met in connection with the Lister Institute. He became permanent secretary of the new Ministry of Health in 1919, but died of influenza in March 1920. *ODNB*.

[47] Roscoe, *Life and Experiences*, 371–382.

[48] Thorpe, *Roscoe*, 197.

[49] Roscoe, *Life and Experiences*, 357–361; "The Celebration of Sir Henry Roscoe's Graduation Jubilee," *Nature* (28 April 1904): 613–614.

Figure 11.2 Woodcote Lodge, near Guildford, Surrey.

at the Royal Albert Hall.[50] The patron of the meeting was King Edward VIII, who had studied chemistry at the University of Edinburgh, and the conference was opened by the vice-patron, the Prince of Wales, whom Roscoe had met at the Athenaeum Club in 1904. Roscoe was the Honorary President of the conference and took part in running it, despite his age. Thousands of chemists from around the world were present and there were several social events, most notably the party given by Ludwig Mond which was attended by no less than 1,700 people. The next conference was to be held in Washington, D.C., in 1912 and the Americans played a leading role, including the U.S. ambassador Whitelaw Reid. This conference thus showed the increasing importance of the United States in chemistry and especially applied chemistry. In October, Roscoe formally opened the new John Morley Laboratories in Manchester, commissioned by William Henry Perkin, Jr.[51] Perhaps not coincidentally, Whitelaw Reid was one of the few people given an honorary degree that day. Roscoe was admitted to the Privy Council in November 1909, one of the relatively few members of this elite body who was not an active politician.[52]

Roscoe was much affected by the sudden death of his beloved wife on 5 July 1910. Roscoe then mostly lived in quiet retirement in his country residence near Leatherhead, Surrey (Figure 11.2). When he first moved to London, he had the idea of having a country retreat. In 1892, he was able to obtain Woodcote Lodge, a cottage with an attached farm of 70 acres in West Horsley near Guildford in Surrey in

[50] "International Congress of Applied Chemistry," *Journal of the Society of Chemical Industry* 28 (1909): 580–584; Charles Baskerville, "The Seventh International Congress of Applied Chemistry," *Science*, new series, 30 (September 17, 1909): 374–384. The conference was well-covered in *The Times*, 3 May, 18; 24 May, 11; 27 May, 12; 28 May, 8 and 9; 29 May, 8; 31 May, 8; 1 June, 8; 2 June, 3, 4, 8, and 11; and 3 June, 6.

[51] *The Times*, 5 October 1909, 7.

[52] *The Times*, 23 November 1909, 13.

the Surrey Hills, about 26 miles by road from Kensington, on the property of another Fellow of the Royal Society, William King-Noel, 1st Earl of Lovelace, the widower of the now celebrated pioneer of computing (and daughter of Lord Byron) Ada Lovelace.[53] Roscoe spent weekends there, and by 1914, he lived there more or less permanently. He would often invite friends down to Woodcote to stay with him; they were almost invariably fellow chemists and former students.[54] Roscoe characteristically threw himself into the running of the rather run-down farm, breeding Jersey cattle, Berkshire pigs, and poultry as well as growing cereals, potatoes and vegetables.[55]

On his 80th birthday in January 1913, his former students donated a bust of Roscoe (Figure 11.3) by Arthur Drury RA to the Chemical Society (which was given to the society in November with Roscoe present).[56] A deputation led by Sir Edward Thorpe, accompanied by Arthur Smithells, Phillips Bedson, Arthur Harden, Arthur Crossley, Charles Keane, John Bevan (of rayon fame), and Watson Smith, visited Woodcote Lodge and gave him an address from 140 of his former students.[57] In July 1914, Roscoe was presented with the medal of the Society of Chemical Industry by Rudolph Messel at the society's Annual General Meeting in Nottingham.[58]

He was presumably saddened, if not surprised, by the outbreak of war between Britain and Germany. For years he had been warning in print of the dangers of a rift between the two countries he loved dearly.[59] He was overwhelmed by the bellicose attitude of his German friends, especially Ostwald who had been one of the few guests from abroad at his Jubilee celebration. Soon after the outbreak of war, Roscoe wrote to Thorpe:

> I have been laid up, more or less, since the war broke out with dyspepsia and gout, but now I am recovering.... What do you say to Ostwald! ... I agree with you that his swelled head is cracked. What horrors! One can scarcely believe that the German, as you and I knew him, could have assumed such brutal characteristics as we read of.[60]

In a letter to Thorpe in February 1915, Roscoe remarked, "The Government have been bamboozled by a want of scientific acumen at the head—a not unusual occurrence *chez nous*."[61] This was an anxiety shared by other scientists.[62] In a letter to *The*

---

[53] Roscoe, *Life and Experiences*, 386. For an obituary of Lord Lovelace, see https://www.icevirtuallibrary.com/doi/pdf/10.1680/imotp.1894.20022.

[54] Roscoe, *Life and Experiences*, 391–394.

[55] Roscoe, *Life and Experiences*, 391; Thorpe, *Roscoe*, 198–200.

[56] *The Times*, 21 November 1913, 6.

[57] "Sir Henry Roscoe's Eightieth Birthday," *Chemical News* 107 (27 January 1913): 31; *Manchester Guardian*, 8 January 1913, 6.

[58] *Manchester Guardian*, 17 July 1914, 10.

[59] Thorpe, *Roscoe*, 179–189.

[60] Letter to Thorpe, dated 20 September 1914, in Thorpe, *Roscoe*, 201.

[61] Letter to Thorpe, dated 19 February 1915, in ibid., 202.

[62] For the background to the exchange of letters between Armstrong and Roscoe, see Iain Varcoe, "Comment: Practical Proposals by Scientists for Reforming the Machinery of Scientific Advice, 1914–17," *British Journal for the History of Science* 33 (2000): 109–114. I wish to thank William H. Brock for this reference. For a broader account, see Andrew Hull, "War of Words: The Public Science of the British Scientific Community and the Origins of the Department of Scientific and Industrial Research, 1914–16," *British Journal for the History of Science* 32 (1999): 461–481.

Figure 11.3 Bust of Harry Roscoe in the Royal Society of Chemistry's rooms at Burlington House, London.
Image courtesy of the Royal Society of Chemistry Library.

Times in July 1915, Henry E. Armstrong noted, "There is no proper organized body of scientific opinion as yet behind our Government."[63] Specific scientific advice was being given to the government by advisory committees set up by the Royal Society and the scientific societies, but there was a general feeling that there should be a broader forum of scientists to assist the war effort. The Royal Society was one obvious candidate to create such a forum, but the leadership of the society had imposed a rule of secrecy on its committees, so members of one committee could not discuss a problem with another committee, nor did it approve of a general meeting. In his letter

[63] Henry E. Armstrong, letter in *The Times*, 15 July 1915, 7.

to *The Times* on 15 July, Armstrong urged the Royal Society to set up grand committees along disciplinary lines. An anonymous FRS criticized part of Armstrong's rather rambling letter and appealed for loyalty in a letter published in *The Times* the next day,[64] and the physicist Ambrose Fleming supported Armstrong in a letter published on 17 July.[65] Roscoe entered the debate over a week later with a letter published on 5 August.[66] He supported the idea of a general meeting of scientists and suggested that the forthcoming meeting of the British Association for the Advancement of Science (BAAS) offered such a forum. On the 9th, Armstrong welcomed Roscoe's support for a meeting, but objected to the BAAS meeting on the grounds that the gentleman enforcing the culture of secrecy at the Royal Society (namely Arthur Schuster, the secretary of the Royal Society) would be chairing the meeting as the new president of the BAAS.[67] Roscoe then protested in the columns of *The Times* that Armstrong had taken his support for a meeting about the war effort at the BAAS as support for Armstrong's plans for the Royal Society.[68] By assuming that Roscoe was supporting him, Armstrong had put Roscoe in a difficult position. Armstrong was at loggerheads with Schuster. As well as being a friend of Schuster since his Owens College days, Roscoe would not have wanted anything to come between them while Schuster was organizing the upcoming meeting of the BAAS. Hence, Roscoe's irritation at Armstrong's letter of 9 August is understandable. This appears to have been Roscoe's last foray into the public arena.

The annual meeting of the BAAS in Manchester between 7 and 11 September 1915 was surely one that Roscoe wanted to attend, because of his debate with Armstrong, because it was in Manchester and because he was a Vice-President of Council. However, a telegram from Roscoe was read out at the beginning of the General Meeting:

> My best wishes for the success of the Meeting. I greatly regret that I cannot be present to support my distinguished friend, the President [Arthur Schuster]. I send my love to Manchester.[69]

Clearly his health was failing fast, but he carried on as usual at Woodcote Lodge until the final evening.[70] Following an attack of angina just after dawn on 18 December 1915, Roscoe died of cardiac failure, three weeks before his 83rd birthday, which he had been looking forward to celebrating with some of his former students.[71] A memorial service for Roscoe was held at Rosslyn Hill Unitarian Chapel, Hampstead, on 22 December, attended by Viscount Iveagh who had supported the Lister Institute,

---

[64] "A Fellow [of the Royal Society]," letter in *The Times*, 16 July 1915, 9.

[65] J. A. Fleming, letter in *The Times*, 17 July 1915, 7.

[66] Henry E. Roscoe, letter in *The Times*, 5 August 1915, 5.

[67] Henry E. Armstrong, letter in *The Times*, 9 August 1915, 9.

[68] Henry E. Roscoe, letter in *The Times*, 13 August 1915, 7.

[69] *Report of the Eighty-fifth Meeting of the British Association for the Advancement of Science, Manchester, 1915, September 7–11* (London: John Murray, 1916), 50.

[70] Thorpe, *Roscoe*, 203.

[71] As stated in the death certificate and thus not a heart attack, as stated by several biographies (e.g., *ODNB*).

his fellow former Liberal MP Sir William Mather, his former students J. J. Thomson and Sir Edward Thorpe, and Professors Dixon and Schuster from the University of Manchester, among others.[72] The funeral took place at Brookwood Cemetery in Woking, Surrey, where he is buried with his wife.[73]

[72] *The Times*, December 23, 1915, 11.
[73] *The Times*, December 21, 1915, 13 and personal communication from Maria Brownsea of Brookwood Cemetery, 15 November 2021.

12
Roscoe's Legacy

The Invisibility of Roscoe

During his lifetime, Harry Roscoe was given many honors. In addition to being a Knight Bachelor and a member of the Privy Council, he was:

D.C.L. of Oxford and LL.D. of Cambridge, Dublin, Glasgow, and Montreal; D.Sc. of Aberdeen, Liverpool, and Victoria. On the occasion of the eighth jubilee of the foundation of Heidelberg University, he was made an honorary M.D. He was an Officer of the French Legion of Honour.[1]

Yet Roscoe is hardly prominent today. There has not been a biography since 1916, and his name rarely appears in other biographies. Only in the history of spectroscopy and photochemistry does his name come up with any regularity.[2] As far as we are aware, there is no statue of Roscoe; though there is a Roscoe building at the University of Manchester, erected in 1964. There is no apparatus named after Roscoe, and there is no Roscoe reaction (mainly because he was not an organic chemist); nor is there a Roscoe test. His work on photochemistry and exotic metals did not have a major impact on later research in these fields. There is the Bunsen-Roscoe Law in photochemistry, which is effectively the same as the Grotthuss-Draper Law. He does not have a lasting reputation as a chemical researcher. However, he did much to popularize spectroscopy. He also developed magnesium-based flash photography. Consequently, Roscoe is not regarded today as one of the leading British scientists of the nineteenth century, alongside Darwin, Faraday, Joule, and Huxley. Nor can he be rehabilitated as a leading scientist as Colin Russell strove to do for Edward Frankland, who had fallen into a similar obscurity.

Roscoe was eager to reform chemical education, but he did not succeed in the short run, although most universities now incorporate some original research into their undergraduate degrees; most notably at Oxford where the fourth-year research Part II was introduced just after Roscoe's death by another professor from Manchester,

---

[1] T.E.T[horpe], "Sir Henry Roscoe," *Proceedings of the Royal Society of London* A93 (1917): i–xxi, on xx. The "eighth jubilee" of the foundation of the University of Heidelberg was held in 1886; see Henry Enfield Roscoe, *The Life and Experiences of Sir Henry Enfield Roscoe DCL, LLD, FRS, Written by Himself* (London: Macmillan, 1906), 75–76; Roscoe says it was 400 years after it was founded, but it was actually 500 years (see Chapter 7). The expression may be Roscoe's own invention, as it does not seem to appear anywhere else. Roscoe could have been made a member of the Order of Merit after it was founded by King Edward VII in June 1902, alongside Lord Kelvin, Lord Lister, Lord Rayleigh, and Sir William Huggins, but he was not. They had been (Kelvin, Lister), were (Huggins), or would become (Rayleigh) President of the Royal Society, one of the few positions in the scientific hierarchy never held by Roscoe.

[2] For example, in Klaus Hentschel, *Mapping the Spectrum: Techniques of Visual Representation in Research and Teaching* (Oxford: Oxford University Press, 2002).

Henry Enfield Roscoe. Peter J.T. Morris and Peter Reed, Oxford University Press. © Oxford University Press 2024.
DOI: 10.1093/oso/9780190844257.003.0012

William Henry Perkin Jr. He supplied professors to other northern universities, but was not a prolific "chemist breeder" in the mold of Justus von Liebig or Robert Bunsen.

Unusually for a chemist, Roscoe was a member of Parliament for 10 years and was an effective backbencher in a period when his party was largely out of power. Yet there is no great act of Parliament associated with his name, and he did not help to found a political party like Baron Kelvin. His plans for compulsory metrication and local temperance came to naught. Unlike Lyon Playfair and Richard Haldane, he did not join the cabinet.

He supported industry in various ways and argued for the value of an academic education for those working in industry. Yet he was not associated with any chemical companies except for the innovative Castner–Kellner Company, which was partly the result of a prejudice in this period (and long afterward) against academics being associated with industry or being paid by firms. Few of his students went into industry, and many of those who did set up their own small firms. When the mighty Imperial Chemical Industries (ICI) was formed in 1926, there were no Roscoe students in its senior management, and the firm drew its scientifically trained managers from Oxford, not Manchester. He strongly believed that pure chemistry had to be the basis of a chemical career, even one in industry, and it was not the task of academic chemists to teach practical skills for industrial work. He later modified his views to the extent of teaching technological chemistry as part of a broader education in chemistry, but he was not one of the founders of chemical engineering. He did, however, employ Watson Smith, who can be regarded as a pioneer in the teaching of technological chemistry, but his departure from Manchester was not Roscoe's fault and indeed was partly the result of Roscoe leaving the university. His greatest weakness in this regard was perhaps his failure to realize that it was not enough to supply chemists for industry; one also had to foster new branches of industry that would employ chemists—he thought in terms of supply, rather than demand.

If all this is true, why is Roscoe a significant figure in the history of science? We would argue that his impact lay in three areas: textbooks, institution building, and the expansion of technical (rather than specifically chemical) education. Before we review these areas, it is important to reflect on the influence of Germany on Roscoe's thinking and outlook.

Roscoe and Germany

Roscoe's attachment to Germany and his high opinion of German scientific education was a vein which ran through his life and career, shaping much of what he did and thought. What was his legacy in this area? At the time of his death the Anglo-German relationship lay in ruins. Even leading German scientists were supporting the aggressive policies of the German government. At the same time, the war could be seen as confirming Roscoe's main argument that German education was preparing the country for the future in a way that British education was not.

Roscoe's error was to admire German education and German universities without considering the path Germany was taking on the international stage. As mentioned at the end of Chapter 4, his Germany was composed of small liberal university towns like

Heidelberg and liberal German academics. This can be considered short-sightedness, but even on these narrow terms, his view was flawed. For he overlooked Heinrich von Treitschke, the ultra-nationalist historian, who was at Heidelberg University between 1867 and 1873. Roscoe claimed he did not know Treitschke, but was he quite as ignorant of Treitschke as he made out? His friend Heinrich Helmholtz was one of Treitschke's few friends in Heidelberg and engineered his appointment in Berlin.[3] If he was unfamiliar with these currents in German academia (and German society as a whole), it could be argued that this was a culpable ignorance on his part. Perhaps he looked forward to the reign of the Emperor Friedrich and his British wife Victoria, which would have perhaps been a liberal period, but we will never know, as he reigned for only 98 days.

Even if we ignore the wider context of German education, was Roscoe's advocacy of German educational methods wise? No country likes to be told that its way of doing things is not as good as another country; and certainly not one as powerful as Britain in the late nineteenth century. It would have been better to advocate reform without pointing out that you are basing your reform on a system used by another country, especially when that country is a major rival. If appeal has to be made to what is happening elsewhere, it surely would be better to refer to a number of countries adopting that approach to education rather than one country, even if these other countries (e.g., the United States) are in fact copying the German system.

To go further, was it a wise move to advocate the German approach to education, even if it worked well in Germany (which is at least debatable)? No two countries are alike, and what works in one is not necessarily effective in another nation. To give credit to Roscoe, he did work within the English system of technical education, setting up a series of public lectures and developing evening classes at Owens College. Would he have been more effective if he had embraced this approach wholeheartedly, for example advocating apprenticeships rather than an academic training? Or building up postgraduate chemical education in Britain rather than sending his students to Germany? Perhaps it was impossible for an academically trained chemist, from a comfortable middle-class background, who was partly trained in Germany before taking up an academic career, to escape from his life experience.

Did Roscoe's Germanophilia leave any tangible legacy? He died in the middle of World War I and thus could not help to restore scientific relations with Germany afterward, when many scientists would have left Germany outside the scientific community altogether. This difficult task was carried out in chemistry by William Pope and Arthur Eddington for physics. Pope had no direct connection with Roscoe and was influenced by Henry Edward Armstrong, who taught him at Finsbury Technical College, London.[4] Eddington had been at Owens College before going to Cambridge,

---

[3] David Cahan, *Helmholtz: A Life in Science* (Chicago: University of Chicago Press, 2018), 342 and 419–420.

[4] For Pope, see *ODNB*; and Arnold Thackray and Mary Ellen Bowden, "The Rise and Fall of the 'Papal State,'" in *The 1702 Chair of Chemistry at Cambridge: Transformation and Change*, ed. Mary Archer and Christopher Haley (Cambridge: Cambridge University Press, 2005), 189–209; and Danielle Fauque, "Reorganizing Chemistry after World War I: The Birth of the International Union of Pure and Applied Chemistry (IUPAC)," *Rendiconti dell'Accademia Nazionale delle Scienze detta dei XL* (2020): 75–86.

but he was taught by Arthur Schuster (who was German) rather than by Roscoe; he was also influenced by his Quaker pacificism.[5] Anglo-German relations in chemistry have been celebrated by taking August Wilhelm Hofmann or Justus Liebig as the link between Britain and Germany, rather than Roscoe.[6]

Roscoe did much to foster academic links between Britain and Germany, sending his best students to Heidelberg and other German universities for their postgraduate training. After the rupture of World War I, this model lost its luster and British chemists took their postgraduate education at home. Eventually, British chemists looked to the United States for inspiration and training, and the brain drain to America got underway in the 1960s. Germany's former preeminence in chemistry gave way to American supremacy and Roscoe's ideal of "Heidelberg du Feine" faded into history.

Textbooks

Roscoe wrote textbooks on three levels: elementary, secondary, and tertiary education. It perhaps frustrated him, as we noted in Chapter 7, that his textbooks were influential in inverse proportion to the level of their teaching (and indeed their bulk), since by far the most read of his books was his small primer of chemistry, which was used in secondary schools as much as in the elementary schools for which it was intended (Figure 12.1). It was translated into many languages, including some Asian languages, and was particularly influential in Japan and Korea. As it encouraged young children to carry out simple experiments, it perhaps fostered an interest in chemistry sets and the setting up of primitive laboratories in sheds (as in the case of one of the authors, as well as Roscoe) for many years afterward. Other later chemists, including Frank Sherwood Taylor, wrote books of home experiments which must have been influenced by Roscoe's primer.[7] Indeed, as the primer taught chemistry through experiments, it can be argued that it even presaged the Nuffield approach to the teaching of chemistry in the 1970s and 1980s. However, its authors were probably more influenced by the ideas of Henry Edward Armstrong and his heuristic method of teaching chemistry.[8] Roscoe and his colleague Schorlemmer expended much effort in the writing of their massive treatise of chemistry, but we suspect it was used more as a reference book than a textbook. All these works were much used and translated into other languages. The treatise did much to bring the periodic table to the attention of chemists at a time when its future was still in doubt. They were read by students and chemists up to World War II, when they were gradually replaced by a later generation of textbooks.

[5] For Eddington, see *ODNB*.

[6] For example, the celebration of the 150th anniversary of the Gesellschaft Deutscher Chemiker by the Royal Society of Chemistry in October 2017; see https://www.rsc.org/events/detail/27201/symposium-commemorating-the-150th-anniversary-of-the-gesellschaft-deutscher-chemiker.

[7] F. Sherwood Taylor, *The Young Chemist* (London: Nelson, 1934).

[8] For example, see Robin Millar, "Training the Mind: Continuity and Change in the Rhetoric of School Science," *Journal of Curriculum Studies* 17 (1985): 369–382.

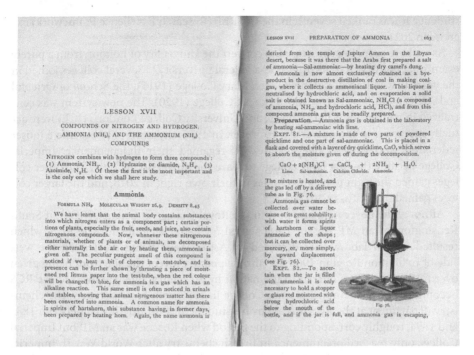

Figure 12.1 Pages (162–163) on ammonia from Sir Henry Roscoe and Joseph Lunt, *Inorganic Chemistry for Beginners* (London: Macmillan, 1893, 1895 reprint), with the sentence about the smell of heated cheese quoted by Beatrix Potter in Figure 4.1.

Institution Building

When Roscoe arrived at Owens College in 1857, it was a small and almost bankrupt institution. Owens College was moved to new buildings in Oxford Road in 1873. By the time he resigned his Chair almost 30 years later, it was part of Victoria University, a federation of Owens and University College Liverpool, largely thanks to his own efforts. Leeds joined Victoria University in 1887, but Liverpool left in 1904 and Leeds a year later. Victoria University merged with its remaining member, Owens College, and became the Victoria University of Manchester, usually just called Manchester University. By this time, thanks to new laboratories built during the professorship of William Henry Perkin Jr., Manchester was one of the most prestigious chemistry departments in the country, sending its professors to Oxford (Perkin, Robert Robinson, and Ewart Jones), Cambridge (Alexander Todd), and Imperial College (Ian Heilbron). In 2004, the Victoria University of Manchester merged with UMIST (University of Manchester Institute of Science and Technology) to form the University of Manchester; Victoria University thus ceased to exist 124 years after it was formed. The new university is now the second largest conventional university in the United Kingdom by total enrollment and also by the number of postgraduate students. The chemistry department has around 600 undergraduate students and 200 postgraduate students. It was ranked seventh in the UK REF (Research Excellence Framework) exercise in 2021.[9]

As we have seen, Roscoe helped to convert the University of London from a purely examining body back to its origins as a federal university.[10] Several colleges joined in the years that followed; notably Queen Mary College (1915), the School of Oriental and African Studies (1916), and Birkbeck College (1920). The new system was centralized in nature, and King's College and University College London (UCL) were actually merged into the university. However, in 1978, the university became a looser federation, and King's College and UCL regained their independence. Imperial College became the first large college to leave the university in 2007. The other colleges gained the right to be universities in their own right in 2018, and the larger colleges gained this status in 2023.

Although he never held any post in the Royal College of Science (RCS) in South Kensington, as an MP Roscoe campaigned for better facilities for the RCS, which resulted in new laboratories designed by Aston Webb being erected on the west side of Exhibition Road in 1906. He also helped to prevent the site from being given over to the art gallery offered to the nation by Sir Henry Tate. He lived to see RCS becoming part of the new Imperial College in 1907 with his assistance. The chemistry department at the RCS was led by his student Sir Edward Thorpe between 1885 and 1894, roughly corresponding to the period when Roscoe was an MP, but Imperial College only reached preeminence in British chemistry after World War II. Imperial College now has around 17,000 students, and the chemistry department has about 900 students.[11]

Roscoe had a long association with the science collections of the South Kensington Museum which were renamed the Science Museum in 1885, although they remained part of the South Kensington Museum. The collections were then housed in the rather unsatisfactory galleries built for the 1862 International Exhibition. With Norman Lockyer (and later Robert Morant at the Board of Education, which was responsible for the South Kensington Museum as the successor of the Science and Art Department), Roscoe campaigned both for the independence of the Science Museum from the South Kensington Museum, which was renamed the Victoria & Albert Museum in 1899, and better accommodation for its collections. Thanks to their efforts, the Science Museum became an independent institution in 1909 and moved into a new building on the west side of Exhibition Road between 1925 and 1928. In 1910, the number of visitors to the Science Museum was 462,000. The number of visitors reached one million in 1929 after it was formally opened in March 1928, two

[9] See entries in *Wikipedia* for the University of Manchester and the Chemistry Department at the University of Manchester, https://en.wikipedia.org/wiki/Victoria_University_of_Manchester; https://en.wikipedia.org/wiki/University_of_Manchester; and https://en.wikipedia.org/wiki/Department_of_Chemistry,_University_of_Manchester.

[10] Negley Harte, *The University of London, 1936–1986: An Illustrated History* (London: Athlone Press, 1986), 162–285.

[11] Hannah Gay and William P. Griffith, *The Chemistry Department at Imperial College London: A History, 1845–2000* (London: World Scientific, 2017); and the entry in *Wikipedia* on the chemistry department at Imperial College, https://en.wikipedia.org/wiki/Department_of_Chemistry,_Imperial_College_London.

million in 1968, three million in 1977, and a peak of 4.2 million in 1980; the last pre-Covid pandemic (2019) figure was 3.3 million.[12]

As we have seen, Roscoe helped to found the Lister Institute of Preventive Medicine in 1891, and he worked hard to ensure its survival when it went through a rocky patch soon afterward. For many years, the institute subsided on its income from making and selling vaccines. By the 1970s, it was continually running annual deficits and its facilities in Chelsea and Elstree were sold in the late 1970s. The Chelsea building was converted into a private hospital. The net funds raised from these sales were used to set up a research fellowship scheme, which was replaced by a prize scheme in 2004. During its existence between 1891 and the 1970s, the Lister Institute made major advances in the development of vaccines, the study of blood, microbiology, immunology, and especially biochemistry. Its most notable achievements include the synthesis of adenosine triphosphate (ATP), the determination of the structure of co-enzyme A, and the development of Factor VIII concentrate for hemophiliacs. Arthur Harden, a student of Roscoe, won a Nobel Prize in 1929 for his work on fermentation at the Lister Institute.[13]

Roscoe was also instrumental in setting up the Society of Chemical Industry (SCI) in 1881 and served as its first President. He ensured that it was both a national society and not one limited to industrial chemists, but open to anyone interested in the chemical industry. The SCI was particularly influential around World War I, when the chemical industry was essential to Britain's survival. It was involved in setting up the International Union of Pure and Applied Chemistry in 1919 and the Institution of Chemical Engineers in 1922.[14] The following year the SCI founded a new journal, *Chemistry and Industry*, which was very broad in its coverage and soon took over the role formerly played by William Crookes's *Chemical News*. After World War II the SCI strengthened its links with agriculture and biotechnology. More recently it reinforced its connections with materials science, environmental science, and energy. In 1955 the SCI moved into its current imposing premises in Belgrave Square—Roscoe would surely be delighted that it is only a stone's throw from the German embassy. From an early membership of 300, the SCI grew to over 5,000 in 1920, reached a peak of just under 6,000 in the early 1950s, and its membership in 2023 was 7,500.[15]

The Association of Public-School Science Masters was founded in 1901 by four science teachers at the leading public school, Eton College. Roscoe had been appointed to the governing body of Eton by the Royal Society in 1889, replacing Thomas Huxley, and he served until 1912.[16] Perhaps surprisingly, given his background, he was proud of his association with Eton. He successfully pushed for the construction of new

[12] Peter J.T. Morris, ed., *Science for the Nation: Perspectives on the History of the Science Museum* (Basingstoke, Hants: Palgrave Macmillan, 2010), appendix 3; and ALVA, 2019 Visitor Figures, https://www.alva.org.uk/details.cfm?p=610.

[13] The former website of the Lister Institute of Preventive Medicine at the Wayback Machine, https://web.archive.org/web/20120214000139/http://www.lister-institute.org.uk/scientificheritage.html.

[14] See the society's website at https://www.soci.org/about-us/history.

[15] Earlier figures from figure 8 in the appendix to Colin A. Russell, Noel G. Coley, and Gerrylynn K. Roberts, *Chemists by Profession: The Origins and Rise of the Royal Institute of Chemistry* (Milton Keynes: Open University Press, 1977); 2023 figure from Ian Stewart, Stakeholder Engagement Manager, SCI, personal communication, 14 February 2023.

[16] Roscoe, *Life and Experiences*, 253–254; Thorpe, *Roscoe*, 165–167.

chemistry and physics laboratories. He also encouraged the school to allow boys to take up science and modern languages at an earlier age than previously allowed—the main curriculum at Eton was based on the classics and remained so for many years. It is quite likely that the changes brought about by Roscoe enabled Eton-educated Henry Moseley (1887–1915) to pursue a scientific career and thus discover the physical basis of the periodic table that Roscoe had done much to promote.[17] Roscoe had chaired a meeting of science teachers organized by the London County Council in January 1900 by virtue of being Vice-Chancellor of the University of London, which led to Eton masters writing a letter advocating the setting up of the association.[18] A year later the founding meeting of the association was held in rooms in the University of London provided by Roscoe, who chaired the meeting. David Layton has suggested that Roscoe at least probably encouraged the Eton masters to proceed when they decided to form the association. The association later became the Science Masters Association, which merged with the Association of Women Science Teachers to form the Association for Science Education in 1963.[19] The association supports science teachers and the teaching of science in schools. It publishes several journals, including *Education in Science* and *School Science Review*. It had over 8,000 members in 2022.[20]

Technical Education

Among the strongest parts of Harry Roscoe's legacy is his role as a Victorian campaigner in the field of education, and technical education in particular. From his regular periods in Heidelberg with Bunsen, his tireless work at Owens College and the Victoria University as professor of chemistry, and through his period as a parliamentarian, Roscoe took a strong advocacy role to advance all levels of science education, from elementary schools through to technical education. Even with his many commitments, Roscoe always found the time and energy as opportunities arose: acting as a member of a Royal Commission or Select Committee, giving oral or written evidence to formal inquiries, or when provided with an appropriate speaking engagement. His outstanding contribution as a member of the Royal Commission of Technical Education (1881) led to his knighthood in 1884.

Looking at the wider context, the nineteenth century is rightly considered an important period in British history for major reforms in voting expansion, local authority powers, abolition of slavery, public health, and environmental regulation, but reforms in education never received the parliamentary attention they warranted. The recommendations of most nineteenth-century Royal Commissions and Select Committees devoted to education remained "on the shelf," and successive governments' general lack of commitment toward education was most markedly reflected

---

[17] For Moseley, see Chapter 7.

[18] David Layton, *Interpreters of Science: A History of the Association for Science Education* (London: John Murray, 1984), 4–5.

[19] Layton, *Interpreters of Science*, and the ASE's website, https://www.ase.org.uk/.

[20] Personal communication from Melanie Bennett, membership officer of the ASE, 7 December 2022.

in the delayed appointment of a Minister of Education to plan and oversee education policy and funding across its many strands.

The 1870 Education Act was important, but did not go far enough in addressing the need for an increasingly educated population that would enable Britain to compete economically with other leading nations. The Technical Instruction Act of 1889 was viewed by Roscoe as a first installment, coming before his 1891 Act that allowed available government funding (the "whisky money") to be more widely applied in support of technical education.

At the time of Roscoe's death in 1915 during the early stage of World War I, little had changed as far as education policy was concerned. During the postwar period and the economic depression of 1926, any changes in education policy were piecemeal and lacked the broad systematic review that was so urgently needed to strengthen Britain's economic position in an increasingly competitive world driven by the United States and Germany. Such policy changes would have to await the 1944 Education Act, though the first framework for a coherent system for higher education only came in 1963 with the Robbins Report.

Before reviewing British education policy changes in the periods to the 1944 Education Act and then to the 1963 Robbins Report, what were the issues that concerned Roscoe and had informed his advocacy for educational advancement? The issues focused on: government funding for universities, the research role for universities and research collaboration with industry, and the role for technical colleges. The overarching priority for Roscoe was the appointment of a Minister for Education who would have oversight of all education provision in Britain.

World War I had confirmed the remarkable advantage Germany had acquired through its huge investment in scientific and industrial research in the period through the second half of the nineteenth century and to the beginning of the war. Britain was forced to catch up on its research activities during the war, marking a sharp lesson that informed its postwar policies toward scientific and industrial research, financial support for universities, and increasing the number of scientists, technologists, and engineers. There was now greater government interest, though implementation continued to lag.

In 1916 the government created the Department for Scientific and Industrial Research (DSIR) to encourage research in industry and educational institutions for the postwar period rather than for the war effort. The DSIR's annual grant was allocated to encourage and support research in universities, technical colleges (and other institutions), and also create research organizations to advance trade and industry.

The years between World War I and World War II saw successive governments dramatically increase state grants for universities (see Table 12.1). From 1919 the review and allocation of these grants became the purview of the government-created Universities Grants Committee (UGC), responsible to the Treasury rather than the Board of Education (which was still acting effectively as the Ministry of Education).[21] The increases in grants reflected the urgent national need to increase the number of students studying science, engineering, and technology. In 1912–1913 there were only

[21] John Carwell, *Government and the Universities in Britain: Programme and Performance 1960–1980* (Cambridge: Cambridge University Press, 1985), 10–15.

Table 12.1 Government Grants to
Universities

Period	Grants (£)
1903–1904	27,000
1919–1920	692,150
1928–1929	1,535,230
1938–1939	2,007,900
1945–1946	5,149,000

Data from Argles, *From South Kensington to Robbins*
(1964), 72.

1,487 full-time students of engineering and technology in universities and colleges in England and Wales, and 1,199 students in technical colleges; this was in sharp contrast to the 11,000 students in the German Technische Hochschulen.[22]

In 1918 the Thomson Committee had highlighted: the need for additional funding for university science (an issue later taken up by the UGC); the requirement of a year of research after a first degree; that the appointment of senior staff should reflect not just research ability but also teaching aptitude; and that heads of technological departments should be allowed time for private practice.[23] But by 1922–1923 the proportion of full-time technology students compared to full-time science students had dropped.[24]

Even with the remarkable increases in UGC grants, the situation through to 1938–1939 had not improved much due to the General Strike of 1926 and the worldwide industrial Depression that followed; a period when there was a move away from the sciences to the arts. There were 5,970 science students and 3,882 technologists in 1922–1923, and 6,061 science and 4,217 technology students in 1938–1939 in English universities and university colleges.[25] But as Michael Argles has highlighted, the World War II years "led to a great amount of rethinking in the field of education."[26]

The key outcome came in 1943 with the White Paper, *Educational Reconstruction*, drafted by the Board of Education (under its then President, R. A. [Rab] Butler) that acknowledged the inadequate provision of education for those up to the age of 15.[27] The White Paper also touched on the inadequacy of further education provision due to the spotty response of Local Authorities; where technical education was provided, it was subpar for an advanced industrial nation.[28] The subsequent 1944 Education

[22] Michael Argles, *South Kensington to Robbins: An Account of English Technical and Scientific Education since 1851* (London: Longmans, Green, 1964).

[23] *Natural Science in Education* (London: HMSO, 1918). See also Argles, *South Kensington to Robbins*, 72–73.

[24] Ibid., 73.

[25] Ibid., 74.

[26] Ibid., 83.

[27] *Educational Reconstruction*, P.P. 1943 (6458). For Butler, see *ODNB*.

[28] *Educational Reconstruction*, 21.

Act proved to be very effective in addressing issues that had remained unresolved for many decades.[29] Key among its provisions were creation of a Ministry of Education and a Minister of Education (for which Roscoe had campaigned so strongly over many years) and strengthening the education provision for all to age 15.[30]

A Ministry of Education booklet, *Further Education: The Scope and Content of Its Opportunities under the Education Act, 1944*, published in 1947, set out the hoped-for outcomes of the Education Act 1944 as far as further education was concerned.[31] The aim was for a more uniform provision with better facilities and equipment, and with cooperation between industry and colleges.[32]

The future provision of higher and technical education largely stood outside the 1944 Education Act, but was addressed by a series of government-appointed committees between 1945 and 1963 when the Robbins Committee set out the first blueprint for a coherent system of higher education in Britain.[33] The first committee was the Percy Committee of 1945. Chaired by Lord Eustace Percy, a former President of the Board of Education with a lifelong commitment to expanding access to scientific and industrial training.[34] The main recommendation of the Committee was the designation of a small number of technical colleges developing university-standard courses with postgraduate study.[35] Unfortunately the Committee's recommendations, like those of most earlier education reports, languished.

The following year the Barlow Committee was asked to review the country's likely future scientific manpower requirement and how it could be achieved.[36] It was chaired by the civil servant Sir Alan Barlow and included such distinguished members as P.M.S. Blackett, Solly Zuckerman, C.P. Snow, and Geoffrey Crowther. Its main findings included: 70,000 graduate-qualified scientists were required by 1950 and 90,000 by 1955, and the annual output of 2,500 should be doubled; the recommendations of the earlier Percy Committee were supported; and university provision should be expanded with government support. By 1950 the number of full-time graduate students in technology had doubled since 1938–1939 (5,288 in 1938–1939; 10,933 in 1949–1950) and postgraduate students had also increased.

In 1953 the House of Commons Select Committee on Estimates issued a blistering indictment on postwar governments for their miserly approach to education funding.[37] In response, David Eccles, the newly appointed Conservative Secretary of State for Education, drew on the recommendations of the Barlow and Percy reports in his 1954 policy plan for education costing some £85m and set to start within five years. Eccles's plan was published in 1956 as a White Paper on technical education

[29] *Education Act 1944*, P.P. 1944 (c31).

[30] Argles, *South Kensington to Robbins*, 84–85.

[31] Ministry of Education, *Further Education: The Scope and Content of Its Opportunities under the Education Act, 1944* (London: HMSO, 1947).

[32] Ibid., 13–14 and 15–16.

[33] John Carswell, *Government and the Universities in Britain: Programme and Performance, 1960–1980* (Cambridge: Cambridge University Press, 1985), 145–146.

[34] For Percy, see *ODNB*. See also Eustace Percy, *Education at the Crossroads* (London: Evans Bros., 1930).

[35] Ministry of Education, *Higher Technological Education* (London: HMSO, 1945).

[36] Lord President of the Council, *Scientific Manpower* (London: HMSO, 1946). For Barlow, see *ODNB*.

[37] Correlli Barnett, "Prelude to Industrial Defeat: From the 1944 Education Act to the 1956 White Paper on Technological Education," *Royal Society of Arts Journal* 149, no. 5501 (2002): 37–39, on 38.

and included a key initiative to develop 25 regional colleges as Colleges of Advanced Technology (CATs) (some 20 years after Barlow had recommended them), where technology would be taught at the university level to increase the number of engineers and technologists; and a 10-year development program was proposed to address the ongoing deficiencies.[38] Sandwich courses that embraced part-time in college and part-time in industry were proving popular and were to be extended.[39] The White Paper also set out the four categories of qualification: university degree; technical college diploma; National Diplomas and Certificates; and City & Guilds.[40]

With government inertia prevailing yet again, it was not until 1957 that the then Minister of Education, Viscount Hailsham, announced the designation of eight Colleges of Advanced Technology (CATs). The CATs were to focus on advanced and postgraduate study and research; have representatives of industry, Local Education Authorities (LEAs), universities, and professional bodies on their managing councils; and appoint qualified staff as in universities.[41]

Quite separately, the UGC had continued to develop the university sector in the postwar period while ensuring "the grant did not exceed reasonable bounds and was properly administered."[42] From 1953 the expansionist approach had accelerated with the appointment of Sir Keith Murray (former Rector of Lincoln College, Oxford) as Chairman for the next 10 years.[43] During his tenure "the annual capital expenditure on universities increased fifteen fold, and seven new universities planned and designated."[44]

The foregoing summarizes the ad hoc manner in which plans were reviewed (and only partly implemented) in the postwar period in Britain. More concerning was the messy and unsystematic framework that had come about by 1960, and which faced the Robbins Committee during their deliberations between 1961 and 1963.

When Conservative Prime Minister Harold Macmillan announced the Robbins Committee, the remit was intentionally broad:

> To review the pattern of full-time education in Great Britain and in the light of national needs and resources to advise Her Majesty's Government on what principles its long-term development should be based. In particular, to advise, in the light of these principles, whether there should be any changes in pattern, whether any new types of institution are desirable and whether any modifications should be made in the present arrangements for planning and coordinating the development of the various types of institution.[45]

The Committee was chaired by Professor Lord (Lionel) Robbins (Chairman of Governors at the London School of Economics) and its members included Anthony

---

[38] Barnett, "Prelude to Industrial Defeat," 39.

[39] H. V. Lowry, "Technical Education in Great Britain," *Nature* 177 (1956): 970–971, on 970.

[40] *White Paper on Technical Education* (London: HMSO, 1956). City & Guilds qualifications were developed for learners who wished to train for a technical occupation.

[41] Ministry of Education, *The Organisation of Technical Colleges, Circular 305/56* (London: HMSO, 1956).

[42] Carswell, *Government and the Universities*, 13.

[43] Geoffrey Caston, "Keith Anderson Hope Murray, Baron Murray of Newhaven (1903–1993)," *ODNB*.

[44] Caston, "Keith Murray," *ODNB*.

[45] *Higher Education. Report of the Committee Appointed by the Prime Minister under the Chairmanship of Lord Robbins 1961–63.* P.P. 1963 (2154), iii.

Chenevix-Trench (Head Master of Eton College), Sir Patrick Linstead (Rector of Imperial College London) and Sir Philip Morris (Vice-Chancellor of Bristol University). The deliberations of the Committee were supported by Professor Claus Moser (Professor of Social Statistics at the London School of Economics) who collated the extensive statistical information about higher education in Britain and overseas (in particular Western Europe, the United States, and the Soviet Union).[46]

Having undertaken a thorough survey of the existing provision of higher education in all its forms in Britain, the Committee structured its final Report around a series of major issues linked to its remit;[47] but not before setting down its position on the aims of higher education:

> We must postulate that what is taught should be taught in such a way as to promote the general powers of the mind. The aim should be to form not mere specialists but rather cultivated men and women. And it is the distinguishing characteristic of a healthy higher education that, even where it is concerned with practical techniques, it imparts them on a plane of generality that makes possible their application to many problems.... It is this that the world of affairs demands of the world of learning.[48]

This is not the place to review all the recommendations, but to highlight those issues that Roscoe had strongly advocated for. He would have particularly welcomed the increased capacity for higher education study at both undergraduate and postgraduate level, the focus on engineering and technological education at CATs and technical colleges, the in-depth analysis of the financial resources required for capital and recurring expenditure and the adoption of a 10-year forward plan, and the closer focus on the education and training of teachers to meet the increasing number of young people studying at school.

The Committee also recommended a Ministry of Art and Science with a Secretary of State for Art and Science who would work with a revised UGC, and have the ability to negotiate with the Treasury on the annual capital and recurring funding as part of the 10-year plan. The more prominent profile of science in government and a place in the cabinet, where most legislative decisions were made, would have been wholeheartedly welcomed by Roscoe.

When the government issued its formal response a month after the Report's publication in October 1963, it broadly welcomed the recommendations. Given the level of detail and its long-term consequences for education and science, such a prompt acceptance was remarkable, surprising, and perhaps even refreshing, given the usual practice was for such reports to "rest on the shelf" for years; or more often than not, to be forgotten. Both main political parties (Conservative and Labour) committed to implementing the recommendations of the Robbins Report and adopting them as a key part of their campaign statements for the next General Election.[49] This was almost unheard of in British parliamentary history and reflected the growing concern among

---

[46] Ibid., iii. For Lord Moser, see *ODNB*.
[47] *Higher Education*, vi–x, and 257.
[48] *Higher Education*, 6.
[49] Carswell, *Government and the Universities*, 26.

the public over adequate opportunities in higher education given the high birth rate after the end of World War II.

The government's subsequent review of the Committee's recommendations led to the creation of the Department of Education and Science (DES) that better reflected the emerging umbilical cord between education and science.[50] With the election of Labour Prime Minister Harold Wilson in the October 1964 General Election, the recommendations were subsequently contained in the 1965 Science and Technology Act due to the leadership of Anthony Crosland as Secretary of State for Education.[51] The Act also created research councils, including the Science Research Council (SRC), which were charged with advising the Secretary of State for Education on research policy.[52]

With education and science brought together in the new department's responsibilities, Roscoe would certainly have welcomed the DES as the culmination of his campaigning beyond all other issues. The creation of the post of Minister of Education would also have chimed with the views of his grandfather William Roscoe on education that his grandson had highlighted during his address to the Liverpool Royal Institution on 21 December 1888.[53] However, Roscoe would have been disappointed, though probably not too surprised, that such an appointment had taken some 75 years from his Liverpool address and almost 50 years after his death in 1915.

Roscoe the Campaigning Chemist

Chemists are often considered to work in ivory towers, carrying out research in laboratories with little or no impact on ordinary life. Roscoe was the antithesis of the ivory tower chemist. He was an energetic academic administrator who had an impact on three modern-day universities, and he helped to found a major research institute—and saved it from closing toward the end of his life. He was also instrumental in establishing two important societies which today have a membership numbering in the thousands. His textbooks were read by thousands of students from primary school to university and were translated into numerous languages. Even more remarkably, he served as the MP for a Manchester constituency for 10 years, campaigning for electoral reform, the reform of the House of Lords, improved working conditions for calico printers, and Irish Home Rule. Above all, he tirelessly campaigned for wider access to education at all levels, and especially the expansion of technical education.

Roscoe sought to create a Britain in which the population was better educated and workers were more highly skilled, which in his view would lead to Britain being a more prosperous and competitive nation. While other scientists are often lauded and remembered for one specific achievement, Roscoe's legacy compasses many achievements across a broad swath of cultural life due to his energetic and tireless campaigning over 60 years of his life.

---

[50] Ibid., 74.
[51] Ibid., 65–66.
[52] *Science and Technology Act 1965*, P.P. 1965 (c.4).
[53] Henry Enfield Roscoe, "An Educational Parallel," *Journal of Education*, new series, 11 (1 February 1889): 118–120.

A Complete Bibliography of Henry Enfield Roscoe's Publications

Only British editions of Roscoe's books are listed here, for reasons of space. Most of them were also published in the United States and translated into many languages. Publications in the *Proceedings of the Royal Institution of Great Britain* are listed under the year the lecture was delivered, not the date of publication. Publications in the *Proceedings of the Manchester Literary and Philosophical Society* have not been listed if they also appear in the *Memoirs*. Regarding the numbering of the volumes of these *Memoirs*, the numbering is that used at the time (series 3, vol no.), but there is an alternative method extant which uses a continuous numbering of all the volumes. This number can be obtained by adding 20 to the series 3 volume number, for example, volume 5 of series 3 can be written as 25.

1854

"Notiz über die Zusammensetzung einiger Gneise." *Justus Liebigs Annalen der Chemie* 91 (1854): 302–306 (with Franz Schönfeld).

1855

"Ueber das Verhalten des Chlors bei der Absorption in Wasser." *Justus Liebigs Annalen der Chemie* 95 (1855): 357–372 (published as A. E. Roscoe [sic]).
"II. Photochemische Untersuchungen, Erste Abhandlung." *Annalen der Physik* 172 (1855): 373–394 (with Robert W. Bunsen). Papers in the *Annalen der Physik* used to be cited as *Poggendorfs Annalen* with a different volume number.

1856

"II. On the Absorption of Chlorine in Water." *Quarterly Journal of the Chemical Society of London* 8 (1856): 14–26.
"X. Photochemical Researches." *Quarterly Journal of the Chemical Society* 8 (1856): 193–211.
"On the Measurement of the Chemical Action of Light." *Proceedings of the Royal Institution of Great Britain* 2 (1854–1858): 223–225. Lecture delivered on 4 April 1856.
"Photochemical Researches with Reference to the Law of the Chemical Action of Light." In *Report of the Twenty-Fifth Meeting of the British Association for the Advancement of Science; Held at Glasgow in September 1855*, 48–49 (with Robert W. Bunsen). London: John Murray, 1856.

1857

Introductory Lecture on the Development of Physical Science. London: Longman, Brown, Green, Longmans, and Roberts, 1857.

"III. Photochemische Untersuchungen. Zweite Abhandlung. Maassbestimmung der chemischen Wirkungen des Lichts." *Annalen der Physik* 176 (1857): 43–88 (with Robert W. Bunsen).

"I. Photochemische Untersuchungen. Dritte Abhandlung. Erscheinungen der photochemischen Induction." *Annalen der Physik* 176 (1857): 481–516 (with Robert W. Bunsen).

"IV. Photochemische Untersuchungen. Vierte Abhandlung. Optische und chemische Extinction der Strahlen." *Annalen der Physik* 177 (1857): 235–263 (with Robert W. Bunsen).

"XVII. Photo-Chemical Researches. Part I. Measurement of the Chemical Action of Light." *Philosophical Transactions of the Royal Society of London* 147 (1857): 355–380 (with Robert W. Bunsen).

"I. Photo-Chemical Researches. Part II. Phenomena of Photo-Chemical Induction." *Proceedings of the Royal Society of London* 8 (1857): 326–330 (with Robert W. Bunsen).

"XVIII. Photo-Chemical Researches. Part II. Phenomena of Photo-Chemical Induction." *Philosophical Transactions of the Royal Society of London* 147 (1857): 381–402 (with Robert W. Bunsen).

"X. Photochemical Researches. 3rd Communication. The Optical and Chemical Extinction of the Chemical Rays." *Proceedings of the Royal Society of London* 8 (1857): 516–520 (with Robert W. Bunsen).

"XXIX. Photo-Chemical Researches. Part III. Optical and Chemical Extinction of the Chemical Rays." *Philosophical Transactions of the Royal Society of London* 147 (1857): 601–620 (with Robert W. Bunsen).

"LVII. On the Influence of Light upon Chlorine: To the Editors of the *Philosophical Magazine and Journal.*" *The London, Edinburgh, and Dublin Philosophical Magazine and Journal of Science* 14 (1857): 504–506.

"Photo-Chemical Researches." In *Report of the Twenty-Sixth Meeting of the British Association for the Advancement of Science; Held at Cheltenham in August 1856,* 62–68 (with Robert W. Bunsen). London: John Murray, 1857.

1858

"XXVI. Some Chemical Facts Respecting the Atmosphere of Dwelling-Houses." *Quarterly Journal of the Chemical Society of London* 10 (1858): 251–269.

1859

"I. Photochemische untersuchungen, Fünfte Abhandlung. Die Sonne." *Annalen der Physik* 184 (1859): 193–273 (with Robert W. Bunsen).

"XXXV. Photo-Chemical Researches. Part IV." *Philosophical Transactions of the Royal Society of London* 149 (1859): 879–926 (with Robert W. Bunsen).

"Ueber die Absorption des Chlorwasserstoffs und des Ammoniaks in Wasser." *Justus Liebigs Annalen der Chemie* 112 (1859): 327–355 (with William Dittmar).

1860

"VIII. Photochemical Researches. Part IV." *Proceedings of the Royal Society of London* 10 (1860): 39–49 (with Robert W. Bunsen).

"XV. On the Absorption of Hydrochloric Acid and Ammonia in Water." *Quarterly Journal of the Chemical Society of London* 12 (1860): 128–151 (with William Dittmar).

"Ueber die Zusammensetzung der wässerigen Säuren von constantem Siedepunkt." *Justus Liebigs Annalen der Chemie* 116 (1860): 203–220.

"On the Alleged Practice of Arsenic-Eating in Styria." *The London, Edinburgh, and Dublin Philosophical Magazine and Journal of Science* 20 (1860): 550–551.

"On the Measurement of the Chemical Action of the Solar Rays." *Proceedings of the Royal Institution of Great Britain* 3 (1858–1862): 210–217. Lecture delivered on 2 March 1860.

1861

"XVIII. On the Composition of the Aqueous Acids of Constant Boiling Point." *Quarterly Journal of the Chemical Society of London* 13 (1861): 146–164.

"On Bunsen and Kirchhoff's Spectrum Observations." *Chemical News* 3 (9 March 1861): 153–155; (16 March 1861): 170–172.

"On the Application of the Induction Coil to Steinheil's Apparatus for Spectrum Analysis." *Chemical News* 4 (31 August 1861): 118–122.

"On Bunsen and Kirchhoff's Spectrum Observations." *Proceedings of the Royal Institution of Great Britain* 3 (1858–1862): 323–328. Lecture delivered on 1 March 1861.

1862

"Note on Perchloric Ether." *Journal of the Chemical Society* 15 (1862): 213–215.

"On Perchloric Acid and Its Hydrates." *Proceedings of the Royal Society of London* 11 (1862): 493–503.

"Ueber die Ueberchlorsäure, deren Hydrate, und einige Salze derselben." *Justus Liebigs Annalen der Chemie* 121 (1862): 346–356.

"XXXVI. On the Composition of the Aqueous Acids of Constant Boiling Point. Second Communication." *Journal of the Chemical Society* 15 (1862): 270–276.

"VIII. On the Solar Spectrum, and the Spectra of the Chemical Elements: To the Editors of the *Philosophical Magazine and Journal*." *The London, Edinburgh, and Dublin Philosophical Magazine and Journal of Science* 23 (1862): 63–64.

"On the Recent Progress and Present Condition of Manufacturing Chemistry in the South Lancashire District." In *Report of the Thirty-Third Meeting of the British Association for the Advancement of Science held in Manchester in 1861*, 108–128 (with E. R. Schunck and R. Angus Smith). London: John Murray, 1862.

"On the Effect of Increased Temperature upon the Nature of the Light Emitted by the Vapour of Certain Metals or Metallic Compounds." *Proceedings of the Literary and Philosophical Society of Manchester* 2 (1862): 227–230 (with Robert Bellamy Clifton).

"On the Alleged Practice of Arsenic-Eating in Styria." *Memoirs of the Literary and Philosophical Society of Manchester*, series 3, 1 (1862): 208–221.

"Course of Three Lectures on Spectrum Analysis." *Chemical News* 5 (19 April 1862): 218–222; (19 May 1862): 261–265; (24 May 1862): 287–293.

1863

"Photochemische untersuchungen. VI. Abhandlung. Meteorologische Lichtmessungen." *Annalen der Physik* 193 (1863): 529–562 (with Robert W. Bunsen).

"III. Photochemical Researches. Part V. On the Measurement of the Chemical Action of Direct and Diffuse Sunlight." *Proceedings of the Royal Society of London* 12 (1863): 306–312 (with Robert W. Bunsen).

"VII. Photo-Chemical Researches. Part V. On the Direct Measurement of the Chemical Action of Sunlight." *Philosophical Transactions of the Royal Society of London* 153 (1863): 139–160 (with Robert W. Bunsen).

"XV. On Perchloric Acid and Its Hydrates." *Journal of the Chemical Society* 16 (1863): 82–88.

"IX. On the Measurement of the Chemical Brightness of Various Portions of the Sun's Disc." *Proceedings of the Royal Society of London* 12 (1863): 648–650.

"Ueber die Zusammensetzung der wässerigen Säuren von constantem Siedepunkt." *Justus Liebigs Annalen der Chemie* 125 (1863): 319–327. [This is a different paper from the one with the same title published in 1860.]

"On the Existence of a Crystallizable Carbon Compound and Free Sulphur in the Alais Meteorite." *The London, Edinburgh, and Dublin Philosophical Magazine and Journal of Science* 25 (1863): 319–320.

"On the Direct Measurement of the Sun's Chemical Action." *Proceedings of the Royal Institution of Great Britain* 4 (1862–1866): 128–134. Lecture delivered 22 May 1863.

"Ueber die Bestimmung der chemischen Helligkeit von verschiedenen Theilen der Sonnenscheibe." *Annalen der Physik* 196 (1863): 331–333.

"Atmosphere." In *A Dictionary of Chemistry and the Allied Branches of Other Sciences*, edited by Henry Watts, vol. 1, 437–440. London: Longmans, Green, 1863. [Given the cross-reference to "Ventilation" it would seem that Roscoe expected to write that entry, but it never appeared.]

1864

"IV. On a Method of Meteorological Registration of the Chemical Action of Total Daylight." *Proceedings of the Royal Society of London* 13 (1864): 555–559.

"The Existence of a Crystallisable Carbon Compound and Free Sulphur in the Alais Meteorite." *Proceedings of the Manchester Literary and Philosophical Society* 3 (1864): 57–59.

"Note on the Amount of Carbonic Acid Contained in the Air of Manchester." *Proceedings of the Literary and Philosophical Society of Manchester* 3 (1864): 219–223.

"On the Measurement of the Chemical Brightness of Various Portions of the Sun's Disk." *Journal of the Franklin Institute* 77 (1864): 106–108.

"Note on the Existence of Lithium, Strontium, and Copper in the Bath Waters." *Chemical News* 10 (1 October 1864): 158.

"On the Metal Indium and Recent Discoveries on Spectrum Analysis." *Proceedings of the Royal Institution of Great Britain* 4 (1862–66): 284–290. Lecture delivered 6 May 1864.

"Gases, Absorption by Liquids and Solids." In *A Dictionary of Chemistry and the Allied Branches of Other Sciences*, edited by Henry Watts, vol. 2, 790–805. London: Longmans, Green, 1864.

1865

"XII. The Bakerian Lecture: On a Method of Meteorological Registration of the Chemical Action of Total Daylight." *Philosophical Transactions of the Royal Society of London* 155 (1865): 605–631.

"Einfaches Instrument zu meteorologischen Lichtmessungen in allgemein vergleichbarem Maasse." *Annalen der Physik* 200 (1865): 353–390.

"Light." In *A Dictionary of Chemistry and the Allied Branches of Other Sciences*, edited by Henry Watts, vol. 3, 589–695. London: Longmans, Green, 1865.

1866

Lessons in Elementary Chemistry. London: Macmillan, 1866.

"XLVII. On the Isomorphism of Thallium-Perchlorate with the Potassium and Ammonium-Perchlorates." *Journal of the Chemical Society* 19 (1866): 504–505.

"Ueber die relativen chemischen Intensitäten des directen und zerstreuten Sonnenlichtes." *Annalen der Physik* 204 (1866): 291–298 (with Joseph Baxendell).

"On the Opalescence of the Atmosphere." *Proceedings of the Royal Institution of Great Britain* 4 (1862–1866): 651–655. Lecture delivered 1 June 1866.

1867

Lessons in Elementary Chemistry: Inorganic and Organic. London: Macmillan, 1867.

"II. Note on the Relative Chemical Intensities of Direct Sunlight and Diffuse Daylight at Different Altitudes of the Sun." *Proceedings of the Royal Society of London* 15 (1867): 20–24 (with Joseph Baxendell).

"XVII. On the Chemical Intensity of Total Daylight at Kew and Pará, 1865, 1866, and 1867." *Philosophical Transactions of the Royal Society of London* 157 (1867): 555–569.

"Ueber die chemische Intensität des gesammten Tageslichtes zu Kew und Pará." *Annalen der Physik* 208 (1867): 404–425.

"On the Isomorphism of Thallium-Perchlorate with Potassium and Ammonium-Perchlorate." *Proceedings of the Manchester Literary and Philosophical Society* 6 (1867): 9–11.

"Ueber den Isomorphismus des überchlorsauren Thalliums mit dem überchlorsauren Kalium und Ammonium." *Justus Liebigs Annalen der Chemie* 144 (1867): 127–128.

1868

"XXXVI. Researches on Vanadium." *Journal of the Chemical Society* 21 (1868): 322–350.

"VIII. On the Chemical Intensity of Total Daylight at Kew and Pará in 1865–67." *Proceedings of the Royal Society of London* 16 (1868): 41–44.

"Bakerian Lecture: Researches on Vanadium. Part I." *Proceedings of the Royal Society of London* 16 (1868): 220–228.

"I. The Bakerian Lecture. Researches on Vanadium." *Philosophical Transactions of the Royal Society of London* 158 (1868): 1–27.

"On Vanadium, One of the Trivalent Group of Elements." *Proceedings of the Royal Institution of Great Britain* 5 (1866–1869): 287–294. Lecture delivered 14 February 1868.

"Spectral Analysis." In *A Dictionary of Chemistry and the Allied Branches of Other Sciences*, edited by Henry Watts, vol. 5, 376–397. London: Longmans, Green, 1868.

1869

Spectrum Analysis: Six Lectures Delivered in 1868 before the Society of Apothecaries of London. London: Macmillan, 1869.

Lessons in Elementary Chemistry: Inorganic and Organic. 2nd edition. London: Macmillan, 1869.

"XXV. Researches on Vanadium. Part II." *Philosophical Transactions of the Royal Society of London* 159 (1869): 679–692.

"Untersuchungen über Vanadium." *Journal für Praktische Chemie* 108 (1869): 303–310.

"Science Education in Germany. I. The German University System." *Nature* 1 (1869): 157–159.

1870

Spectrum Analysis: Six Lectures Delivered in 1868 before the Society of Apothecaries of London. 2nd edition. London: Macmillan, 1870.

"XVIII. Researches on Vanadium. Part III." *Philosophical Transactions of the Royal Society of London* 160 (1870): 317–331.

"XXVI. Researches on Vanadium. Part II." *Journal of the Chemical Society* 23 (1870): 344–358.

"XVII. On the Relation between the Sun's Altitude and the Chemical Intensity of Total Daylight in a Cloudless Sky." *Philosophical Transactions of the Royal Society of London* 160 (1870): 309–316 (with Thomas Edward Thorpe).

"Atomgewicht des Vanadiums." *Zeitschrift für analytische Chemie* 9 (1870): 433–436.
"Vanadiummetall." *Zeitschrift für analytische Chemie* 9 (1870): 386–387.
"XIX. Researches on Vanadium." *Proceedings of the Royal Society of London* 18, no. 114–122 (1870): 37–42.
"On the Artificial Production of Alizarine, the Colouring Principle of Madder." *Proceedings of the Royal Institution of Great Britain* 6 (1870–1872): 120–125. Lecture delivered 1 April 1870.
"Science Education in Germany. II. The Polytechnic Schools." *Nature* 1 (1870): 475–477.
"Women at College." *Owens College Magazine* 2 (May 1870): 129–133.

1871

Lessons in Elementary Chemistry: Inorganic and Organic. 3rd edition. London: Macmillan, 1871.
Course of Four Lectures in Elementary Chemistry. In *Science Lectures for the People*, lecture 2. Manchester: John Heywood, 1871.
"XVII. On the Measurement of the Chemical Intensity of Total Daylight Made at Catania during the Total Eclipse of Dec. 22nd, 1870." *Philosophical Transactions of the Royal Society of London* 161 (1871): 467–476 (with Thomas Edward Thorpe).
"III. Researches on Vanadium. Part III." *Journal of the Chemical Society* 24 (1871): 23 36.
"VI. On the Measurement of the Chemical Intensity of Total Day-Light Made at Catania during the Total Eclipse of Dec. 22, 1870." *Proceedings of the Royal Society of London* 19 (1871): 511–514 (with Thomas Edward Thorpe).
"On the Application of the Spectroscope to the Manufacture of Steel by the Bessemer Process." *Chemical News* 23 (14 April 1871): 174–176 and (21 April 1871): 182–185.
"Zu den Reactionen der orthovanadinsauren Salze." *Zeitschrift für analytische Chemie* 10 (1871): 224–225.
"Ueber die Beziehungen zwischen der Sonnenhöhe und der chemischen Intensität des Gesammt-Tageslichtes bei unbewölktem Himmel." *Annalen der Physik* 218 (1871): 177–192 (with T. E. Thorpe).
"On Measurements of the Chemical Intensity of Total Daylight, Made during the Recent Total Eclipse of the Sun, by Lieut. J. Herschel, RE." *Memoirs of the Literary and Philosophical Society of Manchester*, series 3, 4 (1871): 202–206.
"The People's University." *Nature* 4 (1871): 41.

1872

Chemistry, Science Primers: 2. 1st and 2nd editions. London: Macmillan, 1872.
"On the Study of Some Tungsten Compounds." *Chemical News* 37 (23 February 1872): 90.
"A Study of Certain Tungsten Compounds." *Proceedings of the Manchester Literary and Philosophical Society* 11 (1872): 79–90.
"Ueber einige Wolframverbindungen." *Justus Liebigs Annalen der Chemie* 162 (1872): 349–368.
"Atomgewicht des Wolframs." *Zeitschrift für analytische Chemie* 11 (1872): 363–364.

1873

Spectrum Analysis: Six Lectures Delivered in 1868 before the Society of Apothecaries of London. 3rd edition. London: Macmillan, 1873.

"Justus Liebig." *Nature* 8 (1873): 27–28.

1874

Chemistry, Science Primers: 2. 3rd edition. London: Macmillan, 1874.

"II. On a Self-Recording Method of Measuring the Intensity of the Chemical Action of Total Daylight." *Proceedings of the Royal Society of London* 22 (1874): 158–159.

"XVIII. On a Self-Recording Method of Measuring the Intensity of the Chemical Action of Total Daylight." *Philosophical Transactions of the Royal Society of London* 164 (1874): 655–673.

"XLIV. On a New Chloride of Uranium." *Journal of the Chemical Society* 27 (1874): 933–935.

"I. Note on the Absorption-Spectra of Potassium and Sodium at Low Temperatures." *Proceedings of the Royal Society of London* 22 (1874): 362–364 (with Arthur Schuster).

"Note sur les spectres d'absorption du potassium et du sodium à de basses temperatures." *Journal de Physique Théorique et Appliquée* 3 (1874): 344–346 (with Arthur Schuster and M. Bertholomey).

"Ueber ein selbstregistrirendes Instrument zu meteorologischen Lichtmessungen in allgemein vergleichbarem Maasse." *Annalen der Physik* 227 (1874): 268–285.

"Ueber ein neues Uranchlorid." *Berichte der deutschen chemischen Gesellschaft* 7 (1874): 1131–1133.

"Original Research as a Means of Education." In *Essays and Addresses by Professors and Lecturers of Owens College, Manchester*, 21–57. London: Macmillan, 1874.

1875

Lessons in Elementary Chemistry: Inorganic and Organic. With many alterations and additions. London: Macmillan, 1875.

"XIX. On the Heat of Sunshine at London during the Twenty-four [*sic*] Years 1855 to 1874, as Registered by Campbell's Method." *Proceedings of the Royal Society of London* 23 (1875): 578–582 (with Balfour Stewart).

"A Study of Certain Tungsten Compounds." *Memoirs of the Manchester Literary and Philosophical Society*, series 3, 5 (1875): 76–99.

"On the Corrosion of Leaden Hot-Water Cisterns." *Proceedings of the Manchester Literary and Philosophical Society* 14 (1875): 23.

1876

Chemistry, Science Primers: 2. 5th edition. London: Macmillan, 1876.

Chemistry, Science Primers: 2. 6th edition, with questions. London: Macmillan, 1876.

"What the Earth Is Composed of: Three Lectures." In *Manchester Science Lectures for the People*, series 8. London: Macmillan, 1876.

"Recent Discoveries about Vanadium." *Proceedings of the Royal Institution of Great Britain* 8 (1875–1878): 221–230. Lecture delivered 2 June 1876.

"Notes on a Collection of Apparatus Employed by Dr. Dalton at the Loan Exhibition of Scientific Apparatus at South Kensington." *Proceedings of the Manchester Literary and Philosophical Society* 15 (1876): 77–82.

"Some Remarks on Dalton's First Table of Atomic Weights." *Memoirs of the Manchester Literary and Philosophical Society*, series 3, 5 (1876): 269–275.

"Physical Science in Schools." *Nature* 13 (1876): 386–387.

1877

"IV. On Two New Vanadium Minerals." *Proceedings of the Royal Society of London* 25 (1877): 109–112.

"VI. On the Absorption-Spectra of Bromine and of Iodine Monochloride." *Philosophical Transactions of the Royal Society of London* 167 (1877): 207–212 (with Thomas Edward Thorpe).

1070

A Treatise on Chemistry, vol. 1: *The Non-metallic Elements*. London: Macmillan, 1878 (with Carl Schorlemmer).

A Treatise on Chemistry, vol. 1: *The Non-metallic Elements*. 2nd edition. London: Macmillan, 1878 (with Carl Schorlemmer).

A Treatise on Chemistry, vol. 2: *Metals*, part 1. London: Macmillan, 1878 (with Carl Schorlemmer).

Lessons in Elementary Chemistry: Inorganic and Organic. 4th edition. London: Macmillan, 1878.

"VII. Note on the Specific Gravity of the Vapours of the Chlorides of Thallium and Lead." *Proceedings of the Royal Society of London* 27 (1878): 426–428.

"Note on Metallic Niobium and a New Niobium Chloride." *Chemical News* 37 (18 January 1878): 25–26.

"Note on Metallic Niobium and a New Niobium Chloride." *Memoirs of the Literary and Philosophical Society of Manchester*, series 3, 6 (1878): 186–191.

"Ueber das specifische Gewicht der Dämpfe der Chloride des Thalliums und Bleis." *Berichte der deutschen chemischen Gesellschaft* 11 (1878): 1196–1197.

1879

A Treatise on Chemistry, vol. 2: *Metals*, part 1. 2nd edition. London: Macmillan, 1879 (with Carl Schorlemmer).

"Technical Chemistry." In *Science Lectures at South Kensington*, vol. 2, 299–344. London: Macmillan, 1879.

"Ueber die Absorptionsspectren des Broms und des Jodmonochlorids." *Zeitschrift für analytische Chemie* 18 (1879): 95 (with T. E. Thorpe).

"A New Chemical Industry, Established by M. Camille Vincent." *Proceedings of the Royal Institution of Great Britain* 9 (1879–81): 51–58. Lecture delivered 21 February 1879. [The industry in question was the use of sugar beet vinasses in the chemical industry.]

1880

A Treatise on Chemistry, vol. 2: *Metals*, part 2. London: Macmillan, 1880 (with Carl Schorlemmer).

"V. On the Absence of Potassium in Protagon Prepared by Dr. Gamgee." *Proceedings of the Royal Society of London* 30 (1880): 365.

"Note on the Identity of the Spectra Obtained from the Different Allotropic Forms of Carbon." *Proceedings of the Manchester Literary and Philosophical Society* 19 (1880): 46–49.

1881

Description of the Chemical Laboratories at the Owens College, Manchester. Manchester: J. E. Cornish, 1881.

A Treatise on Chemistry, vol. 3: *The Chemistry of the Hydrocarbons and Their Derivatives, or Organic Chemistry*, part 1. London: Macmillan, 1881 (with Carl Schorlemmer).

A Treatise on Chemistry, vol. 1: *The Non-metallic Elements.* New edition. London: Macmillan, 1881 (with Carl Schorlemmer).

"IV. Note on Protagon." *Proceedings of the Royal Society of London* 32 (1881): 35–36.

"Indigo and Its Artificial Production." *Journal of the Franklin Institute* 112 (1881): 295–311.

"Indigo and Its Artificial Production." *Proceedings of the Royal Institution of Great Britain* 9 (1879–1881): 580–594. Lecture delivered 27 May 1881.

"Scientific Worthies: XVII. Robert Wilhelm Bunsen." *Nature* 23 (1881): 597–600.

1882

"XLII. A Study of Some of the Earth-Metals Contained in Samarskite." *Journal of the Chemical Society, Transactions* 41 (1882): 277–282.

"XLIII. The Spectrum of Terbium." *Journal of the Chemical Society, Transactions* 41 (1882): 283–287 (with A. Schuster).

"Ueber das Spectrum des Terbiums." *Berichte der deutschen chemischen Gesellschaft* 15 (1882): 1280–1284 (with A. Schuster).

"Sur la combustion des diamants du Cap." *Annales de Chimie et de Physique* 26 (1882): 136–141.

"Sur l'équivalent du carbone déterminé par la combustion du diamant." *Comptes Rendus* 94 (1882): 1180.

"Light, Chemical Action of." In *Dictionary of Chemistry*, edited by Henry Watts, vol. 3, 678–695. London: Longmans, Green, 1882.

"Anniversary Report." *Journal of the Chemical Society* 41 (1882): 229–239.
"President's Address." *Journal of the Society of Chemical Industry* 1 (1882): 250–252.

1883

A *Treatise on Chemistry*, vol. 2: *Metals*, part 1. New edition. London: Macmillan, 1883
(with Carl Schorlemmer).
Chemistry, Science Primers: 2. Revised edition. London: Macmillan, 1883.
"Das Aequivalentgewicht des Kohlenstoffs." *Zeitschrift für analytische Chemie* 22 (1883):
306–307.

1884

A *Treatise on Chemistry*, vol. 3: *The Chemistry of the Hydrocarbons and Their Derivatives, or
Organic Chemistry*, part 2. London: Macmillan, 1884 (with Carl Schorlemmer).
A *Treatise on Chemistry*, vol. 1: *The Non-metallic Elements*. New edition. London:
Macmillan, 1884 (with Carl Schorlemmer).
"On a New Variety of Halloysite from Maidenpek, Serbia." *Memoirs of the Manchester Literary
and Philosophical Society*, series 3, 8 (1884): 213–214. [Now called Majdanpek, Serbia.]
"On the Heat of Sunshine at the Kew Observatory, as Registered by Campbell's Method." In
*Report of the Fifty-Third Meeting of the British Association for the Advancement of Science; Held
at Southport in September 1883*, 414–418 (with Balfour Stewart). London: John Murray, 1884.
"Progress of Chemistry Since 1848." *Science* 83 (1884): 206–208.
"Remarks on the Teaching of Chemical Technology." *Journal of the Society of Chemical
Industry* 3 (1884): 592–594.

1885

Spectrum Analysis. 4th edition revised and considerably enlarged, with Arthur Schuster.
London: Macmillan, 1885.
"LXX. Note on the Spontaneous Polymerisation of Volatile Hydrocarbons at the Ordinary
Atmospheric Temperature." *Journal of the Chemical Society, Transactions* 47 (1885):
669–671.
"On the Diamond-Bearing Rocks of South Africa." *Proceedings of the Manchester Literary
and Philosophical Society* 24 (1885): 5–10.
"Progress in Chemistry since 1848." Presidential address to the chemistry section. In *Report
of the Fifty-Fourth Meeting of the British Association for the Advancement of Science; Held
at Montreal in August and September 1884*, 659–669. London: John Murray, 1885.

1886

A *Treatise on Chemistry*, vol. 3: *The Chemistry of the Hydrocarbons and Their Derivatives, or
Organic Chemistry*, part 3. London: Macmillan, 1886 (with Carl Schorlemmer).

Lessons in Elementary Chemistry: Inorganic and Organic. 5th edition. London: Macmillan, 1886.
"Notiz über die freiwillige Polymerisation flüchtiger Kohlenwasserstoffe bei gewöhnlicher Temperatur." *Justus Liebigs Annalen der Chemie* 232 (1886): 348–352.
"On Recent Progress in the Coal-Tar Industry." *Proceedings of the Royal Institution of Great Britain* 11 (1884–1886): 450–466. Lecture delivered 16 April 1886.

1887

Record of Work Done in the Chemical Department of the Owens College, 1857–1887. London: Macmillan, 1887.
"Remarks on the Synthesis of Organic Bodies." *British Medical Journal* 2 (1887): 495–497.

1888

A Treatise on Chemistry, vol.3: *The Chemistry of the Hydrocarbons and Their Derivatives, or Organic Chemistry,* part 4. London: Macmillan, 1888 (with Carl Schorlemmer).
A Treatise on Chemistry, vol. 1: *The Non-metallic Elements.* New and thoroughly revised edition. London: Macmillan, 1888 (with Carl Schorlemmer).
"Presidential Address." In *Report of the Fifty-Seventh Meeting of the British Association for the Advancement of Science; Held at Manchester in August and September 1887,* 3–28. London: John Murray, 1888.

1889

A Treatise on Chemistry, vol. 3: *The Chemistry of the Hydrocarbons and Their Derivatives, or Organic Chemistry,* part 5. London: Macmillan, 1889 (with Carl Schorlemmer).
A Treatise on Chemistry, vol. 2: *Metals,* part 2. New and revised edition. London: Macmillan, 1889 (with Carl Schorlemmer).
"LIII. On Schützenberger's Process for the Estimation of Dissolved Oxygen in Water." *Journal of the Chemical Society, Transactions* 55 (1889): 552–576 (with Joseph Lunt).
"Aluminium." *Proceedings of the Royal Institution of Great Britain* 12 (1887–1889): 451–464. Lecture delivered 3 May 1889.
"The Industrial Value of Technical Training." *The Contemporary Review* 55 (1889): 771–796.
"An Educational Parallel." *Journal of Education,* new series, 11 (1 February 1889): 118–120.
"The Life and Work of a Chemist." *Nature* 40 (1889): 578–583. [The chemist in question was Pasteur.]

1890

A Treatise on Chemistry, vol. 3: *The Chemistry of the Hydrocarbons and Their Derivatives, or Organic Chemistry,* part 2. New and revised edition. London: Macmillan, 1890 (with Carl Schorlemmer).
"An Address on the Advancement of Medicine by Research." *British Medical Journal* 2 (1890): 135.

1891

A Treatise on Chemistry, vol. 3: *The Chemistry of the Hydrocarbons and Their Derivatives, or Organic Chemistry*, part 3. New and revised edition. London: Macmillan, 1891 (with Carl Schorlemmer).

"I. Contributions to the Chemical Bacteriology of Sewage." *Proceedings of the Royal Society of London* 49 (1891): 455–457.

"IX. Contributions to the Chemical Bacteriology of Sewage." *Philosophical Transactions of the Royal Society of London (B)* 182 (1891): 633–664 (with Joseph Lunt).

"Note on the Action of Water Gas on Iron." *Proceedings of the Chemical Society of London* 7 (1891): 126–128 (with Frank Scudder).

"Notiz über Einwirkung von Wassergas auf Eisen." *Berichte der deutschen chemischen Gesellschaft* 24 (1891): 3843–3845 (with Frank Scuder [*sic*]).

"Untersuchung des Wassers." *Zeitschrift für analytische Chemie* 30 (1891): 370–378 (with Adolph F. Jolles, Joseph Lunt, G. Loof, Franz Musset, E. Schmidt, C. Denner, et al.).

1892

A Treatise on Chemistry, vol. 3: *The Chemistry of the Hydrocarbons and Their Derivatives, or Organic Chemistry*, part 6. London: Macmillan, 1892 (with Carl Schorlemmer)

Lessons in Elementary Chemistry: Inorganic and Organic. 6th edition. London: Macmillan, 1892.

"Carl Schorlemmer, LL.D., FRS." *Nature* 46 (1892): 394–395.

"The Manchester Municipal Technical School." *Nature* 47 (1892): 201–204.

"Photochemische Untersuchungen," two volumes, with Robert W. Bunsen. *Ostwald's Klassiker der exacten Wissenschaften*, no. 34. Leipzig: W. Engelmann, 1892.

1893

Inorganic Chemistry for Beginners, assisted by Joseph Lunt. London: Macmillan, 1893.

"The Secondary Education Movement." *Nature* 49 (1893): 203–204.

1894

A Treatise on Chemistry, vol. 1: *The Non-metallic Elements.* 3rd edition, assisted by H. G. Colman and A. Harden. London: Macmillan, 1894.

1895

John Dalton and the Rise of Modern Chemistry. London: Cassell, 1895.

"The "Hermite" Process of Sewage Treatment." *Journal of the Society of Chemical Industry* 14 (1895): 224–233.

1896

A New View of the Origin of Dalton's Atomic Theory: A Contribution to Chemical History, with Arthur Harden. London: Macmillan, 1896.
"Chemical Education in England and Germany." Chemical News 74 (1896): 175–177.

1897

A Treatise on Chemistry, vol. 2: Metals. New edition [3rd], completely revised, assisted by H. G. Colman and A. Harden. London: Macmillan, 1897.
"Die Genesis der Atomtheorie." Zeitschrift für Physikalische Chemie 22 (1897): 241–249 (with Arthur Harden).
"The Genesis of Dalton's Atomic Theory." The London, Edinburgh, and Dublin Philosophical Magazine and Journal of Science 43 (1897): 153–161 (with Arthur Harden).

1899

Inorganic Chemistry for Advanced Students, with Arthur Harden. London: Macmillan, 1899.
"Death of Professor Bunsen." The Times, August 17, 1899, 4.
"Professor Bunsen." Nature 60 (1899): 424–425.

1900

Lessons in Elementary Chemistry: Inorganic and Organic. 7th edition. London: Macmillan, 1900.
"Bunsen Memorial Lecture." Journal of the Chemical Society, Transactions 77 (1900): 513–554.
"Bunsen." Proceedings of the Royal Institution of Great Britain 16 (1899–1901): 437–447. Lecture delivered 1 June 1900.

1905

A Treatise on Chemistry, vol. 1: The Non-metallic Elements. 4th edition, assisted by H. G. Colman and A. Harden. London: Macmillan, 1905.

1906

The Life and Experiences of Sir Henry Enfield Roscoe DCL, LLD, FRS, Written by Himself. London: Macmillan, 1906.

1907

A Treatise on Chemistry, vol. 2: *Metals*. New edition [4th], completely revised, assisted by A. Harden. London: Macmillan, 1907.

1909

"Es soll in Europa kein Krieg mehr sein." *Deutsche Revue* 34, no. 4 (1909): 24–26.

1910

Inorganic Chemistry for Advanced Students, with Arthur Harden. 2nd edition. London: Macmillan, 1910.

1915

"L'industrie chimique en Allemagne depuis trente ans." *Revue de Métallurgie* 12 (1915): 722–728.

Bibliography of Works Consulted

Accumulated Index to the Memoirs and Proceedings, 1780-1989. Manchester: Manchester Literary and Philosophical Society, 1991.

Adunka, Roland, and Mary Virginia Orna. *Carl Auer von Welsbach: Chemist, Inventor, Entrepreneur*. Cham: Springer, 2018.

Alberti, Samuel J.T.M. "Placing Nature: Natural History Collections and Their Owners in Nineteenth-Century Provincial England." *British Journal for the History of Science* 35 (2002): 291–311.

Albini, Angelo. "Some Remarks on the First Law of Photochemistry." *Photochemical and Photobiological Sciences* 15 (2016): 319–324.

Alekperov, Vagit. *Oil of Russia: Past, Present and Future*. Minneapolis, MN: East View Press, 2011.

Alverstone, Viscount. *Recollections of Bar and Bench*. London: Edward Arnold, 1914.

"Amateur Photographic Competition (First Notice)." *Amateur Photographer* 3 (16 April 1889): 189.

Anderson, Robert G.W. "Thomas Charles Hope and the Limiting Legacy of Joseph Black." In *Cradle of Chemistry: The Early Years of Chemistry at the University of Edinburgh*, edited by Robert G.W. Anderson, 147–162. Edinburgh: John Donald, 2015.

Anschütz, Richard. *August Kekulé*, vol. 1: *Leben und Wirken*. Berlin: Verlag Chemie, 1929.

Anschütz, Richard. "Ludwig Claisen. Ein Gedenkblatt." *Berichte der Deutschen Chemischen Gesellschaft* 69, no. 7 (1936): A97–A170.

Argles, Michael. *South Kensington to Robbins: An Account of English Technical and Scientific Education since 1851*. London: Longmans, Green, 1964.

Armytage, W.H.G. *A. J. Mundella 1825-1897: The Liberal Background to the Labour Movement*. London: Ernest Benn, 1951.

Arrowsmith, Peter. "The Population of Manchester from cAD79 to1801." *Greater Manchester Archaeological Journal* 1 (1985): 99–102.

Atkins, W.R.G. "Sydney Young." *Obituary Notices of Fellows of the Royal Society* 2 (1938): 370–379.

Baker, H. B. "Frank Clowes." *Journal of the Chemical Society, Transactions* 125 (1924): 985–987.

Baker, H. B. "Francis Jones." *Journal of the Chemical Society (Resumed)* 129 (1926): 1020–1021.

Baker, H. B., and W. A. Bone. "Harold Baily Dixon." *Journal of the Chemical Society (Resumed)* (1931): 3349–3368.

Balfour, Arthur James. "Reflections Suggested by the New Theory of Matter." *Science* 20 (1904): 257–266.

Balmain, William H. *Lessons on Chemistry, for the Use of Pupils in Schools, Junior Students in the Universities, and Readers Who Wish to Learn the Fundamental Principles, and the Leading Facts*. London: Longman, Brown, Green and Longmans, 1844.

Barnett, Correlli. "Prelude to Industrial Defeat: From the 1944 Education Act to the 1956 White Paper on Technological Education." *Royal Society of Arts Journal* 149, no. 5501 (2002): 37–39.

Barton, Ruth. *The X Club: Power and Authority in Victorian England*. Chicago: University of Chicago Press, 2018.

Baskerville, Charles. "The Seventh International Congress of Applied Chemistry." *Science*, new series, 30 (17 September 1909): 374–384.

Beardsley, Edward H. *The Rise of the American Chemistry Profession, 1850–1900*. Gainesville: University of Florida Press, 1964.

Bebbington, David W. *The Nonconformist Conscience: Chapel and Politics, 1870–1914*. London: George Allen & Unwin, 1982.

Becker, Barbara J. *Unravelling Starlight: William and Margaret Huggins and the Rise of the New Astronomy*. Cambridge: Cambridge University Press, 2011.

Bedson, P. Phillips. "Lothar Meyer Memorial Lecture." *Journal of the Chemical Society, Transactions* 69 (1896): 1403–1439.

Bedson, P. Phillips. "Watson Smith." *Journal of the Chemical Society, Transactions* 117 (1920): 1637–1638.

Bedson, P. Phillips. "Georg Lunge." *Journal of the Chemical Society, Transactions* 123 (1923): 948–950.

Bedson, P. Phillips. "Sir Edward Thorpe." *Journal of the Chemical Society (Resumed)* 129 (1926): 1031–1050.

Bedson, P. Phillips. "William Carleton Williams." *Journal of the Chemical Society (Resumed)* (1927): 3202–3205.

Beer, John Joseph. *The Emergence of the German Dye Industry to 1925*. Urbana: University of Illinois Press, 1956.

Bell, James. *The Analysis and Adulteration of Foods*, 2 vols. London: Chapman and Hall, 1881–1883.

Bell, James. *The Chemistry of Tobacco*. London: n.p., 1887.

Bellot, Hugh Hale. *University College London 1826–1925*. London: University of London Press, 1929.

Ben-David, Joseph. "Scientific Productivity and Academic Organization in Nineteenth Century Medicine." *American Sociological Review* 25 (1960): 828–843.

Ben-David, Joseph. "The Universities and the Growth of Science in Germany and the United States." *Minerva* 7 (1968): 1–35.

Ben-David, Joseph, and Awraham Zloczower. "Universities and Academic Systems in Modern Societies." *European Journal of Sociology* 3 (1962): 45–84.

Bernthsen, August. "Die Heidelberger chemischen Laboratorien für den Universitätsunterricht in den letzten hundert Jahren." *Zeitschrift für Angewandte Chemie* 42 (1929): 382–384.

Berzelius, J. "Ueber das Vanadin und seine Eigenschaften." *Annalen der Physik* 98, no. 5 (1831): 1–67.

Biggs, Norman. "A Tale Untangled: Measuring the Fineness of Yarn." *Textile History* 35 (2004): 120–129.

Black, R.D. Collison, ed. *Papers and Correspondence of William Stanley Jevons*, vol. II: *Correspondence, 1850–1862*. Clifton, NJ: Augustus M. Kelley, 1973.

Borscheid, Peter. *Naturwissenschaften, Staat und Industrie in Baden, 1848–1914*. Stuttgart: Klett, 1976.

Bourne, John Michael. "The East India Company's Military Seminary, Addiscombe, 1809–1858." *Journal of the Society of Army Historical Research* 57 (1979): 206–222.

Bowen, Edmund John. "David Leonard Chapman." *Biographical Memoirs of Fellows of the Royal Society* 4 (1958): 34–44.

Bowers, Brian. *Sir Charles Wheatstone FRS, 1802–1875*. London: HMSO, 1975.

Bradshaw's General Railway Directory, Shareholders' Guide, Manual, and Almanack etc. London; Manchester: W. J. Adams, Bradshaw and Blacklock, 1852.

Brand, John C.D. *Lines of Light: The Sources of Dispersive Spectroscopy, 1800–1930*. Luxemburg City: Gordon and Breach, 1995.

Brennan, E.J.T. "Sidney Webb and the London Technical Education Board." *The Vocational Aspect of Secondary and Further Education* 11, no. 23 (1959): 85–96.

British Association for the Advancement of Science. *Report of the Eighty-Fifth Meeting of the British Association for the Advancement of Science, Manchester, 1915, September 7–11*. London: John Murray, 1916.

"The British Science Guild." *Nature* 72 (12 October 1905): 585–586.

Broadberry, Steve N. *The Productivity Race: British Manufacturing in International Perspective*. Cambridge: Cambridge University Press, 1997.

Brock, William H. *Justus von Liebig: The Chemical Gatekeeper*. Cambridge: Cambridge University Press, 1997.

Brock, William H. *William Crookes (1832–1919) and the Commercialisation of Science*. Aldershot, Hampshire: Ashgate, 2008.

Brock, William H. "The Chemical Origins of Practical Physics." In William H. Brock, *The Case of the Poisonous Socks*, 97–113. Cambridge: RSC, 2011.

Brock, William H. "Bunsen's British Students." *Ambix* 60 (2013): 203–233.

Brooke, Xanthe. "Roscoe's Italian Paintings in the Walker Art Gallery, Liverpool." In *Roscoe and Italy*, edited by Stella Fletcher, 65–93. Farnham, Surrey: Ashgate, 2013.

Brown, A. C. "William Dittmar." *Nature* 45 (24 March 1892): 493–494.

Bud, Robert. "'Infected by the Bacillus of Science': The Explosion of South Kensington." In *Science for the Nation: Perspectives on the History of the Science Museum*, edited by Peter J.T. Morris, 11–40. Basingstoke, Hampshire: Palgrave Macmillan, 2010.

Bud, Robert. "Responding to Stories: The 1876 Loan Collection of Scientific Apparatus and the Science Museum." *Science Museum Journal* 1 (Spring 2014), http://journal.sciencemuseum.ac.uk/browse/2014/responding-to-stories/.

Bud, Robert, and Gerrylynn K. Roberts. *Science versus Practice: Chemistry in Victorian Britain*. Manchester: Manchester University Press, 1984.

Bunsen, Robert. *Gasometrische Methoden*. Brunswick: F. Vieweg, 1857.

Bunsen, Robert. *Gasometry, Comprising the Leading Physical and Chemical Properties of Gases*. Translated by Henry E. Roscoe. London: Walton and Maberly, 1857.

Bunsen, Robert, and Lyon Playfair. "Report on the Gases Evolved from Iron Furnaces, with Reference to the Theory of the Smelting of Iron." In *Report of the Fifteenth Meeting of the British Association for the Advancement of Science*, 142–186. London: John Murray, 1846.

Burgess, Charles Hutchens, and David Leonard Chapman. "CXXXVIII. The Interaction of Chlorine and Hydrogen." *Journal of the Chemical Society, Transactions* 89 (1906): 1399–1434.

Burn, Richard. *Statistics of the Cotton Trade*. London: Simpkin, Marshall, 1847.

Busteed, Mervyn. *The Irish in Manchester, c. 1750–1921: Resistance, Adaptation and Identity*. Manchester: Manchester University Press, 2016.

Butler, Stella V.F. "A Transformation in Training: The Formation of University Medical Faculties in Manchester, Leeds, and Liverpool, 1870–84." *Medical History* 30 (1986): 115–132.

Cahan, David. *Helmholtz: A Life in Science*. Chicago: University of Chicago Press, 2018.

Campbell, John. *Haldane: The Forgotten Statesman Who Shaped Modern Britain*. London: C. Hurst, 2020.

Cannadine, David. *The Victorious Century: The United Kingdom, 1800–1906*. London: Viking, 2017.

Cardwell, Donald S.L. *The Organisation of Science in England*. London: Heinemann, 1972.

Cardwell, Donald S.L. *James Joule: A Biography*. Manchester: Manchester University Press, 1989.

Carswell, John. *Government and the Universities in Britain: Programme and Performance, 1960–1980*. Cambridge: Cambridge University Press, 1985.

"The Castne–Kellner Alkali Company Limited." *Chemical Trade Journal*, 3 April 1897, 238.

"Castner–Kellner Alkali Co." *Chemical Trade Journal*, 4 June 1898, 381.

"Castner–Kellner Alkali Co." *Chemical Trade Journal*, 3 June 1899, 367.

Castner–Kellner Alkali Co. *Fifty Years of Progress. The Story of the Castner-Kellner Alkali Company Told to Celebrate the Fiftieth Anniversary of its Formation, 1895–1945*. Birmingham: Kynoch Press, 1947.

Caswell, Lyman R. "Andrés del Río, Alexander von Humboldt, and the Twice-Discovered Element." *Bulletin for the History of Chemistry* 28 (2003): 35–41.

"The Celebration of Sir Henry Roscoe's Graduation Jubilee." *Nature* 69 (28 April 1904): 613–614.

Chaloner, W. H. *The Movement for the Extension of Owens College, Manchester, 1863–73*. Manchester: Manchester University Press, 1973.

Chapman, David Leonard, and Patrick Sarsfield MacMahon. "XIX. The Interaction of Hydrogen and Chlorine." *Journal of the Chemical Society, Transactions* 95 (1909): 135–138.

Chapman, S. D. *Cotton Industry in the Industrial Revolution*. London: Macmillan, 1972.

Charlton, H. B. *Portrait of a University 1851–1951*. Manchester: Manchester University Press, 1951.

Chemical Society. *The Jubilee Celebration [of the Chemical Society of London]*. London: Chemical Society, 1896.

Chick, Harriette, Margaret Hume, and Marjorie Macfarlane. *War on Disease: A History of the Lister Institute*. London: André Deutsch, 1971.

Chinnici, Ileana. *Decoding the Stars: A Biography of Angelo Secchi, Jesuit and Scientist*. Leiden: Brill, 2019.

Claisen, L., and F. E. Matthews. "On the Action of Haloïd Acids upon Hydrocyanic Acid." *Journal of the Chemical Society, Transactions* 41 (1882): 264–268.

Clapp, Brian W. *John Owens, Manchester Merchant*. Manchester: Manchester University Press, 1965.

Clarke, P. F. *Lancashire and the New Liberalism*. Cambridge: Cambridge University Press, 1971.

Cohen, J. M. *The Life of Ludwig Mond*. London: Methuen, 1956.

Cohen, Julius B. *The Owens College Junior Course of Practical Organic Chemistry*. London: Macmillan, 1887.

Cohen, Julius B. *Practical Organic Chemistry for Advanced Students*. London: Macmillan, 1900.

Coleman, Donald C. "Gentlemen and Players." *Economic History Review*, 2nd series, 26 (1973): 92–116.

Collier, Leslie. *The Lister Institute of Preventive Medicine: A Concise History*. Bushey Heath: Lister Institute of Preventive Medicine, 2000.

"Conference on Technical Education." *Journal of the Society of Arts* 14, no. 793 (31 January 1868): 183–209.

Cook, Andrew. *Cash for Honours: The Story of Maundy Gregory*. Stroud, Gloucestershire: History Press, 2008.

Corson, John W. *Loiterings in Europe*. New York: Harper & Brothers, 1848.

Cotgrove, Stephen. *Technical Education and Social Change*. London: George Allen & Unwin, 1958.

Cronin, Bernard. *Technology, Industrial Conflict and the Development of Technical Education in 19th-Century England*. Aldershot, Hampshire: Ashgate, 2001.

Crosland, Maurice. *Gay-Lussac, Scientist and Bourgeois*. Cambridge: Cambridge University Press, 1978.

Cser, Andreas. *Kleine Geschichte der Stadt Heidelberg und ihrer Universität*. Heidelberg: Kleine Buch Verlag, 2007.

Cunningham, Colin, and Prudence Waterhouse. *Alfred Waterhouse 1830–1905: Biography of a Practice*. Oxford: Clarendon Press, 1992.

Curtius, Theodor, and Johannes Rissom. *Aus der Geschichte des Chemischen Universitätslaboratoriums zu Heidelberg seit der Gründung durch Bunsen*. Heidelberg: Verlag Rochow, 1908.

Dainton, Frederick Sydney, and Brian Arthur Thrush. "Ronald George Wreyford Norrish." *Biographical Memoirs of Fellows of the Royal Society* 27 (1981): 379–424.

Davies, D. I. "Charles Loudon Bloxam." *Analytical Proceedings* 18 (1981): 327–331.

Debré, Patrice. *Louis Pasteur*. Translated by Elborg Forster. Baltimore, MD: Johns Hopkins Press, 1998.

Desmond, Adrian. *Huxley: The Devil's Disciple to Evolution's High Priest*. London: Penguin, 1997.

DeYoung, Ursula. *A Vision of Modern Science: John Tyndall and the Role of the Scientist in Victorian Culture*. New York: Palgrave Macmillan, 2011.

Dickens, Charles. *Hard Times*. London: Bradbury & Evans, 1854.

Dingle, A. E. "The Monster Nuisance of All: Landowners, Alkali Manufacturers, and Air Pollution, 1828–64." *Economic History Review* 35 (1982): 529–548.

Dixon, Harold Baily. "Berthelot Memorial Lecture." *Journal of the Chemical Society, Transactions* 99 (1911): 2353–2371.

Dixon, H. B., and J. N. Collie. "James Campbell Brown." *Journal of the Chemical Society, Transactions* 99 (1911): 1457–1460.

Donnelly, James Francis. "Chemical Education and the Chemical Industry in England from the Mid-Nineteenth to the Early Twentieth Century." PhD dissertation, University of Leeds, West Yorkshire, 1987.

Donnelly, James F. "Consultants, Managers, Testing Slaves: Changing Roles for Chemists in the British Alkali Industry, 1850–1920." *Technology and Culture* 35 (1994): 100–128.

Donnelly, James F. "Defining the Industrial Chemist in the United Kingdom, 1850–1921." *Journal of Social History* 29 (1996): 779–796.

Donnelly, James F. "Getting Technical: The Vicissitudes of Academic Industrial Chemistry in Nineteenth-Century Britain." *History of Education* 26 (1997): 125–143.

Dorpalen, Andreas. *Heinrich von Treitschke*. New Haven, CT: Yale University Press, 1957.

Draper, John W. "XXX. On Some Analogies between the Phenomena of the Chemical Rays, and Those of Radiant Heat." *Philosophical Magazine*, series 3, 19 (1841): 195–210.

Draper, John W. "XLIX. Description of the Tithonometer, an Instrument for Measuring the Chemical Force of the Indigo-Tithonic Rays." *Philosophical Magazine*, series 3, 23 (1843): 401–415.

Duckworth, C.L.D., and G. E. Langmuir. *Railway and Other Steamers*. Glasgow: Shipping Histories, 1948.

Dunston, Frederick W. *Roscoeana, being some account of the kinsfolk of William Roscoe of Liverpool and Jane, née Griffies, his wife*. Donhead St Mary: privately printed, 1910.

Durrant, P. J. *Organic Chemistry*, 9th impression. London: Longmans, Green, 1963.

Durrant, P. J. *General and Inorganic Chemistry*, 3rd edition. London: Longmans, Green, 1964.

Edgerton, David. *Science, Technology, and the British Industrial "Decline," 1870–1970*. Cambridge: Cambridge University Press, 1996.

Edgerton, David. "The Decline of Declinism." *Business History Review* 71 (1997): 201–206.

Edgerton, David. *Warfare State: Britain, 1920–1970*. Cambridge: Cambridge University Press, 2006.

"Editorial." *Journal of the Society of Chemical Industry* 1 (1882): 1.

Edmondson, John. *William Roscoe and Liverpool's First Botanic Garden*. Liverpool: National Museums Liverpool and The University of Liverpool, 2005.

Egerton, A. C. "William Arthur Bone." *Obituary Notices of Fellows of the Royal Society* 2 (1939): 586–611.

Ellison, Thomas. *The Cotton Trade of Great Britain*. London: Effingham Wilson, 1886.

Emy, H. V. *Liberals, Radicals and Social Politics, 1892–1914*. Cambridge: Cambridge University Press, 1973.

Engels, Friedrich. *The Conditions of the Working Class in England in 1844*, translated by F. K. Wischnewetzky. London: Allen and Unwin, 1892.

Ensor, R.C.K. *England, 1870–1914*. Oxford: Clarendon Press, 1968.

Evans, George Eyre. *Vestiges of Protestant Dissent*. Liverpool: F. and E. Gibbons, 1897.

Evans, Richard. *A Short History of Technical Education* (2009), at https://web.archive.org/web/20230426022149/https://technicaleducationmatters.org/category/publications/publications-ashte/.

Fahy, David M. *The Politics of Drink in England, from Gladstone to Lloyd George.* Newcastle-upon-Tyne: Cambridge Scholars, 2022.

Faraday, Michael. *Chemical Manipulation.* London: W. Phillips, 1827.

Farrar, W. V. "The Society for the Promotion of Scientific Industry, 1872–1876." *Annals of Science* 29 (1972): 81–86.

Farrar, W. V. "Synthetic Dyes before 1860." *Endeavour* 32 (September 1974): 149–154.

Fauque, Danielle. "Reorganizing Chemistry after World War I: The Birth of the International Union of Pure and Applied Chemistry (IUPAC)." *Rendiconti dell'Accademia Nazionale delle Scienze detta dei XL* (2020): 75–86.

Felkin, Henry M. *Technical Education in a Saxon Town.* London: Kegan Paul, 1881.

Festing, E. R. *Report of Visits to Chemical Laboratories at Bonn, Berlin, Leipzig, etc.* London: HMSO, 1871.

Fiddes, Edward. *Chapters in the History of Owens College and of Manchester University 1851–1914.* Manchester: Manchester University Press, 1937.

Fiddes, Edward. "Introductory Chapter: The Admission of Women to Owens College." In *The Education of Women at Manchester University 1883 to 1933*, edited by Mabel Tylecote, 1–16. Manchester: Manchester University Press, 1941.

Fitzgerald, George Francis. "Helmholtz Memorial Lecture." *Journal of the Chemical Society, Transactions* 69 (1896): 885–912.

Fletcher, Stella, ed. *Roscoe and Italy: The Reception of Italian Renaissance History and Culture in the Eighteenth and Nineteenth Centuries.* Farnham, Surrey: Ashgate, 2012.

Floud, Roderick. "Technical Education and Economic Performance: Britain, 1850–1914." *Albion: A Quarterly Journal Concerned with British Studies* 14 (1982): 153–171.

Fontani, Marco, Mariagrazia Costa, and Mary Virginia Orna. *The Lost Elements: The Periodic Table's Shadow Side.* New York: Oxford University Press, 2015.

Forgan, Sophie. "The Architecture of Science and the Idea of a University." *Studies in the History and Philosophy of Science* 20 (1989): 405–434.

Forgan, Sophie, and Graeme Gooday. "'A Fungoid Assemblage of Buildings': Diversity and Adversity in the Development of College Architecture and Scientific Education in Nineteenth-Century South Kensington." *History of Universities* 13 (1994): 153–192.

Forgan, Sophie, and Graeme Gooday. "Constructing South Kensington: The Buildings and Politics of T. H. Huxley's Working Environments." *British Journal for the History of Science* 29 (1996): 435–468.

Foster, G. C. "Frederick Guthrie." *Nature* 35 (4 November 1886): 8–10.

Fowler, Alan. *Lancashire Cotton Operatives and Work, 1900–1950.* Aldershot, Hampshire: Ashgate, 2003.

Gáldy, Andrea M. "William Roscoe and Thomas Coke of Holkham." In *Roscoe and Italy*, edited by Stella Fletcher, 204–206. Farnham, Surrey: Ashgate, 2012.

Garber, Elizabeth, ed. *Beyond History of Science: Essays in Honor of Robert E. Schofield.* Bethlehem, PA: Lehigh University Press, 1990.

Gavroglu, Kostas. "The Transmission and Assimilation of Scientific Ideas to the Greek Speaking World ca. 1700–1900: The Case of Chemistry." In *The Making of the Chemist: The Social History of Chemistry in Europe 1789–1914*, edited by David Knight and Helge Kragh, 289–304. Cambridge: Cambridge University Press, 1998.

Gay, Hannah. *The History of Imperial College, 1907–2007: Higher Education and Research in Science, Technology and Medicine.* London: Imperial College Press, 2007.

Gay, Hannah. "Science, Scientific Careers and Social Exchange in London: The Diary of Herbert McLeod, 1885–1900." *History of Science* 46 (2008): 457–496.

Gay, Hannah, and William P. Griffith. *The Chemistry Department at Imperial College London: A History, 1845–2000.* London: World Scientific, 2017.

Gay-Lussac, Joseph-Louis, and Louis-Jacques Thénard. "Des Mémoires lus a l'Institut national, depuis le 7 mars 1808 jusqu'au 27 février 1809." *Mémoires de physique et de chimie, de la Société d'Arcueil* 2 (1809): 295–358.

Geison, Gerald L. *The Private Science of Louis Pasteur.* Princeton, NJ: Princeton University Press, 1995.

Gibbs, F. W. *Joseph Priestley: Adventurer in Science and Champion of Truth.* London: Nelson and Sons, 1965.

Gillin, Edward J. *The Victorian Palace of Science: Scientific Knowledge and the Building of the Houses of Parliament.* Cambridge: Cambridge University Press, 2017.

Gingerich, Owen. "The Nineteenth Century Birth of Astrophysics." In *Physics of Solar and Stellar Coronae: GS Vaiana Memorial Symposium,* edited by J. F. Linsky and S. Serio, 47–58. Dordrecht: Kluwer, 1993.

Glover, E. Otho. "Harry Baker." *Journal of the Chemical Society (Resumed)* (1936): 539.

Gowing, Margaret. "Science, Technology and Education: England in 1870: The Wilkins Lecture, 1976." *Notes and Records of the Royal Society of London* 32 (July 1977): 71–90.

Graebe, Carl. "Marcelin Berthelot." *Berichte der Deutschen Chemischen Gesellschaft* 41 (1908): 4805–4872.

Greenaway, A. John, Jocelyn F. Thorpe, and Robert Robinson. *The Life and Work of Professor William Henry Perkin.* London: Chemical Society, [1932].

Greenaway, Frank. *John Dalton and the Atom.* London: Heinemann, 1966.

Griffiths, John G.A., and Ronald G.W. Norrish. "The Photosensitised Decomposition of Nitrogen Trichloride and the Induction of the Hydrogen-Chlorine Reaction." *Transactions of the Faraday Society* 27 (1931): 451–458.

Griffiths, John G.A., and Ronald G.W. Norrish. "The Induction Period of the Photochemical Reaction between Hydrogen and Chlorine." *Proceedings of the Royal Society of London, A. Mathematical and Physical Sciences* 147 (1934): 140–151.

Guagnini, Anna. "The Fashioning of Higher Technical Education in Britain: The Case of Manchester, 1851–1914." In *Industrial Training and Technical Innovation, A Comparative and Historical Study,* edited by Howard F. Gospel, 69–92. London: Routledge, 2010.

Guinot, Eugène. *A Summer in Baden-Baden.* London: J. Mitchell, n.d.

Hamer, D. A. *Liberal Politics in the Age of Gladstone and Rosebery: A Study in Leadership and Policy.* Oxford: Clarendon Press, 1972.

Hamlin, Christopher. "William Dibdin and the Idea of Biological Sewage Treatment." *Technology and Culture* 29 (1988): 189–218.

Hammond, Peter W., and Harold Egan. *Weighed in the Balance: A History of the Laboratory of the Government Chemist.* London: HMSO, 1992.

Hankins, Thomas. "In Defence of Biography: The Use of Biography in the History of Science." *History of Science* 17 (1979): 1–16.

Hannaway, Owen. "The German Model of Chemical Education in America: Ira Remsen at Johns Hopkins (1876–1913)." *Ambix* 23 (1976): 145–164.

Harden, A. "Carl Schorlemmer." *Journal of the Chemical Society, Transactions* 63 (1893): 756–763.

Harden, Arthur. *Alcoholic Fermentation*. London: Longmans, Green, 1911.

Hardie, D.W.F. *A History of the Chemical Industry in Widnes*. [Liverpool]: Imperial Chemical Industries, 1950.

Hare, Clark, and Martin H. Boyè. "On the Perchlorate of the Oxide of Ethule or Perchloric Ether." *Transactions of the American Philosophical Society* 8 (1843): 73–76.

Harris, J., and W. H. Brock. "From Giessen to Gower Street: Towards a Biography of Alexander William Williamson (1824–1904)." *Annals of Science* 31 (1974): 95–130.

Harte, Negley. *The University of London, 1936–1986: An Illustrated History*. London: Athlone Press, 1986.

Hartog, Philip J., ed. *The Owens College Manchester, Founded 1851: A Brief History of the College and Description of Its Various Departments*. Manchester: J. E. Cornish, 1900.

Harvey, A. D. "The Ministry of All the Talents: The Whigs in Office, February 1806 to March 1807." *The Historical Journal* 15 (1972): 619–648.

Hassan, John. *A History of Water in Modern England and Wales*. Manchester: Manchester University Press, 1998.

Hatchard, C. G., and C. A. Parker. "A New Sensitive Chemical Actinometer: II. Potassium Ferrioxalate as a Standard Chemical Actinometer." *Proceedings of the Royal Society of London, A. Mathematical and Physical Sciences* 235 (1956): 518–536.

Hawes, James. *Englanders and Huns: How Five Decades of Enmity Led to the First World War*. London: Simon & Schuster, 2014.

Hearnshaw, John B. *The Analysis of Starlight: Two Centuries of Astronomical Spectroscopy*. Cambridge: Cambridge University Press, 2014.

Heidelberg, University of. *Addressbuch. Sommer-halbjahr 1843*. Heidelberg: Karl Winter, 1843.

Heidelberg, University of. *Addressbuch. Sommer-halbjahr 1853*. Heidelberg: Karl Winter, 1853.

Heidelberg, University of. *Addressbuch. Winter-hulbjahr 1852–1853*. Heidelberg: Karl Winter, 1852.

Heidelberg, University of. *Addressbuch, Winter-halbjahr 1853–1854*. Heidelberg: Karl Winter, 1853.

Heidelberg, University of. *Addressbuch. Winter-halbjahr 1855–1856*. Heidelberg: Julius Groos, 1855.

Heilbron, J. L. *H. G. J. Moseley: The Life and Letters of an English Physicist, 1887–1915*. Berkeley: University of California Press, 1974.

Heimann, P. M. "Moseley and Celtium: The Search for a Missing Element." *Annals of Science* 23 (1967): 249–260.

Heine, Jens Ulrich. *Verstand und Schicksal*. Weinheim: VCH, 1990.

Heinig, Karl. *Carl Schorlemmer, Chemiker und Kommunist ersten Ranges*. Leipzig: BSB B. G. Teubner, 1974.

Hennig, Jochen. *Der Spektralapparat Kirchhoffs und Bunsens*. Berlin: Verlag für Geschichte der Naturwissenschaften und Technik, 2003.

Hennig, Jochen. "Die spektroskopischen Arbeiten von Gustav Kirchhoff und Robert Bunsen." In *Kanonische Experimente der Physik: Fachliche Grundlagen und historischer Kontext*, edited by Peter Heering, 153–168. Berlin: Springer, 2022.

Hentschel, Klaus. *Mapping the Spectrum: Techniques of Visual Representation in Research and Teaching*. Oxford: Oxford University Press, 2002.

Hentschel, Klaus. "Why Not One More Imponderable?: John William Draper and his Tithonic Rays." *Foundations of Chemistry* 4 (2002): 5–59.

Hervel, John. *Joseph Fourier, the Man and the Physicist*. Oxford: Clarendon Press, 1975.

Heyrovsky, J. "Prof. Bohuslav Brauner." *Nature* 135 (1935): 497–498.

Hingley, Peter D. "The First Photographic Eclipse?" *Astronomy & Geophysics* 42, no. 1 (2001): 1–18.

Hodgkinson, W. R. "Francis Edward Matthews." *Journal of the Chemical Society (Resumed)* (1929): 2970–2972.

Hoffman, Ross J. S. *Great Britain and the German Trade Rivalry, 1875–1914*. Philadelphia: University of Pennsylvania Press, 1933.

Hofmann, A. W. von. *Erinnerung an vorangegangene Freunde*, vol. 3. Brunswick: Vieweg, 1888.

Hofmann, A. W. von. "Heinrich Will. Ein Gedenkblatt." *Berichte der Deutschen Chemischen Gesellschaft* 23 (1890): 852–899.

Holloway, Sydney W.F. *Royal Pharmaceutical Society of Great Britain, 1841–1991: A Political and Social History*. London: Pharmaceutical Press, 1991.

Holmes, Rachel. *Sylvia Pankhurst: Natural Born Rebel*. London: Bloomsbury, 2020.

Holt, Anne. *Walking Together: A Study in Liverpool Nonconformity, 1688–1938*. London: George Allen & Unwin, 1938.

Homburg, Ernst. "The Rise of Analytical Chemistry and Its Consequences for the Development of the German Chemical Profession (1780–1860)." *Ambix* 46 (1999): 1–32.

Hopkins, F. Gowland. "John Wade." *Journal of the Chemical Society, Transactions* 103 (1913): 767–774.

Hopkins, F. Gowland, and C. J. Martin. "Arthur Harden." *Obituary Notices of Fellows of the Royal Society* 4 (1942): 2–14.

Hopkins, James. "The Disciplinary Development of University Buildings: Medicine and Manchester." *Baltic Journal of Art History* 16 (2018): 101–114.

Hopkins, James. "The (Dis)assembling of Form: Revealing the Ideas Built into Manchester's Medical School." *Journal of the History of Medicine and Allied Sciences* 75 (2020): 24–53.

Hughes, David. "Dr Robert West Pearson: The Scandalous Life of the Rev. Robert West Pearson: Over-reaching Ambition or 'a Bad Egg'?" (June 2018), Blackburn, https://www.cottontown.org/Names%20of%20Note/Pages/Rev.-Dr.-Robert-West-Pearson.aspx.

Hull, Andrew. "War of Words: The Public Science of the British Scientific Community and the Origins of the Department of Scientific and Industrial Research, 1914–16." *British Journal for the History of Science* 32 (1999): 461–481.

Hyde, Francis E. *Liverpool and the Mersey: An Economic History of a Port 1700–1970*. Newton Abbot, Devon: David and Charles, 1971.

"International Congress of Applied Chemistry." *Journal of the Society of Chemical Industry* 28 (1909): 580–584.

Inuzuka, Takaaki. *Alexander Williamson: A Victorian Chemist and the Making of Modern Japan*, translated by Haruko Laurie. London: UCL Press, 2021.

Jackson, Catherine M. "Analysis and Synthesis in Nineteenth-century Organic Chemistry." PhD dissertation, University College London, 2009.

Jackson, Myles W.W. *Spectrum of Belief: Joseph von Fraunhofer and the Craft of Precision Optics*. Cambridge, MA: MIT Press, 2000.

Jackson, Roland. *The Ascent of John Tyndall: Victorian Scientist, Mountaineer and Public Intellectual*. Oxford: Oxford University Press, 2018.

James, Frank A.J.L. "The Establishment of Spectro-Chemical Analysis as a Practical Method of Qualitative Analysis, 1854–1861." *Ambix* 30 (1983): 30–53.

James, Frank A.J.L. "The Creation of a Victorian Myth: The Historiography of Spectroscopy." *History of Science* 13 (1985): 1–24.

James, Frank, ed. *The Correspondence of Michael Faraday*, vol. 1. London: Institution of Electrical Engineers, 1991.

James, Frank A.J.L. "Science as a Cultural Ornament: Bunsen, Kirchhoff and Helmholtz in Mid-Nineteenth-Century Baden." *Ambix* 42 (1995): 1–9.

Japp, Francis R. "Kekulé Memorial Lecture." *Journal of the Chemical Society Transactions* 73 (1898): 97–138.

Jensen, W. B. "The Origin of the Bunsen Burner." *Journal of Chemical Education* 82 (2005): 518.

Jevons, Harriet. *Letters and Journal of W. Stanley Jevons*. London: Macmillan, 1886.

Johnson, Jeffrey A. "Academic Chemistry in Imperial Germany." *Isis* 76 (1985): 500–524.

Johnston, James F.W. "Report on the Recent Progress and Present State of Chemical Science." In *Report of the First and Second Meetings of the British Association for the Advancement of Science at York in 1831 and Oxford in 1832*, 414–529. London: John Murray, 1833.

Johnston, Neil. "The History of the Parliamentary Franchise." *Research Papers of the House of Commons Library* 13, no. 14 (2013): 8.

Jones, Alan W. *Lyulph Stanley: A Study in Educational Politics*. Waterloo, Ontario: Wilfrid Laurier Press, 1979.

Jones, Francis. *The Owens College Junior Course of Practical Chemistry*. London: Macmillan, 1872.

Jordan, Ellen. "'The Great Principle of English Fair-Play': Male Champions, the English Women's Movement and the Admission of Women to the Pharmaceutical Society in 1879." *Women's History Review* 7 (1998): 381–410.

Kargon, Robert H. *Science in Victorian Manchester: Enterprise and Expertise*. Manchester: Manchester University Press, 1977.

Kelly, Thomas. *A History of Adult Education*. Liverpool: Liverpool University Press, 1970.

Kikuchi, Yoshiyuki. *Anglo-American Connections in Japanese Chemistry: The Lab as Contact Zone*. New York: Palgrave Macmillan, 2013.

Kingszett, C. T. "Calcic Hypochlorite." *Chemical News* 46 (15 September 1882): 120.

Knobel, E. B. "Pierre Jules Cesar Janssen." *Monthly Notices of the Royal Astronomical Society* 68 (1908): 245–249.

Kohler, Robert E. "The Background to Arthur Harden's Discovery of Cozymase." *Bulletin of the History of Medicine* 48 (1974): 22–40.

König, Wolfgang. "Technical Education and Industrial Performance in Germany: A Triumph of Heterogeneity." In *Education, Technology and Industrial Performance in Europe, 1850–1939*, edited by Robert Fox and Anna Guagnini, 65–87. Cambridge: Cambridge University Press, 1993.

Landolt, Hans Heinrich. "Carl Löwig." *Berichte der Deutschen Chemischen Gesellschaft* 23 (1890): 905–909.

Lawes, John Bennet. *Reply to Baron Liebig's Principles of Agricultural Chemistry.* London: W. Clowes, 1855.

Layton, David. *Interpreters of Science: A History of the Association for Science Education.* London: John Murray, 1984.

Le Conte, David. "Two Guernseymen and Two Eclipses." *Antiquarian Astronomer* 4 (January 2008): 55–68.

Le Conte, David. "Warren de la Rue: Pioneer Astronomical Photographer." *Antiquarian Astronomer* 5 (February 2011): 14–35.

Leapman, Michael. *The World for a Shilling: How the Great Exhibition of 1851 Shaped a Nation.* London: Headline, 2001.

Lear, Linda. *Beatrix Potter: A Life in Nature.* London: Allen Lane, 2007.

Lees, Colin, and Alex Robertson. "Owens College: A. J. Scott and the Struggle against Prodigious Antagonistic Forces." *Bulletin of the John Rylands Library* 78 (1996): 155–172.

Lees, Colin, and Alex Robertson. "Early Students and the 'University of the Busy': The Quay Street Years of Owens College 1851–1870." *Bulletin of the John Rylands Library* 7 (1997): 161–194.

Lees, Colin, and Alex Robertson. "Community Access to Owens College, Manchester: A Neglected Aspect of University History." *Bulletin of the John Rylands Library* 80 (1998): 125–152.

Leitch, Diana, and Alfred G. Williamson. *The Dalton Tradition.* Manchester: John Rylands University Library, 1991.

Levinstein, I. "Address by the Chairman." *Journal of the Society of Chemical Industry* 3 (1884): 69–73.

Levy, S. I. "Brauner Memorial Lecture." *Journal of the Chemical Society (Resumed)* (1935): 1876–1890.

Lewis, William Draper. "John Innes Clark Hare." *The American Law Register*, new series, 45 (December 1906): 711–717.

Linder, Leslie, ed. *The Journal of Beatrix Potter, 1881–1897.* London: Frederick Warne, 1989.

"List of Exhibitors." *The Photographic Journal, Including the Transactions of the Photographic Society of Great Britain*, new series, 11 (1877–1878): 19.

Lloyd-Jones, Naomi. "The 1892 General Election in England: Home Rule, the Newcastle Programme and Positive Unionism." *Historical Research* 93 (2020): 73–104.

Lockemann, Georg. *Robert Wilhelm Bunsen: Lebensbild eines deutschen Naturforschers.* Stuttgart: Wissenschaftliche Verlagsgesellschaft, 1949.

Lockemann, Georg. "The Centenary of the Bunsen Burner." *Journal of Chemical Education* 33 (January 1956): 20–22.

Longmate, Norman. *The Waterdrinkers: A History of Temperance.* London: Hamish Hamilton, 1968.

Lord President of the Council. *Scientific Manpower.* London: HMSO, 1946.

Lowry, H. V. "Technical Education in Great Britain." *Nature* 177 (1956): 970–971.

Lund, Valerie K. "The Admission of Religious Nonconformists to the Universities of Oxford and Cambridge, and to Degrees in those Universities, 1828–1871." MA thesis, College of William and Mary, Williamsburg, Virginia, 1978.

Lundgren, Anders, and Bernadette Bensaude-Vincent, ed. *Communicating Chemistry: Textbooks and Their Audiences, 1789–1939*. Canton, MA: Science History Publications/ USA, 2000.

Macdonald, Lee T. "'Solar Spot Mania': The Origins and Early Years of Solar Research at Kew Observatory, 1852–1860." *Journal for the History of Astronomy* 46 (2015): 469–490.

Macdonald, Lee T. *Kew Observatory and the Evolution of Victorian Science, 1840–1910*. Pittsburgh: University of Pittsburgh University, 2018.

MacLeod, Roy M. "The Alkali Acts Administration, 1863–1884: The Emergence of the Civil Scientist." *Victorian Studies* 9 (1965): 85–112.

MacLeod, Roy M. "The Support of Victorian Science: The Endowment of Research Movement in Great Britain, 1868–1900." *Minerva* 9 (1971): 197–230.

MacLeod, Roy. "Science for Imperial Efficiency and Social Change: Reflections on the British Science Guild, 1905–1936." *Public Understanding of Science* 3 (1994): 155–193.

MacLeod, Roy. *Archibald Liversidge, FRS: Imperial Science under the Southern Cross*. Sydney: Sydney University Press, 2009.

MacLeod, Roy, Russell G. Egdell, and Elizabeth Bruton, eds. *For Science, King and Country: The Life and Legacy of Henry Moseley*. London: Uniform Press, 2018.

Maddison, Angus. *Contours of the World Economy, 1–2030 AD: Essays in Macro-Economic History*. Oxford: Oxford University Press, 2007.

Magnus, Philip. *Gladstone; A Biography*. London: John Murray, 1970.

Mann, F. G., and B. C. Saunders. *Practical Organic Chemistry*. London: Longmans, Green, 1936.

Mansergh, Nicholas. *The Irish Question, 1840–1921*, 3rd edition. London: George Allen and Unwin, 1975.

Marie, D., J. Burrows, R. Meller, and G. K. Moortgat. "A Study of the UV-Visible Absorption Spectrum of Molecular Chlorine." *Journal of Photochemistry and Photobiology A: Chemistry* 70 (1993): 205–214.

Marriner, Sheila. *The Economic and Social Development of Merseyside*. London: Croom Helm, 1982.

Marsh, J. E. "William Odling." *Journal of the Chemical Society, Transactions* 119 (1921): 553–564.

Martin, H. "Present Safeguards in Great Britain against Pesticide Residues and Hazards." *Residue Reviews* 4 (1963): 17–32.

Matsumi, Yutaka, Masahiro Kawasaki, Tetsuya Sato, Tohoru Kinugawa, and Tatsuo Arikawa. "Photodissociation of Chlorine Molecule in the UV Region." *Chemical Physics Letters* 155 (1989): 486–490.

Matthew, H.C.G. *Gladstone, 1809–1898*. Oxford: Oxford University Press, 1999.

Matthews, M. H. "The Development of the Synthetic Alkali Industry in Great Britain by 1823." *Annals of Science* 33 (1976): 371–382.

McCartney, Mark, Andrew Whitaker, and Alastair Wood, ed. *George Gabriel Stokes: Life, Science and Faith*. Oxford: Oxford University Press, 2019.

McGucken, William. *Nineteenth-Century Spectroscopy: Development of the Understanding of Spectra, 1802–1897*. Baltimore, MD: Johns Hopkins Press, 1969.

Meadows, A. J. *Early Solar Physics*. Oxford: Pergamon Press, 1971.

Meadows, A. J. *Science and Controversy: A Biography of Sir Norman Lockyer, Founder of Nature*, 2nd edition. London: Macmillan, 2008.

Meinel, Christoph. "Artibus Academicis Inserenda: Chemistry's Place in Eighteenth and Early Nineteenth Century Universities." *History of Universities* 7 (1988): 89–115.

Meldola, Raphael. "The Scientific Development of the Coal-Tar Colour Industry." *Journal of the Society of Arts* 34, no. 1749 (28 May 1886): 759–771.

Mendenhall, T. C. "The Metric System in England." *Science* 2, no. 31 (1895): 119–120.

Mercelis, Joris, Gabriel Galvez-Behar, and Anna Guagnini. "Commercializing Science: Nineteenth- and Twentieth-Century Academic Scientists as Consultants, Patentees, and Entrepreneurs." *History and Technology* 33 (2017): 4–22.

Meyer, Lothar. "Leopold von Pebal." *Berichte der Deutschen Chemischen Gesellschaft* 20 (1887): 997–1015.

Miers, H. A. "Rammelsberg Memorial Lecture." *Journal of the Chemical Society, Transactions* 79 (1901): 1–43.

Millar, I. T. "Frederick George Mann." *Biographical Memoirs of Fellows of the Royal Society* 30 (1984): 408–441.

Millar, Robin. "Training the Mind: Continuity and Change in the Rhetoric of School Science." *Journal of Curriculum Studies* 17 (1985): 369–382.

Ministry of Education. *Higher Technological Education*. London: HMSO, 1945.

Ministry of Education. *Further Education: The Scope and Content of Its Opportunities under the Education Act, 1944*. London: HMSO, 1947.

Ministry of Education. *The Organisation of Technical Colleges*. London: HMSO, 1956.

Mond, Ludwig. "On the Manufacture of Sulphur from Alkali Waste in Great Britain." In *Report of the Thirty-Eighth Meeting of the British Association for the Advancement of Science Held in Norwich in 1868*, 40. London: John Murray, 1869.

Moore, James R. *The Transformation of Urban Liberalism: Party Politics and Urban Government in Late Nineteenth Century England*. Aldershot, Hampshire: Ashgate, 2006.

Morgan, Charles. *The House of Macmillan (1843–1943)*. London: Macmillan, 1944.

Morrell, Jack B. "The Chemist Breeders: The Research Schools of Liebig and Thomas Thomson." *Ambix* 19 (1972): 1–46.

Morrell, Jack. "W. H. Perkin, Jr., at Manchester and Oxford: From Irwell to Isis." *Osiris*, 2nd series, vol. VIII, *Research Schools: Historical Reappraisals* (1993): 104–126.

Morrell, Jack. "Research as the Thing: Oxford Chemistry 1912–1939." In *Chemistry at Oxford: A History from 1600–2005*, edited by Robert J.P. Williams, Allan Chapman, and John S. Rowlinson, 131–186. Cambridge: RSC, 2009.

Morrell, R. S. "M. M. Pattison Muir." *Journal of the Chemical Society (Resumed)* (1932): 1330–1334.

Morris, Peter J.T., ed. *Science for the Nation: Perspectives on the History of the Science Museum*. Basingstoke, Hampshire: Palgrave Macmillan, 2010.

Morris, Peter J.T. *The Matter Factory: A History of the Chemistry Laboratory*. London: Reaktion, 2015.

Morris, Peter J.T. "Who Are the Chemists in the Picture of Henry Roscoe and Dimitri Mendeleev?" *RSC Historical Group Newsletter* 79 (Winter 2021): 29–37 (electronic version), https://www.rsc.org/globalassets/03-membership-community/connect-with-others/through-interests/interest-groups/historical/newsletters/historical-group-newsletter---winter-2021.pdf.

Morris, Peter J.T. "Aspects of the Social Organization of the Chemical Laboratory in Heidelberg and Imperial College, London." In *The Laboratory Revolution and the Creation of the Modern University, 1830–1940*, edited by Klaas van Berkel and Ernst Homburg, 225–239. Amsterdam: Amsterdam University Press, 2023.

Morris, Peter J.T., and Colin A. Russell. *Archives of the British Chemical Industry, 1750–1914*. Stanford in the Vale, Oxon: British Society for the History of Science, 1988.

Mosley, Stephen. *The Chimney of the World: A History of Smoke Pollution in Victorian and Edwardian Manchester*. Cambridge: White Horse, 2001.

Muspratt, E. K. *My Life and Work*. London: Bodley Head, 1917.

Nash, Leonard K. "The Origin of Dalton's Chemical Atomic Theory." *Isis* 47 (1956): 101–116.

Nath, Biman B. *The Story of Helium and the Birth of Astrophysics*. New York: Springer, 2013.

Natural Science in Education. London: HMSO, 1918.

Nawa, Christine. *Robert Wilhelm Bunsen und sein Heidelberger Laboratorium, Heidelberg, 12. Oktober 2011*. Frankfurt am Main: Gesellschaft Deutscher Chemiker, [2011].

Nawa, Christine. "A Refuge for Inorganic Chemistry: Bunsen's Heidelberg Laboratory." *Ambix* 61 (2014): 115–140.

Nawa, Christine, and Christoph Meinel, ed. *Von der Forschung gezeichnete Instrumente und Apparaturen in Heidelberger Laboratorien skizziert von Friedrich Veith (1817–1907)*, Heidelberg: heiBOOKS, Universitätsbibliothek Heidelberg, 2020. https://books.ub.uni-heidelberg.de/heibooks/catalog/book/793.

Newman, John Henry. *The Idea of a University*. London: Longmans, Green, 1919.

Newth, G. S. *A Text-Book of Inorganic Chemistry*. London: Longmans, Green, 1894.

Noyes, William Albert, and James Flack Norris. "Biographical Memoir of Ira Remsen (1846–1927)." *National Academy of Sciences of the United States Biographical Memoirs* 14 (1931): 207–257.

Nye, Mary Jo. "Scientific Biography: History of Science by Another Means." *Isis* 97 (2006): 322–329.

O'Day, Alan. *Irish Home Rule, 1867–1921*. Manchester: Manchester University Press, 1998.

Oesper, R. E. *The Human Side of Scientists*. Cincinnati, OH: University of Cincinnati, 1975.

Owens College. *The Calendar of Owens College, Manchester, Session 1864–5*. Manchester: Thomas Sowler and Sons, 1864.

Owens College. *The Calendar of Owens College, Manchester, Session 1872–3*. Manchester: Thomas Sowler and Sons, 1872.

Owens College. *The Owens College, Manchester, Calendar for the Session 1876–7*. Manchester: J. E. Cornish and Thomas Sowler, 1876.

Owens College. *The Owens College, Manchester: The Calendar for the Session 1878–9*. J. E. Cornish and Thos. Sowler, 1878.

Owens College. *The Owens College, Manchester, Calendar for the Session 1880–1*. Manchester: J. E. Cornish and T. Sowler [1880].

Pang, Alex Soojung-Kim. *Empire and the Sun: Victorian Solar Eclipse Expeditions*. Stanford, CA: Stanford University Press, 2002.

Park, Jongseok, and Byung-Hoon Chung. "British Chemist Henry E. Roscoe's Unintended Contribution to Korean Chemistry in 1907." *Journal of Chemical Education* 92 (2015): 593–594.

Partington, J. R. *History of Chemistry*, vol. 4. London: Macmillan, 1964.

Pennsylvania, University of. *University of Pennsylvania: Biographical Catalogue of the Matriculates of the College ... 1749–1893*. Philadelphia: Society of the Alumni, 1894.

Percy, Eustace. *Education at the Crossroads*. London: Evans Bros., 1930.

Perkin, Harold. "The Historical Perspective." In *Perspectives of Higher Education*, edited by Burton R. Clark, 17–55. Berkeley: University of California Press, 1984.

Perkin, William Henry. "Baeyer Memorial Lecture." *Journal of the Chemical Society, Transactions* 123 (1923): 1520–1546.

Plumpe, Werner, Alexander Nutzenadel, and Catherine Schenk. *Deutsche Bank: The Global Hausbank, 1870–2020*. London: Bloomsbury Business, 2020.

Pollard, Sidney. *Britain's Prime and Britain's Decline: The British Economy, 1870–1914*. London: Edward Arnold, 1989.

Pope, Alexander, and William Roscoe. *Works of Alexander Pope, with notes and illustrations by himself and others. To which are added a new life of the author ... and occasional remarks by W. Roscoe*. London: C. and J. Rivington, 1824.

Pratt, Mary Louise. "Arts of the Contact Zone." *Profession* 91 (1991): 33–40.

Pratt, Mary Louise. *Imperial Eyes: Travel Writing and Transculturation*, 2nd edition. New York: Routledge, 2008.

Pribram, Richard. "Hans Heinrich Landolt." *Berichte der Deutschen Chemischen Gesellschaft* 44 (1911): 3337–3394.

"Proceedings of the First General Meeting." *Journal of the Society of Chemical Industry* 1 (1881): 3–4.

"Professor Roscoe's Lectures on Spectrum Analysis: Lecture One." *British Medical Journal* 1, no. 384 (May 9, 1868): 460; "Lecture Two." *British Medical Journal* 1, no. 385 (May 16, 1868): 488; "Lecture Three." *British Medical Journal* 1, no. 387 (May 30, 1868): 541; "Lecture Four." *British Medical Journal* 1, no. 389 (June 13, 1868): 594.

Pye, Ken. *Discover Liverpool*. Liverpool: Trinity Mirror Media, 2011.

Quirke, Viviane, and Peter Reed. "Chemistry, Consultants, and Companies, c. 1850–2000: Introduction." *Ambix* 67 (2020): 207–213.

Rae, Ian D. "Spectrum Analysis: The Priority Claims of Stokes and Kirchhoff." *Ambix* 44 (1997): 131–144.

Raper, H. S. "Julius Berend Cohen." *Obituary Notices of Fellows of the Royal Society* 1 (1935): 502–513.

Raper, H. S. "Arthur Smithells." *Obituary Notices of Fellows of the Royal Society* 3 (1940): 96–107.

Reader, W. J. *Imperial Chemical Industries: A History*, vol. 1: *The Forerunners 1870–1926*. London: Oxford University Press, 1970.

Reed, Peter. "The British Chemical Industry and the Indigo Trade." *British Journal for the History of Science* 25 (1992): 113–125.

Reed, Peter. *Acid Rain and the Rise of the Environmental Chemist in Nineteenth-Century Britain. The Life and Work of Robert Angus Smith*. Farnham, Surrey: Ashgate, 2014.

Reed, Peter. *Entrepreneurial Ventures in Chemistry: The Muspratts of Liverpool, 1793–1934*. Farnham, Surrey: Ashgate, 2015.

Reed, Peter. "Making War Work for Industry: The United Alkali Company's Central Laboratory during World War One." *Ambix* 62 (2015): 72–93.

Reed, Peter. "John Fletcher Moulton and the Transforming Aftermath of the Chemists' War." *The International Journal for the History of Engineering and Technology* 87 (2017): 1–19.

Reed, Peter. "George E. Davis (1850–1907): Transition from Consultant Chemist to Consultant Chemical Engineer in a Period of Economic Pressure." *Ambix* 67 (2020): 252–270.

Reed, Peter. "Alfred Fletcher's Campaign for Control of Black Smoke in Britain, 1864–1896: Anticipating the 1956 Clean Air Act." *The International Journal for the History of Engineering and Technology* 91 (2021): 27–48.

Reed, Peter. "Learning and Institutions: Emergence of Laboratory-Based Learning, Research Schools and Professionalization." In *A Cultural History of Chemistry in the Nineteenth Century*, edited by Peter Ramberg, 191–215. London: Bloomsbury Academic, 2022.

Reinhardt, Carsten, and Anthony S. Travis. *Heinrich Caro and the Creation of Modern Chemical Industry*. Dordrecht: Kluwer, 2000.

Remsen, Ira. "Sir Henry Roscoe (1833–1915)." *Proceedings of the American Academy of Arts and Sciences* 51 (1916): 923–925.

"The Report of the Gresham University Commission." *Nature* 49 (1 March 1894): 405–409.

"Report of the Royal Commission on a University for London." *Nature* 40 (6 June 1889): 121–122.

Richardson, H. W, Chapter 9, "Chemicals." In *The Development of British Industry and Foreign Competition 1875–1914*, edited by Derek H. Aldcroft, 274–306. London: George Allen and Unwin, 1968.

Richson, Charles. *Educational Facts and Statistics of Manchester and Salford*. London: Longman, Brown, Green and Longmans, 1852.

Roberts, Andrew. *Salisbury: Victorian Titan*. London: Weidenfeld and Nicolson, 1999.

Roberts, Gerrylynn K. "The Establishment of the Royal College of Chemistry: An Investigation of the Social Context of Early-Victorian Chemistry." *Historical Studies in the Physical Sciences* 7 (1976): 437–485.

Roberts, Gerrylynn K. "The Liberally-Educated Chemist: Chemistry in the Cambridge Natural Sciences Tripos, 1851–1914." *Historical Studies in the Physical Sciences* 11 (1980): 157–183.

Roberts, Gerrylynn K. "'A Plea for Pure Science': The Ascendancy of Academia in the Making of the British Chemist, 1841–1914." In *The Making of the Chemist: The Social History of Chemistry in Europe, 1789–1914*, edited by David Knight and Helge Kragh, 107–120. Cambridge: Cambridge University Press, 1998.

Robertson, Alex, and Colin Lees. "Owens College and the Victoria University, 1851–1903." In *A Portrait of the University of Manchester*, edited by Brian Pullan, 10–16. London: Third Millennium, 2007.

Rocke, Alan J. *Chemical Atomism in the Nineteenth Century: From Dalton to Cannizzaro*. Columbus: Ohio State University Press, 1984.

Rocke, Alan J. *The Quiet Revolution: Hermann Kolbe and the Science of Organic Chemistry*. Berkeley: University of California Press, 1993.

Rocke, Alan J. *Nationalizing Science: Adolphe Wurtz and the Battle for French Chemistry*. Cambridge, MA: MIT Press, 2001.

Rocke, Alan J. "Origins and Spread of the 'Giessen Model' in University Science." *Ambix* 50 (2003): 90–115.

Rocke, Alan J. "In Search of El Dorado: John Dalton and the Origins of the Atomic Theory." *Social Research: An International Quarterly* 72 (2005): 125–158.

Rocke, Alan J. *Image and Reality: Kekulé, Kopp, and the Scientific Imagination.* Chicago: University of Chicago Press, 2010.

Roderick, Gordon W., and Michael D. Stephens. *Scientific and Technical Education in Nineteenth-Century England.* Newton Abbot, Devon: David & Charles, 1972.

Roderick, Gordon W., and Michael D. Stephens. *Education and Industry in the Nineteenth Century.* London: Longman, 1978.

Roscoe, Henry. *A Digest of Law of Evidence on the Trial of Actions at Nisi Prius.* London: Saunders and Benning, 1827.

Roscoe, Henry. *The Life of William Roscoe*, 2 volumes. London: Thomas Cadell, 1833.

Roscoe, Henry. *A Digest of the Law of Evidence in Criminal Cases.* London: Benning and Saunders, 1835.

Roscoe, Henry, and Thomas Roscoe. *Westminster Hall, or, Professional Relics and Anecdotes of the Bar, Bench, and Woolsack.* London: J. Knight and H. Lacey, 1825.

Roscoe, William. *Life of Lorenzo de' Medici.* London: A. Strahan, 1796.

Roscoe, William. *The Life of Pope Leo X, Son of Lorenzo de' Medici.* Liverpool: Cadell and Davies, 1805.

Roscoe, William. *On the Origin and Vicissitudes of Literature, Science and Art and Their Influence on the Present State of Society.* Liverpool: n.p., 1817.

Roscoe, William. *The Poetical Works of William Roscoe.* London: Ward and Lock, 1857.

Roussanova, Elena. *Friedrich Konrad Beilstein: Chemiker zweier Nationen. Sein Leben und Werk sowie einige Aspekte der deutsch-russischen Beziehungen in der Chemie in der zweiten Hälfte des 19. Jahrhunderts im Spiegel seines brieflichen Nachlasses.* Hamburg: Norderstedt, 2007.

Rowlinson, John S. "Chemistry Comes of Age: The 19th Century." In *Chemistry at Oxford: A History from 1600 to 2005*, edited by Robert J.P. Williams, Allan Chapman, and John S. Rowlinson, 79–130. Cambridge: RSC, 2009.

Rubinstein, W. D. *Capitalism, Culture, and Decline in Britain, 1750–1990.* London: Routledge, 1993.

Russell, Colin A. *Edward Frankland: Chemistry, Controversy and Conspiracy in Victorian England.* Cambridge: Cambridge University Press, 1996.

Russell, Colin A., Noel G. Coley, and Gerrylynn K. Roberts. *Chemists by Profession: The Origins and Rise of the Royal Institute of Chemistry.* Milton Keynes, Bucks: Open University Press, 1977.

Russell, Colin A., and John A. Hudson. *Early Railway Chemistry and Its Legacy.* Cambridge: RSC Publishing, 2012.

Sakurai, Joji. "Shigetake Sugiura." *Journal of the Chemical Society (Resumed)* 129 (1926): 3246–3248.

Saltzman, Martin. "Academia and Industry: What Should Their Relationship Be? The Levinstein-Roscoe Dialog." *Bulletin of the History of Chemistry* 23 (1999): 34–41.

Sanders, A. J. "A Long Life's Relationship: Archibald Geikie, Alexander Macmillan and His Publishing House." *Geological Society, London, Special Publications* 480 (2018): 139–148.

Sanderson, Michael. *The Universities and British Industry, 1850–1970*. London: Routledge and Kegan Paul, 1972.

Sanderson, Michael. *The Universities in the Nineteenth Century*. London: Routledge and Kegan Paul, 1975.

Sanderson, Michael. *Education and Economic Decline in Britain, 1870 to the 1990s*. Cambridge: Cambridge University Press, 1999.

Sandison, A. T. "Sir Marc Armand Ruffer (1859–1917) Pioneer of Palaeopathology." *Medical History* 11 (1967): 150–156.

Saul, S. B. *The Myth of the Great Depression, 1873–1896*. London: Macmillan, 1969.

Scerri, Eric R. *A Tale of Seven Elements*. Oxford: Oxford University Press, 2013.

Scerri, Eric R. *The Periodic Table: Its Story and Its Significance*, 2nd edition. New York: Oxford University Press, 2020.

Schabas, Margaret. *A World Ruled by Number: William Stanley Jevons and the Rise of Mathematical Economics*. Princeton, NJ: Princeton University Press, 1990.

Schmitz, Rudolf. *Die deutschen pharmazeutisch-chemischen Hochschulinstitute*. Ingelheim am Rhein: C. H. Boehringer Sohn, 1969.

Schoenefeldt, Henrik. "The Historic Ventilation System of the House of Commons, 1840–52: Re-visiting David Boswell Reid's Environmental Legacy." *The Antiquaries Journal* 98 (2018): 245–295.

Schofield, Robert E. *The Enlightenment of Joseph Priestley. A Study of His Life and Work from 1733 to 1773*, University Park: Pennsylvania State University Press, 1997.

Schorlemmer, Carl. *The Rise and Development of Organic Chemistry*. Manchester: J. E. Cornish, 1879.

Schröder, Michael. *The Argand Burner: Its Origin and Development in France and England, 1780–1800*. Odense, Denmark: Odense University Press, 1969.

Schütt, Hans-Werner. *Eilhard Mitscherlich. Baumeister am Fundament der Chemie*. Munich: Oldenbourg, 1992.

Science Lectures for the People: Science Lectures Delivered in Manchester, 1866–1880, Series 1–11. Manchester: John Heywood, 1866–1879.

Seaman, William H. "The Discussion in the British Parliament on the Metric Bill." *Science* 21, no. 524 (1905): 72–75.

Searle, G. R. *The Quest for National Efficiency: A Study in British Politics and Political Thought, 1899–1914*. Berkeley: University of California Press, 1971.

Searle, G. R. *The Liberal Party: Triumph and Disintegration, 1886–1929*. Basingstoke, Hampshire: Macmillan, 1992.

Searle, G. R. *A New England? Peace and War, 1886–1918*. Oxford: Oxford University Press, 2005.

Seubert, Karl. "Lothar Meyer." *Berichte der Deutschen Chemischen Gesellschaft* 28 (1918): 1109–1146.

Shannon, Richard. *Gladstone: Heroic Minister, 1865–1898*. Harmondsworth, Middlesex: Allen Lane, 1999.

Sharp, P. R. "The Entry of County Councils into English Educational Administration, 1889." *Journal of Educational Administration and History* 1 (1968): 14–22.

Sharp, P. R. "'Whiskey Money' and the Development of Technical and Secondary Education." *Journal of Educational Administration and History* 4 (1971): 31–36.

Short, P. J. "The Municipal School of Technology and the University, 1890–1914." In *Artisan to Graduate*, edited by D.S.L Cardwell, 157–164. Manchester: Manchester University Press, 1974.

Siderer, Yona. "Translations of Roscoe's Chemistry Books into Japanese and Hebrew: Historical, Cultural and Linguistic Aspects." *Substantia* 5, no. 2 (2021): 41–54.

Simmons, Anna. "The Chemical and Pharmaceutical Trading Activities of the Society of Apothecaries, 1822–1922." PhD dissertation, Open University, Milton Keynes, Buckinghamshire, 2004.

Simpson, Renate. *How the PhD Came to Britain: A Century of Struggle for Postgraduate Education.* Guildford, Surrey: Society for Research into Higher Education, 1983.

"Sir Henry Roscoe's Eightieth Birthday." *Chemical News* 107 (27 January 1913): 31.

Smith, Crosbie, and M. Norton Wise. *Energy and Empire: A Biographical Study of Lord Kelvin.* Cambridge: Cambridge University Press, 1989.

Smith, Edgar Fahs. "Martin Hans Boyè, Chemist, 1812–1909." *Journal of Chemical Education* 21 (1944): 7–11.

Smith, Robert Angus. "On the Air of Towns." *Journal of the Chemical Society* 9 (1859): 196–235.

Smith, Robert Angus. "A Centenary of Science in Manchester, for the 100th Year of the Literary and Philosophical Society of Manchester." *Memoirs of the Manchester Literary and Philosophical Society*, series 3, 9 (1883): 1–487.

Smith, Watson, and T. Takamatsu. "On Pentathionic Acid." *Journal of the Chemical Society, Transactions* 37 (1880): 592–608.

Smith, Watson, and T. Takamatsu. "On Phenylnaphthalene." *Journal of the Chemical Society, Transactions* 39 (1881): 546–551.

Smith, Watson, and T. Takamatsu. "Sulphonic Acids Derived from Isodinaphthyl (ββ-dinaphthyl)." *Journal of the Chemical Society, Transactions* 39 (1881): 551–554.

Smith, Watson, and T. Takamatsu. "On Pentathionic Acid (Part II)." *Journal of the Chemical Society, Transactions* 41 (1882): 162–167.

Smythe, J. A. "Peter Phillips Bedson." *Journal of the Chemical Society (Resumed)* (1944): 40–41.

"The Society of Chemical Industry." *Journal of the Society of Chemical Industry*, Jubilee Number (July 1931): 9–23.

Söderqvist, Thomas. "Introduction: A New Look at the Genre of Scientific Biography." In *The History and Poetics of Scientific Biography*, edited by Thomas Söderqvist, 1–15. Aldershot, Hampshire: Ashgate, 2007.

Solov'ev, Yu I. "DI Mendeleev and the English Chemists." *Journal of Chemical Education* 61 (1984): 1069–1071.

"Some Scientific Centres: VI. The Heidelberg Physical Laboratory." *Nature* 65 (24 April 1902): 587–590.

Soukup, Rudolf Werner, and Roland Zenz. *Eine Bibliothek als beredte Zeugin eines umfassenden Wandels des wissenschaftlichen Weltbilds. Teil II: Ansätze einer Rekonstruktion des wissenschaftlichen Netzwerks Bunsens unter besonderer Berücksichtigung von Bunsens Privatbibliothek*, http://www.rudolf-werner-soukup.at/Publikationen/Dokumente/Bunsenbibliothek_Teil_2_Rekonstruktion.pdf.

St John, Ian. *Gladstone and the Logic of Victorian Politics.* London: Anthem Press, 2010.

Stephens, Michael D., and Gordon W. Roderick. "Science, the Working Classes and Mechanics' Institutes." *Annals of Science* 29 (1972): 349–360.

Stewart, Balfour. *Primer in Physics*. London: Macmillan, 1872.

Stewart, Balfour. *The Conservation of Energy: Being an Elementary Treatise on Energy and Its Laws*. London: Henry S. King, 1873.

Stock, Christine, ed. *Robert Wilhelm Bunsens Korrespondenz vor dem Antritt der Heidelberger Professur (1852): kritische Edition*. Stuttgart: Wissenschaftliche Verlagsgesellschaft, 2007.

Štrbáňová, Soňa. "Nationalism and the Process of Reception and Appropriation of the Periodic System in Europe and the Czech Lands." In *Early Responses to the Periodic System*, edited by Masanori Kaji, Helge Kragh, and Gabor Palló, 121–149. New York: Oxford University Press, 2015.

Stumm, Petra. *Leopold Gmelin (1788–1853): Leben und Werk eines Heidelberger Chemikers*. Herbolzheim, Baden-Württemberg: Centaurus, 2012.

Survey of London, vol. 38: *South Kensington Museums Area*. London: Athlone Press, 1975. https://www.british-history.ac.uk/survey-london/vol38/.

Sutton, M. A. "Sir John Herschel and the Development of Spectroscopy in Britain." *British Journal for the History of Science* 7 (1974): 42–60.

Sutton, M. A. "Spectroscopy and the Chemists: A Neglected Opportunity." *Ambix* 23 (1976): 16–26.

Sutton, M. A. "Spectroscopy, Historiography and Myth: The Victorians Vindicated." *History of Science* 24 (1986): 425–432.

Tayler, J. J. "Mr. Tayler on Religion in Germany." *The Christian Reformer; or, Unitarian Magazine and Review*, new series, 12 (November 1856): 651–660.

Taylor, F. Sherwood. *The Young Chemist*. London: Nelson, 1934.

Taylor, Judy. *Beatrix Potter, 1866–1943: The Artist and Her World*. London: F. Warne and the National Trust, 1987.

Thackray, Arnold W. "The Origin of Dalton's Chemical Atomic Theory: Daltonian Doubts Resolved." *Isis* 57 (1966): 35–55.

Thackray, Arnold. *John Dalton: Critical Assessments of His Life and Science*. Cambridge, MA: Harvard University Press, 1972.

Thackray, Arnold, and Mary Ellen Bowden. "The Rise and Fall of the 'Papal State.'" In *The 1702 Chair of Chemistry at Cambridge: Transformation and Change*, edited by Mary Archer and Christopher Haley, 189–209. Cambridge: Cambridge University Press, 2005.

Thom, D. "Liverpool Churches and Chapels; Their Destruction, Removal, or Alteration: With Notices of Clergymen, Ministers, and Others [Pt. 2]." *Proceedings and Papers of the Historic Society of Lancashire and Cheshire* 5 (1852–1853): 3–56.

Thompson, Joseph. *The Owens College: Its Foundation and Growth*. Manchester: J. E. Cornish, 1886.

Thomson, Sir J. J. *Recollections and Reflections*. London: Bell, 1936.

Thorpe, J. F. "John Cannell Cain." *Journal of the Chemical Society, Transactions* 119 (1921): 533–537.

Thorpe, T. E. "Kopp Memorial Lecture." *Journal of the Chemical Society, Transactions* 63 (1893): 775–815.

Thorpe, Thomas Edward. "Dr. James Bell, C.B., F.R.S." *Nature* 77 (1908): 539–540.

Thorpe, Sir [Thomas] Edward. *The Right Honourable Sir Henry Enfield Roscoe, PC, DCL, FRS: A Biographical Sketch.* London: Longmans, Green, 1916.

Thorpe, T. E. "The Right Honourable Sir Henry Enfield Roscoe." *Journal of the Chemical Society, Transactions* (1916): 395–424.

Thorpe, T. E. "Sir Henry Roscoe." *Proceedings of the Royal Society of London* A93 (1917): i–xxi.

Tilden, William A. *Famous Chemists: The Men and their Work.* London: George Routledge & Sons, 1921.

Travers, Morris. *The Life of Sir William Ramsay.* London: Edward Arnold, 1956.

Travis, Anthony S. *The Rainbow Makers: The Origins of the Synthetic Dyestuffs Industry in Western Europe.* Bethlehem, PA: Lehigh University Press, 1993.

Tuchman, Arleen Marcia. *Science, Medicine and the State in Germany; The Case of Baden, 1815–1871.* New York: Oxford University Press, 1993.

Tucker, S. Horwood. "A Lost Centenary: Lassaigne's Test for Nitrogen. The Identification of Nitrogen, Sulfur, and Halogens in Organic Compounds." *Journal of Chemical Education* 22 (1945): 212–215.

Uchida, Jun. "From Island Nation to Oceanic Empire: A Vision of Japanese Expansion from the Periphery." *The Journal of Japanese Studies* 42 (2016): 57–90.

University College London. *The University College London, Calendar for the Session 1853–54.* London: Walton and Maberly, 1853.

University College London. *The University College London Calendar for 1854–55.* London: Walton and Maberly, [1854].

"University of Manchester, I." *Nature* 14 (13 July 1876): 225–226.

"University of Manchester, II." *Nature* 14 (20 July 1876): 245–246.

"University of Manchester, III." *Nature* 14 (27 July 1876): 265–266.

Usselman, Melvyn C. *Pure Intelligence: The Life of William Hyde Wollaston.* Chicago: University of Chicago Press, 2015.

van't Hoff, J. H. *Etudes de Dynamique Chimique.* Amsterdam: Frederik Muller, 1884.

Varcoe, Iain. "Comment: Practical Proposals by Scientists for Reforming the Machinery of Scientific Advice, 1914–17." *British Journal for the History of Science* 33 (2000): 109–114.

Victoria University. *The Victoria University Calendar for the Session 1882–3.* Manchester: J. E. Cornish, 1882.

"Visits to Chemical Works." *Chemical News* 15 (3 February 1867): 71.

Vogel, Arthur I. *A Textbook of Qualitative Inorganic Analysis.* London: Longmans, Green, 1937.

Vogel, Arthur I. *A Textbook of Practical Organic Chemistry Including Organic Analysis.* London: Longmans, Green, 1948.

Vorländer, Daniel. "Jacob Volhard." *Berichte der Deutschen Chemischen Gesellschaft* 46 (1912): 1855–1902.

Wade, John, and Richard William Merriman. "Influence of Water on the Boiling Point of Ethyl Alcohol at Pressures above and below the Atmospheric Pressure." *Journal of the Chemical Society, Transactions* 99 (1911): 997–1011.

Wallace, Alfred Russel. "Government Aid to Science." *Nature* 1 (13 January 1870): 288–289.

Walton, John K. "Mad Dogs and Englishmen: The Conflict over Rabies in Late Victorian England." *Journal of Social History* 13 (1979): 219–239.

Warren, Kenneth. *Chemical Foundation: The Alkali Industry in Britain to 1926*. Oxford: Clarendon Press, 1980.

Warrington, G. "The Copper Mines of Alderley Edge and Mottram St. Andrew, Cheshire." *Journal of the Chester Archaeology Society* 64 (1981): 47–73.

Watts, W. Marshall. *An Introduction to the Study of Spectrum Analysis*. London: Longmans, Green, 1904.

Weeks, Mary Elvira. *Discovery of the Elements*, 6th edition, enlarged and revised. Easton, PA: Journal of Chemical Education, 1960.

Wells, David A. *Wells's Principles and Applications of Chemistry; For the Use of Academies, High-Schools and Colleges*. Chicago: Ivison and Phinney, 1859.

Wentrup, Curt. "Bunsen the Geochemist: Icelandic Volcanism, Geyser Theory, and Gas, Rock and Mineral Analyses." *Angewandte Chemie International Edition* 60 (2021): 1066–1081.

Westheimer, F. H. "Louis Plack Hammett." *Biographical Memoirs of the National Academy of Sciences* 72 (1997): 136–149.

Wheeler, Michael. *The Athenaeum: More than Just Another London Club*. New Haven, CT: Yale University Press, 2020.

White Paper on Technical Education. London: HMSO, 1956.

Williams, Ernst Edwin. *Made in Germany,* 4th edition. London: Heinemann, 1896.

Williamson, Alexander W. *A Plea for Pure Science: Being the Inaugural Lecture at the Opening of the Faculty of Science in University College, London*. London: Taylor & Francis, 1870.

Wilson, F.M.G. *The University of London, 1858–1900: The Politics of Senate and Convocation*. Woodbridge, Suffolk: Boydell Press, 2004.

Wilson, Arline. *William Roscoe: Commerce and Culture*. Liverpool: Liverpool University Press, 2008.

Wisniak, Jaime. "Thomas Carnelley." *Educación Química* 23 (2012): 465–473.

Wisniak, Jaime. "Henry Enfield Roscoe." *Educación Química* 27 (2016): 240–248.

Witt, Otto N. "Friedrich Konrad Beilstein." *Journal of the Chemical Society, Transactions* 99 (1911): 1646–1649.

Wittwer, W. C. "Ueber die Einwirkung des Lichts auf Chlorwasser." *Annalen der Physik* 170 (1855): 597–612.

Wittwer, W. C. "XI. Ueber die Einwirkung des Lichtes auf Chlorwasser." *Annalen der Physik* 173 (1856): 304–310.

Wittwer, W. C. "Ueber die Einwirkung des Lichtes auf Chlorwasser." *Annalen der Physik* 182 (1859): 266–289.

Wöbke, Bernd. "Das Portrait: Leopold Gmelin (1788–1853)." *Chemie in Unserer Zeit* 22 (1988): 208–216.

Wriedt, B., and D. Ziegenbalg. "Application Limits of the Ferrioxalate Actinometer." *ChemRxiv*. Cambridge: Cambridge Open Engage, 2021, https://chemrxiv.org/engage/chemrxiv/article-details/60c7597af96a00617128901c.

Index